COMMERCIAL
AVIATION SAFETY

Other McGraw-Hill Books of Interest

Wells • AIRPORT PLANNING & MANAGEMENT, Fifth Edition
Hessburg • AIR CARRIER MRO HANDBOOK
Davies • STANDARD HANDBOOK FOR AERONAUTICAL AND ASTRONAUTICAL ENGINEERS
de Neufville • AIRPORT SYSTEMS: PLANNING, DESIGN, AND MANAGEMENT

COMMERCIAL AVIATION SAFETY

ALEXANDER T. WELLS
CLARENCE C. RODRIGUES

FOURTH EDITION

McGraw-Hill
New York   Chicago   San Francisco   Lisbon   London   Madrid
Mexico City   Milan   New Delhi   San Juan   Seoul
Singapore   Sydney   Toronto

The McGraw-Hill Companies

Cataloging-in-Publication Data is on file with the Library of Congress

Wells, Alexander T.
    Commercial aviation safety / Alexander T. Wells, Clarence C. Rodrigues.—4th ed.
        p.   cm.
    Includes bibliographical references and index.
    ISBN 0-07-141742-7
        1. Aeronautics—United States—Safety measures.   2. Aeronautics, Commercial—United States—Safety measures.   I. Rodrigues, Clarence C.   II. Title.

TL553.5.W46   2003
363.12'4'0973—dc22

2003066631

2 3 4 5 6 7 8 9 0   DOC/DOC   0 1 0 9 8 7 6 5 4

ISBN 0-07-141742-7

The sponsoring editor for this book was Scott Grillo, the editing supervisor was David E. Fogarty, and the production supervisor was Pamela A. Pelton. It was set in Fairfield by Wayne A. Palmer and Deirdre Sheean of McGraw-Hill Professional's Hightstown, N.J., composition unit. The art director for the cover was Anthony Landi.

Printed and bound by RR Donnelley.

 This book was printed on recycled, acid-free paper containing a minimum of 50% recycled, de-inked fiber.

McGraw-Hill books are available at special quantity discounts to use as premiums and sales promotions, or for use in corporate training programs. For more information, please write to the Director of Special Sales, McGraw-Hill Professional, Two Penn Plaza, New York, NY 10121-2298. Or contact your local bookstore.

CONTENTS

CHAPTER FIVE. REVIEW OF SAFETY STATISTICS   101

CHAPTER SIX. ACCIDENT CAUSATION MODELS   133

CHAPTER SEVEN. HUMAN FACTORS IN AVIATION SAFETY   157

CHAPTER EIGHT. AIR TRAFFIC SAFETY SYSTEMS   187

CHAPTER NINE. AIRCRAFT SAFETY SYSTEMS   219

CHAPTER TWELVE. AIRLINE SAFETY   333

PREFACE

This text is dedicated to studying the principles and regulatory practices of commercial aviation safety in the United States in the twenty-first century. In addition to being a major rewrite of previous editions, this fourth edition introduces co-author Dr. Clarence Rodrigues, who is a member of Embry-Riddle Aeronautical University's College of Aviation faculty at Daytona Beach, Florida.

This edition updates, revises, and makes current the aviation safety and security information contained in previous editions; establishes new changes in the format, content, and order of the chapters to make the flow of information progressive and logical; and broadens the field of study to include regulatory information on OSHA and the EPA that is essential to the practicing aviation safety professional.

Today's aviation safety practitioner has to contend with more than just the safety dictates of the FAA and the NTSB. OSHA and EPA also have regulatory jurisdiction over the aviation sector. It is therefore important that today's aviation safety professional gain a broad understanding of relevant OSHA and EPA regulations. Failure to do so could lead to unsafe operating conditions and regulatory violations that could result in millions of dollars in fines.

This text is intended for individuals with a limited background in the subject and serves as a foundation for further study in the field of aviation safety and security. This book will benefit students enrolled in aviation safety and/or management programs and college or university flight programs. This text also serves as a resource for the practicing safety professional. Course instructors should readily be able to supplement text topics based on specific course needs, as several references are provided throughout the text.

FEATURES OF THE NEW FOURTH EDITION

1. Chapter 1, covering the regulatory framework, has been expanded to include the background and chronology on OSHA and EPA. Additional information on the major functions and activities of the FAA has also been added.

2. Chapter 2 discusses the organizational structure and the rulemaking process of the FAA, OSHA, and EPA. All relevant information from previous editions has been updated. The organizational structure of various departments within the FAA and their missions and responsibilities have been added. Next, the organizational structure, mission, and functions of the Flight Standards Service are detailed together with an in-depth coverage of the FAA rulemaking process. Additional information

on the organizational structure, rulemaking process, and selected regulatory issues of OSHA and EPA has been added to this edition.

3. Chapter 3 discusses the structure, function, and workings of the NTSB. It was deemed appropriate to review the organizations involved in safety before any discussions on safety measures and statistics occurred. To this chapter was added information on how to query and locate accident reports from NTSB databases through the web. This enabled the elimination of App. B contained in the previous edition.

4. Chapter 4 deals with the reporting and recording of safety data. Information from previous editions has been revised and updated, and additional information on FOQA and ASAP has been added. Also, all NTSB and FAA accident and incident definitions are consolidated here. Finally, information on OSHA and EPA accident and incident definitions and reporting requirements has been added.

5. Chapter 5 of this edition contains revised and updated information on safety statistics. Safety statistics of accident rates, fatalities, and other information on commercial jet accidents as compiled by the Boeing Company are included. A discussion on OSHA injury/illness statistics has been added to this chapter.

6. Chapter 6 on accident causation models reformulates some of the information on this topic from previous editions. In this connection a new model of accident causation has been established (the OTA model). Information on Reason's model of accident causation, a widely accepted model in aviation, has been added. Also included here are reasons why we model accidents.

7. Chapters 5 and 10 of the third edition have been consolidated and revised to form the new Chapter 7 on human factors in aviation safety. New information on human performance and human error has been added. The management of human error has been recategorized and discussed separately under engineering and administrative strategies.

8. Information on air traffic control (ATC) safety systems and the NAS scattered throughout previous editions is updated and consolidated here in Chapter 8 on air traffic safety systems. Additional information on Free Flight has been added.

9. Chapter 9 on aircraft safety systems contains revised and updated information. In addition, information on aircraft safety scattered throughout previous editions is consolidated here.

10. Chapter 10 on airport safety is new. The chapter discusses airport certification and compliance with Part 139. In addition, it describes the safety and regulatory issues of terminal buildings, hangars and maintenance shops, ramp operations, fuel handling, aircraft rescue and fire fighting (ARFF), and deicing. This chapter also discusses one of the FAA's top five safety priorities—runway incursions. Several aspects of runway incursions are discussed, including definitions, incident/accident statistics, causes and categories of runway incursions, and strategies that are either in place or are being developed to control runway incursions.

11. This edition has been written after the terrorist events of September 11, 2001. The aviation security world has changed drastically since then, and this fact is reflected in Chapter 11 on aviation security, which is a complete rewrite of the previous edition's chapter on security and safety. This chapter reviews the regulatory movement of aviation security from its inception in the airport and air carrier security acts, through the Air Marshal program, to the Aviation Transportation Security Act of 2001 and formation of the Transportation Security Administration. Additionally, this chapter contains information on security technologies that includes imaging and trace detection technologies, explosive and metal detection systems, biometrics, computer-assisted passenger screening, and strengthening of cockpit doors, aircraft, and baggage containers. Information on nontechnological approaches to security and the role of intelligence and international influences on security has been retained.

12. Chapter 12 on airline safety is a complete rewrite of the previous edition's chapter on this subject. A major section on safety program management has been added which includes extended discussions on management leadership and commitment; employee involvement, responsibility, and accountability; trends analysis and risk assessment; accident investigation and auditing; hazard prevention and control programs; and communications and training. As part of prevention and control programs, the internal evaluation program (IEP) and orders 8300.10 and 8400.10 are discussed at length. This chapter concludes with the safety organizational structure and functional job descriptions of a major air carrier.

13. Tables, figures, statistics, key terms, review questions, and references contained in this text have been updated and added to.

14. Numerous web sites have been included to help students and instructors utilize the vast amount of information available on the World Wide Web.

15. Each chapter contains a number of features that are designed to facilitate student learning. These features include:

- *Chapter outlines.* Each chapter opens with an outline of the major topics.
- *Chapter objectives.* The objectives of the chapter are included so students know exactly what is to be accomplished after completing the material.
- *Relevance.* All examples, applications, and theories are current as of this writing.
- *Figures and tables.* Figures and tables are drawn from sources, such as the FAA, NTSB, Boeing, and other current web sites.
- *Logical organization and frequent headings.* Itemized bulleted lists are used as frequently as possible to enhance reading.
- *Key terms.* Each chapter concludes with a list of key terms used in the text.
- *Review questions.* Review questions at the end of each chapter cover all of the important points.
- *References.* A list of references is included at the end of each chapter for students who wish to pursue the material in greater depth.

ACKNOWLEDGMENTS

We are sincerely appreciative of the many public and private institutions that have provided resource material from which we were able to shape this edition of *Commercial Aviation Safety*. In this regard, we are particularly indebted to the Federal Aviation Administration, National Transportation Safety Board, Occupational Safety and Health Administration, Environmental Protection Agency, International Civil Aviation Organization, the Flight Safety Foundation, Alaska Airlines, and the Boeing Company for their numerous publications. A special thanks to Embry-Riddle Aeronautical University for the use of its excellent library facilites.

We also wish to thank several individuals who were instrumental in making this edition a success. In particular, we wish to thank Professor Bill Martin from Embry-Riddle Aeronautical University's College of Aviation for his insights and for constantly encouraging and challenging us throughout this undertaking. We were able to draw from his past experiences as a military pilot and a College of Aviation Technology dean who also previously served as the University's Chief Flight Instructor and Director of Flight Standards. Among other things, he was instrumental in convincing us to develop the chapter on airport safety. Thanks, Bill.

We would also like to thank our colleague Professor John Griffin from Embry-Riddle for reviewing the chapter on air traffic safety systems. Many thanks to Angel M. Millan, our graduate student in safety, who helped with the research and other administrative details.

We are also grateful to the editor(s) at McGraw-Hill, for their perseverance in pursuing the need for a fourth edition. Also, a debt of gratitude is owed the following members of the McGraw-Hill organization: Scott Grillo, David Fogarty, Mary Ellen Haramis, Larry Hager, and Judith Reiss.

And finally and most importantly, we would be in big trouble if we did not thank our wives. So, to Mary and Nicola, thanks and cheers!

ALEXANDER T. WELLS, ED.D.
Embry-Riddle Aeronautical University
CLARENCE RODRIGUES, PH.D.
Embry-Riddle Aeronautical University

THE REGULATORY FRAMEWORK

The Federal Aviation Administration (FAA)
  Overview
  Background
  Chronology of FAA regulations
The Environmental Protection Agency (EPA)
  Overview
  Background
  Chronology of major environmental laws affecting aviation operations
Occupational Safety and Health Administration (OSHA)
  Overview
  Background
  Chronology of major OSHA standards affecting aviation operations
Key Terms
Review Questions
References

LEARNING OBJECTIVES

After completing this chapter, you should be able to

- Describe some of the early federal legislation that helped shape the airline industry in its formative years.
- Discuss some of the factors that led to the passage of the Federal Aviation Act of 1958.
- Identify some of the safety provisions of the 1958 FAA Act.
- Recognize the important legislation that followed airline deregulation in 1978.
- Distinguish between FAR Part 121 and 135 air carriers.
- Highlight the important features of the Commuter Safety Initiative of 1995.
- Discuss the evolution of EPA.
- List and discuss major EPA laws that are of importance to aviation operations.
- Discuss the evolution of OSHA.
- List and discuss major OSHA standards that are of importance to aviation operations.

This chapter reviews the development and regulatory framework of the three major safety-related agencies that regulate today's commercial aviation sector. The Federal Aviation Administration is discussed first, followed by the Environmental Protection Agency and the Occupational Safety and Health Administration.

THE FEDERAL AVIATION ADMINISTRATION (FAA)

OVERVIEW

The Federal Aviation Administration is a U.S. government agency with primary responsibility for the safety of civil aviation. The FAA was established by the Federal Aviation Act of 1958 and was originally known as the Federal Aviation Agency. It underwent a name change (Agency changed to Administration) in 1967 when the FAA became part of the Department of Transportation. The FAA's major functions as mentioned on its web site include

- Regulating civil aviation to promote safety and fulfill the requirements of national defense
- Encouraging and developing civil aeronautics, including new aviation technology
- Developing and operating a common system of air traffic control and navigation for both civil and military aircraft
- Performing research and development with respect to the National Airspace System and civil aeronautics
- Developing and implementing programs to control aircraft noise and other environmental effects of civil aviation
- Regulating U.S. commercial space transportation

BACKGROUND

The FAA is responsible for performing several activities that are in support of its above-mentioned functions. These activities as reproduced from the FAA web site include

- *Safety regulation.* The FAA issues and enforces regulations and minimum standards relating to the manufacture, operation, and maintenance of aircraft. The agency is responsible for the rating and certification of airmen and for certification of airports serving air carriers.
- *Airspace and air traffic management.* The safe and efficient utilization of the navigable airspace is a primary objective of the FAA. The agency operates a network of airport towers, air route traffic control centers, and flight service stations. It

develops air traffic rules, allocates the use of airspace, and provides for the security control of air traffic to meet national defense requirements.

- *Air navigation facilities.* The FAA is responsible for the construction or installation of visual and electronic aids to air navigation, and for the maintenance, operation, and quality assurance of these facilities. Other systems maintained in support of air navigation and air traffic control include voice/data communications equipment, radar facilities, computer systems, and visual display equipment at flight service stations.

- *Civil aviation abroad.* As mandated by legislation, the FAA promotes aviation safety and encourages civil aviation abroad. Activities include exchanging aeronautical information with foreign authorities; certifying foreign repair shops, airmen, and mechanics; providing technical assistance and training; negotiating bilateral airworthiness agreements; and providing technical representation at international conferences.

- *Commercial space transportation.* The agency regulates and encourages the U.S. commercial space transportation industry. It licenses commercial space launch facilities and private sector launching of space payloads on expendable launch vehicles.

- *Research, engineering, and development.* The FAA engages in research, engineering, and development aimed at providing the systems and procedures needed for a safe and efficient system of air navigation and air traffic control. The agency performs an aeromedical research function and supports development of improved aircraft, engines, and equipment. It also conducts tests and evaluations of specified items such as aviation systems, devices, materials, and procedures.

- *Other programs.* The FAA provides a system for registering aircraft and recording documents affecting title or interest in aircraft and their components. Among other activities, the agency administers an aviation insurance program; develops specifications for aeronautical charts; and publishes information on airways and airport services as well as on technical subjects relating to aeronautics.

CHRONOLOGY OF FAA REGULATIONS

The roots of today's aviation safety programs extend back to the early days of commercial aviation following World War I. Many returning pilots bought surplus war aircraft and went into business. These happy-go-lucky barnstormers toured the country, putting on shows and giving rides to local townsfolk. By the mid-1920s, uses of aircraft included advertising, aerial photography, crop dusting, and carrying illegal shipments of liquor during prohibition. Initial efforts to establish scheduled passenger service were short-lived, as service catered primarily to wealthy east coast tourists. This service was expensive compared to the country's well-developed rail and water travel networks.

AIR MAIL SERVICE.   Growth of commercial aviation was greatly stimulated by the establishment of the U.S. Air Mail Service in the early 1920s. Regulations established by the Post Office Department required its pilots to be tested and to have at least 500 hours of flying experience. The Post Office set up aircraft inspection and preventive maintenance programs for the pilots. These early regulatory requirements improved air mail carrier safety. In 1924, commercial flyers experienced one fatality every 13,500 miles, while the Air Mail Service had one fatality every 463,000 miles.

In 1925, Congress enacted the *Air Mail Act* of 1925, authorizing the Post Office Department to transfer air mail service to private operators. Twelve carriers, some of which evolved into today's major airlines, began air mail operations in 1926 and 1927. These carriers offered limited passenger service, which was much less profitable than carrying mail. Initially, air mail contractors were paid a percentage of postage revenues. In 1926, however, an amendment to the Air Mail Act required payment by weight carried. Small independent operators, using Ford and Fokker trimotor airplanes, handled most of the passenger service in the late 1920s, the forerunners of today's commuter airlines and air taxis.

EARLY SAFETY LEGISLATION.   No federal safety program existed, which prompted a number of states to pass legislation requiring aircraft licensing and registration. In addition, local governments enacted ordinances regulating flight operations and pilots, creating a patchwork of safety-related requirements and layers of authority. Despite strong industry support for federal legislation, Congress was unable to reach agreement on the scope and substance of a statute until 1926, when the *Air Commerce Act* was passed. Key issues debated by Congress included whether to separate military and civil aviation activities, what responsibilities should be left to state and local governments, and how to provide federal support for airports.

The new law gave the Department of Commerce regulatory authority over commercial aviation and responsibilities aimed at promoting the fledgling industry. The major provisions of the act authorized the regulation of aircraft and pilots in interstate and foreign commerce; provided federal support for charting and lighting airways, maintaining emergency fields, and making weather information available to pilots; authorized aeronautical research and development programs; and provided for the investigation of aviation accidents. Local governments were left with jurisdiction over airport control.

Within the Department of Commerce, a new Aeronautics Branch, comprising existing offices already engaged in aviation activities, was formed to oversee the implementation of the new law. Nine district offices of the Regulatory Division of the Aeronautics Branch were established to conduct inspections and checks of aircraft, pilots, mechanics, and facilities. District offices also shared licensing and certification responsibilities with the Washington, D.C., office. The basic allocation of responsibilities survives to this day, although the Department of Commerce responsibilities now rest with the Department of Transportation (DOT) and its branch, the FAA.

The first set of regulations was drafted with substantial input from aircraft manufacturers, air transport operators, and the insurance industry. Compared with current standards, pilot requirements were minimal. In addition to written and flight tests, transport pilots were required to have 100 hours of solo flight experience, while industrial pilots needed only 50 hours. Current procedures for certifying air-craft and engines also originated under these early regulatory programs. Aircraft manufacturers were required to comply with minimum engineering standards issued by the Department of Commerce in 1927, and one aircraft of each type was subject to flight testing to obtain an airworthiness certificate.

The Aeronautics Branch also collected and analyzed data from aircraft inspection reports, pilot records, and accident investigations. Data were made accessible to the insurance industry, which allowed for the development of actuarial statistics. A direct consequence of this step was a significant reduction in insurance rates for many carriers. However, the Department of Commerce, cognizant of its role to promote the aviation industry, was reluctant to make public disclosures about the results of individual accident investigations despite a provision in the Air Commerce Act directing it to do so. Eventually, in 1934, the Air Commerce Act was amended, giving the Secretary of Commerce extensive powers to investigate accidents, including a mandate to issue public reports of its findings. This congressional policy put safety considerations ahead of protecting the industry's image.

As additional regulations to improve safety were implemented, accidents involving passenger carriers and private aircraft decreased significantly. Between 1930 and 1932, the fatality rate per 100 million passenger-miles declined by 50 percent. Updated regulations established more stringent requirements for pilots flying aircraft in scheduled interstate passenger service, including flight-time limitations.

Pilots were restricted to flying 100 hours per month, 1000 hours during any 12-month period, 30 hours for any 7-day period, and 8 hours for any 24-hour period. A 24-hour rest period was also required for every 7-day period. These requirements, which were established in 1934 and are virtually the same today, upgraded earlier restrictions that limited pilots to 110 hours of flight time per month. In addition, a waiver of the 8-hour limitation for a 24-hour period could be granted by the Department of Commerce. The 8-hour waiver rule was ultimately eliminated following a fatal accident involving a pilot who had exceeded 8 hours of flight and pressure from the Air Line Pilots Association.

Other requirements specified the composition of flightcrews, established standards for flight schools, improved takeoff and landing procedures, set minimum flight altitudes and weather restrictions, and required multiengine aircraft to be capable of flying with one inoperative engine. In addition, certification of carriers providing scheduled passenger service in interstate commerce began in 1930. Although financial data were not examined by the Department of Commerce, standards for key personnel, ground organization of a carrier, maintenance procedures, and aircraft equipment and instruments had to be met.

EARLY ECONOMIC LEGISLATION.   During the 1930s, industry expansion and the development of aircraft and communication technologies required continuous improvements of regulations, airways, and airports. However, budget constraints prevented the Department of Commerce from conducting sufficient inspections and keeping up with airway development needs. Moreover, a series of fatal accidents in 1935, 1936, and 1937, including one in New Mexico that killed a New Mexico senator, called into question the adequacy of existing regulations. The fatality rate rose from 4.78 per 100 million passenger-miles in 1935 to 10.1 per 100 million passenger-miles in 1936.

The *Civil Aeronautics Act* of 1938 marked the beginning of economic regulation. It required airlines with or without mail contracts to obtain certificates authorizing service on specified routes if the routes passed a test of public convenience and necessity.

The Civil Aeronautics Act created the *Civil Aeronautics Authority* (CAA), which was responsible for safety programs and economic regulations that included route certificates, airline tariffs, and air mail rates. Within the CAA, a separate Administrator's Office, answering directly to the president, was responsible for civil airways, navigation facilities, and air traffic control. Increasing air traffic between Newark, Cleveland, and Chicago prompted a group of airlines to establish an air traffic control system in 1934. By 1936, however, the Department of Commerce assumed control of the system and issued new regulations for instrument flight.

However, in June 1940, under the Reorganization Act of 1939, the CAA was transferred back to the Department of Commerce. The *Civil Aeronautics Board* (CAB) was created. The CAB was responsible for regulatory and investigatory matters.

Federal responsibilities for airway and airport development grew tremendously during World War II, leading to passage of the *Federal Airport Act* of 1946. Federal financial assistance to states and municipalities was also initiated at this time. The federal government assumed responsibility for air traffic control (ATC). However, the inspector force could not keep pace with the rapidly increasing numbers of new airplanes, pilots, and aviation-related facilities. As early as 1940, the CAA had designated certain parts of the certification process to industry. For example, flight instructors were permitted to certificate pilots, and a certificated airplane repaired by an approved mechanic could fly for 30 days until it was checked by an available CAA inspector. After the war, the CAA limited its aircraft certification and inspection role to planes, engines, and propellers. Manufacturers were responsible for ensuring that other aircraft parts met CAA standards.

Regulatory and organizational changes also took place during and after the war. Regional offices of the CAA, reduced in number to seven in 1938, became more autonomous in 1945. Regional officials became directly responsible for operations in their regions, although technical standards and policies were still developed in Washington, D.C. Except for a brief return to more centralized management in the late 1950s, regional autonomy with the FAA has persisted to this day.

Fatal crashes in the late 1940s and early 1950s prompted revised standards setting minimum acceptable performance requirements that were designed to ensure continued safe flight and landing in the event of failure of key aircraft components. These standards also distinguished small and large airplanes based on existing airplane and power plant design considerations. Small airplanes were those with a maximum certificated takeoff weight of 12,500 pounds or less; airplanes above 12,500 pounds were defined as large. This distinction is still applied by the FAA today despite significant changes in aircraft design.

INDUSTRY GROWTH AFTER WORLD WAR II. Surplus war transport airplanes and a new supply of pilots led to the development of the nonscheduled operator, or air taxi. Exempt from economic regulation by the Civil Aeronautics Act of 1938, these operators transported people or property over short distances in small airplanes, often to locations not serviced by the certificated airlines. The CAA, at the time sympathetic to private and small operators, applied less-stringent safety regulations to air taxis. In 1952, exemption from economic regulation became permanent, even for carriers using small aircraft to provide scheduled service. The Civil Aeronautics Board adopted Economic Regulation Part 298, designating an exempt class of small air carriers known as air taxis.

The decade following World War II witnessed enormous industry growth. Pressurized aircraft traveling at greater speeds and carrying more passengers were introduced. Initially, Lockheed produced the Constellation, which carried 60 passengers and was 70 miles per hour faster than the DC-4. To compete with Lockheed, Douglas developed the DC-6. Subsequently, upgraded versions of each aircraft, the DC-7 and the Super Constellation, were introduced.

In addition to scheduled passenger service, air freight operations expanded when the CAB granted temporary certificates of public convenience and necessity to four all-cargo airlines in 1949. The four carriers were Air News, Flying Tigers, Slick, and U.S. Airlines. Certification and operating rules for commercial operators—those offering air service for compensation or hire—were also adopted in 1949.

However, despite continuing increases in air traffic and the need for better airports to accommodate larger and faster aircraft, federal support for ATC facilities, airport development, and airway modernization was insufficient. The CAA, faced with budget reductions in the early 1950s, was forced to abandon control towers in 18 small cities and numerous communications facilities, postpone jet development and navigation improvements, and curtail research efforts. The federal airport development program, championed by cities and smaller municipalities, was embroiled in controversy. In addition, the number of CAA regional offices was reduced from seven to four, 13 safety inspection field offices were eliminated, and the industry designee program was expanded.

THE FEDERAL AVIATION AGENCY. The impending introduction of jet aircraft and a 1956 midair collision over the Grand Canyon involving a DC-7 and a Super Constellation helped promote congressional authorization of increased levels of safety-related research and more federal inspectors. In 1958, Congress passed the *Federal Aviation Act,* which established a new aviation organization, the Federal Aviation Agency. Assuming many of the duties and functions of the CAA and the CAB, the agency was responsible for fostering air commerce, regulating safety, all future ATC and navigation systems, and airspace allocation and policy. The CAB was continued as a separate agency responsible for economic regulation and accident investigations. However, the Federal Aviation Agency Administrator was authorized to play an appropriate role in accident investigations. In practice, the Federal Aviation Agency routinely checked into accidents for rule violations, equipment failures, and pilot errors. Moreover, the

Civil Aeronautics Board delegated the responsibility to investigate nonfatal accidents involving fixed-wing aircraft weighing less than 12,500 pounds to the Federal Aviation Agency.

The safety provisions of the 1958 act, restating earlier aviation statutes, empowered the Agency to promote flight safety of civil aircraft commerce by prescribing

- Minimum standards for the design, materials, workmanship, construction, and performance of aircraft, aircraft engines, propellers, and appliances.
- Reasonable rules and regulations and minimum standards for inspections, servicing, and overhauls of aircraft, aircraft engines, propellers, and appliances, including equipment and facilities used for such activities. The agency was also authorized to specify the timing and manner of inspections, servicing, and overhauls and to allow qualified private persons to conduct examinations and make reports in lieu of agency officers and employees.
- Reasonable rules and regulations governing the reserve supply of aircraft, aircraft engines, propellers, appliances, and aircraft fuel and oil, including fuel and oil supplies carried in flight.
- Reasonable rules and regulations for maximum hours or periods of service of pilots and other employees of air carriers.
- Other reasonable rules, regulations, or minimum standards governing other practices, methods, and procedures necessary to provide adequately for national security and safety of air commerce.

In addition, the act explicitly provided for certification of pilots, aircraft, air carriers, air navigation facilities, flying schools, maintenance and repair facilities, and airports. In the years following creation of the agency, federal safety regulations governing training and equipment were strengthened despite intense opposition from industry organizations. The number of staff members also grew in the early 1960s, and inspection activities were stepped up, including en route pilot checks and reviews of carrier maintenance operations and organizations. The FAA staff grew from 30,000 in 1959 to 40,000 in 1961.

In 1966, the Federal Aviation Agency became the *Federal Aviation Administration (FAA)*, when it was transferred to the newly formed Department of Transportation (DOT). The *National Transportation Safety Board (NTSB)* was also established to determine and report the cause of transportation accidents and conduct special studies related to safety and accident prevention. Accident investigation responsibilities of the CAB were moved to the NTSB.

Renewed support for improvements to airports, ATC, and navigation systems was also provided by the *Airport and Airway Development Act of 1970*. The act established the Airport and Airway Trust Fund, which was financed in part by taxes imposed on airline tickets and aviation fuel. This act has been reauthorized in subsequent years.

AIRLINE DEREGULATION.   Prompted by widespread dissatisfaction with CAB policies and the belief that increased competition would enhance passenger service and reduce commercial airline fares, Congress enacted the *Airline Deregulation Act of 1978*. Congress believed that fares would drop based on the record of intrastate airlines, where fares were 50 to 70 percent of the Civil Aeronautics Board–regulated fares over the same distance. In addition, the Civil Aeronautics Board had already reduced restrictions on fare competition in 1976 and 1977 and allowed more air-lines to operate in many city-pair markets.

Specifically, in a 6-year period, the act phased out CAB control over carrier entry and exit, routes, and fares. In 1984, the remaining functions of the CAB were transferred to DOT. These functions include performing carrier fitness evaluations and issuing operating certificates, collecting and disseminating financial data on carriers, and providing consumer protection against unfair and deceptive practices.

During the 60-year history of federal oversight, federal regulatory and safety surveillance functions have been frequently reorganized and redefined. Moreover, public concerns about how the FAA carries out its basic functions have remained remarkably constant despite a steadily improving aviation safety record.

The *Airport and Airway Improvement Act of 1982* reestablished the operation of the Airport and Airway Trust Fund with a slightly revised schedule of user taxes. The act authorized a new capital grant program, called the Airport Improvement Program (AIP). In basic philosophy, the AIP was similar to the previous Airport Development Aid Program (ADAP). It was intended to support a national system of integrated airports that recognizes the role of large and small airports together in a national air transportation system. Maximized joint use of underutilized, nonstrategic U.S. military fields was also encouraged. The 1982 act also contained a provision to make funds available for noise compatibility planning and to carry out noise compatibility programs as authorized by the *Noise Abatement Act* of 1979.

The *Aviation Safety and Capacity Expansion Act of 1990* authorized a passenger facility charge (PFC) program to provide funds to finance airport-related projects that preserve or enhance safety, capacity, or security; reduce noise from an airport that is part of such a system; or furnish opportunities for enhanced competition between or among air carriers by local imposition of a charge per enplaned passenger. This act also established a Military Airport Program for current and former military airfields, which should help improve the capacity of the national transportation system by enhancement of airport and air traffic control systems in major metropolitan areas.

COMMERCIAL AVIATION DEFINED.   An *air carrier* is a commercial operator or company that has been certificated by the FAA under *FAR Part 121* or *FAR Part 135* to provide air transportation of passengers or cargo. These operators possess operations specifica-tions and an air carrier certificate, which is a document that describes the conditions, authorizations, and limitations under which the air carrier operates.

Before December 14, 1995, FAR Part 121 included the regulations that govern air carriers in multiengine airplanes with more than 30 seats or 7500-pound pay-load. As authorized by the operations specifications, FAR Part 121 operations can be *domestic air carrier* (scheduled passenger service, generally within the United States), *flag air carrier* (scheduled passenger service in international operations), or *supplemental air carrier* (all cargo and charter operations).

Until December 14, 1995, FAR Part 135 governed air carrier operations in airplanes with 30 or fewer seats and payloads of 7500 pounds or less, including both single- and multiengine aircraft and all rotorcraft. The types of FAR Part 135 operations, per the operations specifications, were *commuter air carrier* (scheduled passenger service), *air taxi operators* (on demand and all cargo and charter operations), and *split certificate* (a carrier that operates aircraft under both FAR Parts 121 and 135, dependent on aircraft size, type, and seating capacity). An industry-coined term, *regional air carrier* refers to a short-haul, scheduled carrier that services small and midsize communities, generally using turboprop and small turbojet aircraft and operating under FAR Part 121 or 135.

Many "commuters" actually operated aircraft with more than 30 seats but considerably fewer than on a jumbo jet. Although these air carriers were called commuters, they were actually already certificated under FAR Part 121. Some commuter air carriers did operate newer versions of older models of aircraft that originally had fewer than 30 seats. This fact, the existence of split certificates, as well as two sets of operating rules, engendered some confusion among the public and the media.

In the 15 years between 1980 and 1995, hours flown by commuters more than doubled, from just above 1,000,000 hours in 1980 to nearly 2,600,000 hours in 1995. For the same period, FAR Part 121 carriers hovered just above or just below 2,000,000 hours. In 1980, commuters carried 9,520,000 passengers; by the end of 1995, they carried nearly 28,000,000.

THE COMMUTER SAFETY INITIATIVE. As of December 14, 1995, all airplanes with 10 or more passenger seats and all turbojets operated in scheduled passenger service must operate under FAR Part 121. Commuter operations with 9 or fewer seats and on-demand air taxi airplanes with 30 or fewer seats and all rotorcraft still operate under FAR Part 135. To operate under FAR Part 121, the aircraft have to meet additional standards involving operational and airplane certification and equipment and performance upgrades.

The *Commuter Safety Initiative,* or the Commuter Rule as it came to be called over the year of its drafting and enactment, requires the 10-or-more-seat aircraft to comply with all FAR Part 121 operational requirements, which include the following:

- Dispatch requirements and certificated dispatchers.
- Retirement at age 60. Pilots operating under FAR Part 135 were not required to stop flying at 60. The Commuter Rule extends the rule to pilots who had been flying for former FAR Part 135 operators. Former FAR Part 135 pilots now over age 60 have 4 years of continued eligibility before they must retire.
- New flight and duty time.
- Manuals and procedures for both flight and ground personnel.

- All cabin safety and flight attendant requirements for 20- to 30-seat airplanes (for 19 or fewer seats, no flight attendant is required).

- Maintenance duty limits.

- A new training rule. The Air Carrier Training Rule increases training requirements for all pilots of scheduled passenger operations in airplanes with 10 seats or more. Training and qualifications are comparable to those for crewmembers of the larger air carriers. The new rule mandates crew resource management (CRM) training for both crewmembers and flight dispatchers.

Also issued with the Commuter Rule is *FAR Part 119*, which consolidates air carrier certification procedures, provides new definitions, and requires new management and safety officer positions for FAR Part 121 operators.

All new type certifications after March 1995 for aircraft with 10 to 19 seats must meet *FAR Part 25* transport category standards. Airplanes in production can be manufactured in the commuter category certification basis with no production time limit, but the airplanes must meet the upgraded equipment requirements. The existing fleet can continue to operate but must eventually meet upgraded equipment requirements.

With some limited exceptions for 10- to 19-seat airplanes, the Commuter Rule requires compliance with the following equipment standards:

- Exterior emergency exit markings
- First aid kits and emergency medical kits
- Wing ice lights
- Weather radar
- Protective breathing equipment
- Locking cockpit doors (20- to 30-seat aircraft only)
- Flight attendant portable and first aid oxygen
- Distance-measuring equipment
- Lavatory fire protection
- Pitot heat indication system
- Landing gear aural warning system
- Additional life rafts
- Additional flashlights

For 10- to 19-seat aircraft, some equipment was excepted if the systems were "functionally equivalent" to FAR Part 121. For example, in 10- to 19-seat aircraft, passengers are no more than 4 feet away from any exit; consequently, floor proximity lighting is not required. Because no flight attendant is required on 10- to 19-seat aircraft, the requirement for a locking cockpit door was excepted so that the crew can easily conduct the safety briefing and see to the safety of the passengers. Other exceptions included no crash ax and other cabin safety equipment and aircraft certi-fication items that would have required a redesign of the aircraft or extensive engi-neering to retrofit.

Operators of 10- to 30-seat aircraft were given 15 months to be recertified under FAR Part 121, but there are extended compliance schedules for certain items, notably installing passenger seat cushions that are not flammable, pitot heat protec-tion systems, lavatory fire protection, and a third altitude indicator. Airplanes currently in production have to be fitted with single-point inertial pilot shoulder harnesses.

Airplanes with 20 to 30 seats currently meet all FAR Part 25 and Part 121 transport category performance requirements, as do all 10- to 19-seat commuter category airplanes. For older 10- to 19-seat airplanes, there was a 15-year phase-in for certain performance requirements. Older airplanes that could not meet the FAR Part 121 performance requirements must be phased out of service at the end of that time period.

THE ENVIRONMENTAL PROTECTION AGENCY (EPA)

OVERVIEW

On December 2, 1970, the *Environmental Protection Agency (EPA)* was established in the executive branch of government as an independent agency pursuant to President Nixon's Reorganization Plan No. 3 of July 9, 1970. The EPA was created to enable coordinated and effective government action on behalf of the environment. The agency strives to abate and control pollution systematically, by integrating a variety of research, monitoring, standard-setting, and enforcement activities. To complement its other activities, EPA coordinates and supports research and antipollution activities by state and local governments, private and public groups, individuals, and educational institutions. The EPA also reinforces efforts among other federal agencies with respect to the impact of their operations on the environment. The EPA is specifically charged with making public its written comments on envi-ronmental impact assessment proposals and lobbying against proposal/projects that are unsatisfactory from the standpoint of public health, community welfare, or environmental quality. The EPA's mandate is to serve as the nation's advocate for a quality livable environment.

BACKGROUND

Before the establishment of the EPA there was no single integrated agency to system-atically address air, water, and land pollution, an issue that affected the daily lives of every person in the United States. The agencies that existed at that time were segregated primarily along pollution medium domains of air, water, and land. The sources of air, water, and land pollution, however, are interrelated and interchangeable and overlap more than one medium. For example, a single source such as a power generating station may pollute the air with smoke and chemicals, the land with solid wastes, and a river or lake with chemical and other wastes. Also, controlling air pollution may produce solid wastes, which then pollute the land or water. Conversely, treating wastewater effluents may generate solid wastes, which must be disposed of on land or burned in an incinerator, which could cause air pollution problems. Finally, some pollutants such as radiation and pesticides can appear in all three media at once. Controlling this multifaceted problem required the coordinated efforts of a variety of separate agencies and departments which often resulted in ineffective regulatory action. Thus, the need to integrate the various departments into a single entity, the EPA, was established. Per the dictates of Reorganization Plan No. 3, the following agencies' functions were consolidated in the Environmental Protection Agency:

- The National Air Pollution Control Administration which resided within the Department of Health, Education, and Welfare.
- The Department of the Interior's Federal Water Quality Administration.
- The Council on Environmental Quality, which has the authority to perform studies relating to ecological systems.
- The Food and Drug Administration's pesticides division, which was part of the Department of Health, Education, and Welfare.
- The Department of the Interior's pesticide studies' division.
- The Department of Agriculture's Research Service group dealing with pesticides registration and related activities.
- The Bureau of Solid Waste Management and the Bureau of Water Hygiene. Also included here were duties carried out by the Bureau of Radiological Health of the Environmental Control Administration, which was part of the Department of Health, Education, and Welfare.
- The Atomic Energy Commission and the Federal Radiation Council's radiation criteria and standards division.

CHRONOLOGY OF MAJOR ENVIRONMENTAL LAWS AFFECTING AVIATION OPERATIONS

NATIONAL ENVIRONMENTAL POLICY ACT (NEPA). The *National Environmental Policy Act of 1969* was significant in that it was among the first laws to establish a broad national framework for protecting the environment. NEPA's fundamental policy was to make certain that all branches of government gave careful consideration to the environment before any major federal undertaking such as airports, military complexes, highways, parkland purchases, and other federal activities that are proposed. The Act requires *Environmental Assessments (EAs)* and *Environmental Impact Statements (EISs)* on the impact of all major undertakings and alternative courses of action on the environment.

CLEAN AIR ACT (CAA). *The Clean Air Act of 1970* regulates air emissions from area, stationary, and mobile sources. By virtue of this law the Environmental Protection Agency is authorized to establish National Ambient Air Quality Standards (NAAQSs) to protect public health and the environment. The goal of the Act was to achieve NAAQSs in every state in the nation by 1975. The Act set upper limits for pollution levels and at the same time required the states to develop specific plans (also called SIPs) for controlling pollution from each industrial pollution source within the state. The act was amended in 1977 to give the states more time to comply with the NAAQS requirements. In 1990, the Clean Air Act was amended again to address special pollution problems of acid rain, ground-level ozone, stratospheric ozone depletion, and air toxics. Sample aviation operations coming under the jurisdiction of this law would include exhaust emissions of smoke from aircraft engines and venting of fuel emissions into the atmosphere.

CLEAN WATER ACT (CWA).   First established as the Federal Water Pollution Control Act Amendments of 1972 to address public concerns for controlling water pollution, this act was amended in 1977 and came to be known as the *Clean Water Act*. The Act gave the EPA the authority to implement pollution control programs such as setting wastewater standards for industry and water quality standards for surface waters. Under the Act, it was unlawful for anyone to discharge pollutants into navigable waters, unless a permit was obtained. It also established grants to fund the construction of sewage treatment plants. In 1981 the municipal construction grants process was streamlined, and treatment plants were required to have enhanced capabilities if funded under the program. In 1987 the funding strategy changed to build on the partnerships developed between EPA and the state. This caused the construction grants program to be phased out and replaced with the State Water Pollution Control Revolving Fund, more commonly known as the Clean Water State Revolving Fund.

RESOURCE CONSERVATION AND RECOVERY ACT (RCRA).   This Act by far is the most far-reaching of all and is of major importance to the aviation industry. By virtue of the RCRA of 1976, the EPA has the authority to control hazardous waste from "cradle to grave." This cradle-to-grave approach governs all phases of the waste from generation and transportation through treatment, storage, and disposal. Hazardous wastes can be solids, liquids, or contained gaseous materials that could pollute the environment. For the purposes of this regulation wastes are categorized into two groups, listed wastes and characteristics wastes. Over 400 specific substances are regulated under the "listed" category. Wastes that are ignitable, corrosive, reactive, or toxic are regulated under the "characteristics" category. Sources of hazardous wastes in aviation operations include these:

• Painting, degreasing, and cleaning of aircraft generate paint wastes, phenols, organic solvents, acids, and alkalis.
• Plating, stripping, rust prevention, and stain removal generate cyanides, chromium, and other toxic metals.

- Spills and leaks from fuel systems and storage tanks generate fuels, oils, and grease.
- Spent or leaking batteries from aircraft, air traffic control (ATC) tower backup, and other power supply sources generate toxic (lead, lithium, nickel, and cadmium) and reactive (acid) wastes.
- Miscellaneous wastes include glycol used for deicing and other detergents.

RCRA focuses only on active and future facilities and does not address abandoned or historical sites. Historical and abandoned sites are addressed by CERCLA (see below).

HAZARDOUS AND SOLID WASTE AMENDMENTS (HSWA). The federal *Hazardous and Solid Waste Amendments of 1984* added major changes to RCRA. These amendments required the phasing out of land disposal of hazardous waste, tightened restrictions on waste recycling and pollutant releases from old abandoned facilities, increased enforcement authority for EPA, and required more stringent hazardous waste management standards. HSWA, under Subtitle I of RCRA, also comprehensively addressed environmental issues of underground storage tanks (USTs) that stored petroleum and other hazardous substances. The greatest potential hazard of USTs is that their contents could leak, seep into the soil, and contaminate groundwater. Leaking USTs can cause fire and explosions in addition to posing other health and environmental risks. HSWA also addressed management of hazardous wastes in aboveground storage tanks and their transportation. In addition, RCRA references the Department of Transportation's Hazardous Materials Transportation Act (HMTA), which governs the packaging, labeling, and transportation of hazardous materials by air, water, rail, or highway. Fuel storage facilities and ramp fueling operations are sample aviation operations governed by this regulation.

TOXIC SUBSTANCES CONTROL ACT (TSCA). In 1976, the *Toxic Substances Control Act* was enacted by Congress to give the EPA the ability to track industrial chemicals produced or imported into the United States and to ban their manufacture and import if those chemicals pose an unreasonable risk. In addition, EPA incorporated the mandate and infrastructure to track the thousands of new chemicals developed by industry each year that may have unknown or dangerous characteristics. With this ability the EPA could monitor and control these chemicals as needed to protect human health and the environment. TSCA supplements the Clean Air Act and the Toxic Release Inventory under EPCRA (see below). Aircraft manufacturing and assembly processes, discharges, and effluents are controlled by this regulation.

COMPREHENSIVE ENVIRONMENTAL RESPONSE, COMPENSATION, AND LIABILITY ACT (CERCLA). CERCLA of 1980 gave to federal authority broad powers to respond directly to releases or potential releases of hazardous chemicals that posed risks to public health or the environment. The Act established requirements concerning closed and abandoned hazardous waste sites and provided for liability of individuals responsible for contaminating sites with hazardous wastes. The Act also gave EPA the power to seek out the parties that were responsible for the pollution and force them to clean up. Finally, this Act created a tax on the chemical and petroleum industries to generate funds to provide for cleanup when a responsible party could not be identified. Hence this Act is also known as the *Superfund.* This fund generated about $1.6 billion in its first 5 years of existence and was mainly used to clean up abandoned hazardous waste sites. Sample applications of this law would be to aircraft burial (retirement) sites and old (discontinued) fuel dumping/storage facilities involving underground fuel storage.

SUPERFUND AMENDMENTS AND REAUTHORIZATION ACT (SARA).   As a result of the complex administering requirements of Superfund, changes and additions were made to the program which resulted in the enactment of the *Superfund Amendments and Reauthorization Act of 1986.* Several site-specific amendments, definitions, clarifications, and technical requirements were added to the legislation, including additional enforcement authorities. The size of the fund was increased to $8.5 billion to pay for cleanup activities around the country. Some of the other salient features of the Act were that it required states to get involved in all phases of the Superfund program, stressed human health concerns posed by hazardous waste sites, and encouraged greater community involvement in the site cleanup decision-making process. Title III of SARA also authorized the Emergency Planning and Community Right-to-Know Act (EPCRA) and is discussed next. Most businesses in the United States are affected by SARA Title III.

EMERGENCY PLANNING AND COMMUNITY RIGHT-TO-KNOW ACT (EPCRA).   Also known as Title III of SARA, EPCRA was passed by Congress in 1986. This law was enacted to help local communities protect public health, safety, and the environment from chemical hazards. As required by the Act, each state was required to appoint a state emergency response commission (SERC), divide the state into emergency planning districts, and name a local emergency planning committee (LEPC) for each district. These planning committees were required to have representation from all groups that were considered essential to managing emergencies. Groups to be given due consideration include firefighters, health officials, government and media representatives, community groups, industrial facilities, and emergency managers. Aircraft rescue and fire fighting (ARFF) and airport fires unrelated to aircraft are governed by this regulation.

THE OIL POLLUTION ACT (OPA).   The *Oil Pollution Act (OPA) of 1990* was passed to strengthen EPA's ability to prevent and respond to catastrophic oil spills. A trust fund, financed by a tax on oil, was established to clean up spills when the responsible party was unable or unwilling to do so. Oil storage facilities and transport vessel operators are required to submit detailed plans to the federal government on how they will manage large-scale inadvertent discharges. The EPA also has specific requirements for aboveground storage facilities. This regulation applies to aviation fueling and storage operations.

THE NOISE CONTROL ACT.   By virtue of the *Noise Control Act of 1972,* which created the Office of Noise Abatement and Control (ONAC), the EPA was required to submit to the FAA proposed aircraft noise control regulations that promoted and protected public health and welfare. The FAA would then be required to publish the proposed regulations in a notice of proposed rulemaking and invite community comments through public hearings. The Quiet Communities Act of 1978 expanded the scope of the Noise Control Act of 1972 to include setting specific decibel limits for civil aircraft; tightening noise emission standards; coordinating federal noise abatement research; working with industry, state, and local regulators to develop consensus standards; and sponsoring research on the effects of noise and abatement strategies. The Act also established a nationwide Quiet Communities Program. The Aviation Safety and Noise Abatement Act of 1979 authorized the FAA, under the Airport Improvement Program (AIP), to award noise mitigation and control grants to state and local governments. In 1981, however, Congress, while keeping the Noise Control Act intact, ceased funding for the ONAC. As a result the EPA was unable to stay on top of its noise abatement and control regulatory effort despite the advancement of relevant science and technology that demonstrated a better understanding on how noise affected people. This led to outdated EPA emission and labeling standards. The Federal Aviation Reauthorization Act of 1996 required the FAA to appoint an aviation noise ombudsman to serve as a liaison with the public on issues regarding aircraft noise and to be consulted when the FAA administrator proposed changes in aircraft routes that included flying over populated areas.

Legislative interest in noise abatement and control of aircraft noise resurfaced in the 106th Congress through the introduction of several bills. Three bills of particular interest included the mitigation of noise levels from aircraft flights over national parks, the banning of commercial operation of certain types of supersonic aircraft due to the high noise levels, and the reduction of noise resulting from take-offs and landings at airports in metropolitan areas. The Reform Act for the 21st Century proposed an increase in the amount of funding set aside for noise mitigation from 31 to 34 percent. Public Law 106-81 set aside nearly $207 million for fiscal year 2000 for AIP noise mitigation. Among other things, this law directed the FAA to develop more stringent aircraft noise standards and develop (with the National Parks Service) air tour management plans for national parks and monuments, to minimize the effects of aircraft noise on the natural environment. In addition to the above legislation, a bill has been introduced to reestablish EPA's Office of Noise Abatement and Control and support its activities with an annual funding of $21 million for fiscal years 2000 through 2004. The important functions of this reestablished office would be to provide states and local communities with technical assistance and grants to develop noise control programs and to fund research on the health effects of noise on humans.

OCCUPATIONAL SAFETY AND HEALTH ADMINISTRATION (OSHA)

OVERVIEW

The *Occupational Safety and Health Administration Act (OSHA Act)* was signed into law by President Nixon on January 29, 1970 (took effect April 28, 1971), to ensure safe and healthful working conditions for working men and women. This was to be accomplished by authorizing enforcement of the standards developed under the Act, assisting and encouraging the states in their efforts to ensure safe and healthful working conditions, and providing for research, information, education, and training in the field of occupational safety and health. The Act established the Occupational Safety and Health Administration, the regulatory body that promulgates and enforces safety and health regulations in the United States. The Act also created the *National Institute for Occupational Safety and Health (NIOSH)* and the *Occupational Safety and Health Review Commission (OSHRC)*. NIOSH is responsible for conducting research and making recommendations for the prevention of work-related disease and injury. The function of the OSHRC, which is comprised of three judges appointed by the President and confirmed by the Senate, is to resolve disputes between OSHA and industry before the issues run through the judicial system. Although NIOSH and OSHA were created by the same Act of Congress, they are two distinct agencies with separate responsibilities. NIOSH is in the U.S. Department of Health and Human Services and is a research agency. OSHA is in the U.S. Department of Labor and is responsible for creating and enforcing workplace safety and health regulations. NIOSH and OSHA often work together toward the common goal of protecting worker safety and health.

The OSHA Act covers all employers and employees who do business in the United States except workplaces already protected by other federal agencies under other federal statutes. This means that OSHA's jurisdiction does not extend into the aircraft, but would apply to all ground, ramp, and airport operations. While state and federal agencies are not covered by the Act, Order 3900.19B requires the FAA to establish and maintain an agencywide occupational safety and health program that is consistent with the OSHA Act of 1970. In a separate agreement, OSHA and FAA signed a *memorandum of understanding (MOU)* on August 7, 2000, in which they pledge to work together to improve the working conditions of flight attendants while aircraft are in operation. By virtue of this MOU, a joint FAA/OSHA team will review the application of OSHA standards to flight attendant safety and health on issues of recordkeeping, bloodborne pathogens, noise, sanitation, hazard communication, and access to employee exposure and medical records. The team will also address whistleblower protections. FAA will issue a new policy statement and request public comment on the team's recommendations. OSHA, for its part, will consult with the FAA before proposing a standard that could compromise aviation safety. Under the MOU, OSHA will continue to enforce its standards for other aviation industry employees, such as maintenance and ground support personnel, while the FAA will continue to cover the flight deck crew, pilots, and copilots.

BACKGROUND

Before OSHA was established, the responsibility for occupational safety and health rested mainly with individual states. Regulations and enforcement were sporadic, varied by state, and were ineffective for the most part. The few federal laws that did exist at that time were limited in scope and application. For example, the Walsh-Healey

Public Contract Act of 1936, the forerunner of the OSHA Act, applied only to federal contracts in excess of $10,000. The industrial sector was rampant with exposures ranging from exposed equipment and unguarded moving parts that maimed or killed workers to electrocutions and blocked/locked exits that trapped individuals during emergencies. Most notable of all was New York City's Triangle Factory fire of 1911 that killed 146 textile workers as they were prevented from escaping because of exits that had been locked to prevent theft. The Pittsburgh Survey of 1907, a one-year study of industrial accidents in Allegheny County, Pennsylvania, found that there were almost 2 deaths for each day of the entire year of the study. The above are two of the many commonly cited examples that mobilized the nation to call for federal intervention to protect citizens in the workplace. Common law doctrine and lack of any worker's compensation laws left few incentives for employers to improve working conditions as survivors and/or their families had the burden of carrying medical costs and losses due to wages. As a result of this failed system, the annual occupational death rate leading up to the Act reached 14,300 with disabling injuries totaling 2.2 million.

CHRONOLOGY OF MAJOR OSHA STANDARDS AFFECTING AVIATION OPERATIONS

According to OSHA's 30-year milestone history, employment in the United States has approximately doubled over the 30 years since the Act went into effect in 1971. Workplace employment increased from 56 million workers at 3.5 million work sites to 105 million workers at nearly 6.9 million sites. Despite this increase, workplace fatalities have declined by 50 percent and occupational injury and illness rates have been reduced by about 40 percent. OSHA's regulatory development and enforcement process has been credited with this feat. What follows is an aviation-related subset of OSHA's regulations that were promulgated over the past 30 years to improve working conditions for employees in the United States.

- *May 29, 1971.* Comprehensive standards were first adopted to provide a baseline for safety and health protection in occupational environments. A majority of standards were incorporated from other standard-setting organizations such as the American National Standards Institute (ANSI), National Fire Protection Agency (NFPA), and American Conference of Governmental Industrial Hygienists (ACGIH). All private sector employees are covered by the OSHA Act. Sample aviation operations that are regulated include aircraft manufacturing and assembly, hangar and other maintenance shop operations, painting and stripping, ramp and flight line operations, baggage handling, cleaning crew activities, and airport operations.

- *November 14, 1978.* The lead standard was introduced to protect workers from occupational exposure to lead, an element that is known to cause damage to human nervous, urinary, and reproductive systems. It was estimated that about 835,000 workers were exposed to lead in its various forms and that this standard would reduce lead exposure risk by 75 percent. Aviation applications of this standard include battery maintenance and aircraft painting and stripping.

- *May 23, 1980.* By virtue of the medical and exposure records standard being finalized, workers and OSHA would be able to access employer-maintained medical and toxic exposure records. Records may include, but are not limited to, personal and area exposure monitoring, baseline measurements, preplacement physicals, and postexposure testing. This regulation would apply to baseline monitoring and audiometric testing for noise exposures for ramp operators and manufacturing and assembly workers, and testing for blood and other potentially infectious material (OPIM) exposures for flight attendants and aircraft cleaning crew.

- *September 12, 1980.* The fire protection standard was updated to include specific procedures for fire brigades that were responsible for extinguishing major workplace fires. Typical applications in aviation would be aircraft manufacturing and maintenance facilities where compressed gases, flammables, and solvents are stored; where welding and brazing operations occur; and where testing and maintenance of sprinkler systems are required.

- *January 16, 1981.* The electrical standards were updated to make compliance easier and adopt a performance-based approach. These standards apply to most businesses. Applications in aviation would include aircraft manufacturing and assembly, and hangar and other maintenance shop activities.

- *November 25, 1983.* The hazard communication standard was passed to require information, training, and labeling of toxic materials handled by employers and employees. Any employee who has the potential of coming in contact with a chemical used at the workplace must be trained in its safe use, handling, and decontamination procedures, if exposed to the chemical. All chemicals must have a material safety data sheet (MSDS) on site within easy access, and all containers must be labeled with their contents. Aviation applications would include aircraft manufacturing and assembly jobs, cleaning crew tasks, and hangar and other maintenance shop activities.

- *March 6, 1989.* The hazardous waste operations and emergency response standard (HAZWOPER) was promulgated to protect 1.75 million public and private sector workers exposed to toxic wastes from spills or at hazardous waste sites. Aircraft refueling, battery maintenance and disposal, deicing operations, and manufacturing process discharges are covered by this standard.

- *September 1, 1989.* The control of hazardous energy sources standard, also known as the *lockout/tagout* (*LOTO*) standard, was promulgated to protect workers (especially during maintenance) from unexpected energizing or start-up of equipment. It was estimated that by implementing this standard 120 deaths and 50,000 injuries would be prevented annually. Individuals who maintain baggage and cargo handling equipment and manufacturing and assembly equipment are covered by this standard.

- *December 6, 1991.* The bloodborne pathogens standard was introduced to prevent occupational exposure to AIDS, hepatitis B, and other infectious diseases. Flight attendants, the members of the go-team, baggage and cargo handlers, aircraft cleanup crew, and high-exposure-potential manufacturing and assembly jobs are covered by this standard. At the very least, employees in these positions require basic training on exposure prevention to blood and OPIM.

- *January 14, 1993.* The confined spaces (and permit-required confined spaces) standard was promulgated to prevent more than 50 deaths and more than 5000 serious injuries annually. Employees who perform maintenance and fabrication work in elevators, bulkheads, and cargo holds are covered by this standard. Manufacturing facilities requiring employees to enter and work in spaces that have a limited means of entry, can engulf the employee, and are not designed for normal continuous work are covered by this standard.

- *November 14, 2000.* The ergonomics program standard was initiated to prevent a painful and debilitating category of musculoskeletal injuries that affect more than 102 million workers. These injuries develop from jobs requiring excessive repetitive motion and/or high forceful applications and/or awkward postures. Sample aviation jobs that could lead to repetitive-motion injuries are those performed by flight attendants, baggage handlers, data entry personnel, and aircraft assembly workers. This standard is still in its development as of this writing.

KEY TERMS

Air carrier

Air Mail Act of 1925

Air Commerce Act of 1926

Air taxi operators

Civil Aeronautics Act of 1938

Civil Aeronautics Authority (CAA)

Civil Aeronautics Board (CAB)

Federal Airport Act of 1946

Federal Aviation Act of 1958

Federal Aviation Administration (FAA)

Department of Transportation (DOT)

National Transportation Safety Board (NTSB)

Airport and Airway Development Act of 1970

Airline Deregulation Act of 1978

Airport and Airway Improvement Act of 1982

Noise Abatement Act of 1979

Aviation Safety and Capacity Expansion Act of 1990

Commuter Safety Initiative of 1995 (Commuter Rule)

FAR Part 25

FAR Part 119

FAR Part 121

FAR Part 135

Domestic air carrier

Flag air carrier

Supplemental air carrier

Commuter air carrier

Regional air carrier

Split certificate

National Environmental Policy Act

Clean Air Act (CAA)

Clean Water Act (CWA)

Resource Conservation and Recovery Act (RCRA)

Hazardous and Solid Waste Amendments (HSWA)

Toxic Substances Control Act (TSCA)

Comprehensive Environmental Response, Compensation, and Liability Act (Superfund)

Superfund Amendments and Reauthorization Act (SARA)

Emergency Planning and Community Right-to-Know Act (EPCRA)

Oil Pollution Act

Noise Control Act

OSHA Act

OSHA the Agency

National Institute for Occupational Safety and Health (NIOSH)

Occupational Safety and Health Review Commission (OSHRC)

Memorandum of Understanding (MOU)

REVIEW QUESTIONS

1. List three major functions of the FAA and discuss some of the activities that support these functions.

2. What was the significance of the Air Mail Act of 1925?

3. Describe the major provisions of the Air Commerce Act of 1926, and discuss the role of the aeronautics branch.

4. Distinguish between the Civil Aeronautics Authority (CAA) and the Civil Aeronautics Board (CAB).

5. Discuss some of the factors that led to the passage of the Federal Aviation Act of 1958, and identify several of the safety provisions of this Act.

6. What was the primary reason for the passage of the Airline Deregulation Act of 1978?

7. What were the important features of the Airport and Airway Improvement Act of 1982?

8. Discuss the highlights of the Airport Safety and Capacity Expansion Act of 1990.

9. Distinguish between FAR Part 121 and Part 135 air carriers.

10. What was the purpose of the Commuter Safety Initiative of 1995?

11. Identify some of the operational requirements imposed on commuter air carriers operating aircraft with 10 or more passenger seats.

12. Explain the evolution of the EPA.

13. Highlight the major environmental acts relevant to aviation, giving examples of each.

14. Explain the evolution of OSHA.

15. Discuss the major OSHA acts that are relevant to aviation, giving examples of each.

REFERENCES

Briddon, Arnold E., Ellmore A. Champie, and Peter A. Marraine. 1974. *FAA Historical Fact Book: A Chronology 1926–1971*. DOT/FAA. Washington, D.C.: U.S. Government Printing Office.

Davies, R. E. G. 1972. *Airlines of the United States Since 1914*. Washington, D.C.: Smithsonian Institution Press.

Jenkins, Darryl (ed.). 1995. ALPA's One Level of Safety (Chapter 73). *Handbook of Airline Economics*. New York: McGraw-Hill.

Komons, Nick A. 1978. *Bonfires to Beacons*. DOT/FAA. Washington, D.C.: U.S. Government Printing Office.

Krieger, G. R., and J. F. Montgomery, 1997. *Accident Prevention Manual for Business and Industry—Engineering and Technology*, 11th ed. Itasca, Ill.: National Safety Council.

Rochester, Stuart I. 1976. *Takeoff at Mid-Century: Federal Aviation Policy in the Eisenhower Years, 1953–1961*. Washington, D.C.: U.S. Government Printing Office.

U.S. Congress, Office of Technology Assessment (1988). *Safe Skies for Tomorrow: Aviation Safety in a Competitive Environment*, OTA-SET-38. Washington, DC: U.S. Government Printing Office.

Wells, Alexander T. 2003. *Air Transportation: A Management Perspective*, 5th. ed. Belmont, Calif.: Wadsworth Publishing Company.

———. 2004. *Airport Planning and Management*, 5th ed. New York: McGraw-Hill.

Wilson, John R. M. 1979. *Turbulence Aloft: The Civil Aeronautics Administration amid Wars and Rumors of Wars, 1938–1953*. DOT/FAA. Washington, D.C.: U.S. Government Printing Office.

WEB REFERENCES

http://www.epa.gov/epahome/laws.htm
http://www.epa.gov/oerrpage/superfund/programs/er/triggers/haztrigs/rqover.htm
http://www.osha.gov
http://www1.faa.gov/index.cfm/apa/1078 (FAA overview)

REGULATORY ORGANIZATION AND RULEMAKING

FAA Organization and Rulemaking
    FAA organization
    Aviation Flight Standards Service (AFS)
    FAA rulemaking
OSHA Organization Structure and Rulemaking
    OSHA organization
    OSHA rulemaking
EPA Organization Structure and Rulemaking
    EPA organization
    EPA rulemaking
Key Terms
Review Questions
References

LEARNING OBJECTIVES

After completing this chapter, you should be able to

- Recognize the organizational structure of the FAA and describe the functions of the six major functional offices.
- Describe the primary mission of the Flight Standards Service.
- Identify at least 10 functional areas that fall within the responsibility of the Flight Standards Service.
- Describe the organizational structure of Flight Standards Service.
- Discuss the responsibilities of air carriers under the FA Act.
- Summarize the types of safety inspections performed by the flight standards district offices (FSDOs).
- Explain why inspector workload has been greatly affected since airline deregulation.
- Describe the Air Transportation Oversight System (ATOS) and explain why it is a significant shift in the way in which the FAA oversees airlines.
- Explain the rulemaking process of the FAA, OSHA, and EPA.
- Recognize the organizational structure of OSHA and describe the functions of its eight directorate offices.

- Recognize the organizational structure of the EPA and describe the functions of its 12 major offices.

To carry out their diverse mandates, regulatory organizations have evolved into organizational structures that best support their missions. The FAA's organization is reviewed first together with detailed discussions on offices that are relevant to enforcement and rulemaking. This topic is followed by discussions on the organizational structure and rulemaking process of OSHA and the EPA.

FAA ORGANIZATION AND RULEMAKING

The FAA is organized into six functional offices spread over nine regions and three centers across the United States, and a headquarters structure that consists of nine staff offices. An overview of FAA's organizational structure is outlined, followed by more detailed discussions on office functions that are of relevance to the focus of this text. This section then concludes with the rulemaking process of FAA.

FAA ORGANIZATION

AIRPORTS (ARP).   The mission of this office is "to provide leadership in planning and developing a safe and efficient national airport system to satisfy the needs of aviation interests of the United States, with due consideration for economics, environmental compatibility, local proprietary rights, and safeguarding the public investment." Two major departments within this office are

- Airport Planning and Programming (APP)
- Airport Safety and Standards (AAS)

AIR TRAFFIC SERVICES (ATS).   The mission of this office is to serve the flying public by directing and coordinating efforts and assuming responsibility for ensuring that the national airspace system is used safely and efficiently. Departments within ATS include

- Air Traffic (AAT)
- Air Traffic System Requirements Service (ARS)
- Airway Facilities (AAF)
- System Capacity (ASC)
- International Flight Inspection Office (IFIO)
- Independent Operational Test and Evaluation (ATQ)

- NAS Configuration Management and Evaluation Staff (ACM)
- Runway Safety Programs
- Terminal Business Service (ATB)

CIVIL AVIATION SECURITY (ACS). This office ensures and promotes a secure and safe civil aviation system by identifying and countering aviation-related threats to U.S. citizens worldwide. The FAA's Civil Aviation Security organization has been integrated into the newly formed Transportation Security Administration (TSA) of the Department of Transportation (DOT).

COMMERCIAL SPACE TRANSPORTATION (AST). The Office of Commercial Space Transportation (AST) was established in 1984 within the Department of Transportation. In 1995 the AST was transferred to the FAA and is currently the only space-related department within the FAA. Under Title 49, U.S. Code, Subtitle IX, Sections 70101–70119, AST is given the responsibility to

- Regulate the commercial space transportation industry to ensure compliance with international obligations of the United States.
- Protect the public health and safety, of property, and national security and foreign policy interest of the United States.
- Encourage, facilitate, and promote commercial space launches by the private sector, and strengthen the overall space transportation infrastructure.
- Recommend appropriate changes in federal regulations and policies that will help support and promote space commercialization.

REGULATION AND CERTIFICATION (AVR). This office is responsible for the certification, production approval, and continued airworthiness of aircraft. It certifies pilots, mechanics, and others in safety-related positions. In addition, the office is responsible for certifying all operational and maintenance activities of U.S. civil aviation organizations, regulating civil flight operations, and certifying and maintaining safety oversight of some 7300 U.S. commercial airlines and air operators. This office employs a workforce of approximately 6000 employees in the FAA's Washington headquarters, 9 regional offices, and more than 125 field offices throughout the world. The FAA's annual regulation budget for this office is in excess of $700 million. Departments within this office include

- Accident Investigation (AAI)
- Aircraft Certification Service (AIR)
- Aviation Flight Standards Service (AFS)
- Aerospace Medicine (AAM)

- Rulemaking (ARM)
- Suspected Unapproved Parts Program Office (SUP)

RESEARCH AND ACQUISITIONS (ARA).   This office provides direction and guidance to the FAA on issues related to acquisition policies, research, system prototyping, and information resource management. Offices within this department include

- Acquisitions (ASU)
- Information Technology (ASU-500)
- Air Traffic Systems Development (AUA)
- Office of Aviation Research (AAR)
- Communications, Navigation, and Surveillance Systems (AND)
- System Architecture and Investment Analysis (ASD)
- Office of Business Management (ABZ)
- The FAA Technical Center

FAA HEADQUARTERS STAFF OFFICES

- Office of Government and Industry Affairs (AGI)
- Office of the Chief Counsel (AGC)
  Office of Dispute Resolution for Acquisition (ODRA)
  FAA Civil Penalty Adjudication
- Assistant Administrator for Financial Services (ABA)
- Office of Policy, Planning, and International Aviation (API)
  Office of Aviation Policy and Plans
  Office of Environment and Energy (AEE)
  International Aviation
- Office of Civil Rights (ACR)
- Office of Human Resource Management (AHR)
- Office of Public Affairs (APA)
- Office of System Safety (ASY)
- Office of Information Services/CIO

FAA REGIONAL OFFICES AND CENTERS

- Alaskan region (AK)
- Central region (KS, MO, IA, NE)

- Eastern region (NY, PA, NJ, MD, DE, VA, WV)
- Great Lakes region (IL, IN, MI, MN, ND, SD, OH, WI)
- New England region (CT, ME, MA, NH, RI, VT)
- Northwest mountain region (CO, MT, OR, WA)
- Southern region (GA, FL, AL, NC, SC, MS, KY, TN)
- Southwest region (NM, TX, OK, LA, AR)
- Western Pacific region (CA, AZ, HI, NV)
- Mike Monroney Aeronautical Center (OK)
- William J. Hughes Technical Center (NJ)
- Center for Management Development (FL)

Additional information on each of the offices together with their contacts is available on the FAA web site. Some of the office functions that are important to the scope and intent of this text will be discussed in greater detail throughout the book.

AVIATION FLIGHT STANDARDS SERVICE (AFS)

When the FAA was created in 1958, the Bureau of Flight Standards was established as one of five operating bureaus within the FAA. This bureau had the responsibility for most of the safety functions of the earlier Office of Aviation Safety at the Department of Commerce. In 1967, the name *Bureau of Flight Standards* was changed to *Flight Standards Service*. The director of this service reported directly to the FAA administrator. The Flight Standards Service was later assigned as one of several offices within the Office of Associate Administrator for Aviation Standards, which had been established in January 1979. In July 1979, three new offices—Flight Operations, Airworthiness, and Aviation Safety—absorbed the safety functions previously assigned to the Flight Standards Service. Most headquarters' flight standards functions were performed by the Office of Flight Operations and the maintenance division of the Office of Airworthiness. In November 1984, the Office of Aviation Safety was reassigned as a staff office reporting directly to the Office of the Administrator. In November 1986, the Office of Flight Standards was created at FAA headquarters by combining the Office of Flight Operations and the maintenance division from the Office of Airworthiness. With this change, flight standards safety responsibilities were aligned at the three flight standards organizational levels (headquarters, regional, and district offices). In 1988, the Office of Flight Standards was redesignated as the Flight Standards Service (AFS). Flight Standards Service resides within the office of Regulation and Certification (AVR).

FLIGHT STANDARDS SERVICE MISSION.   The primary mission of the Flight Standards Service is to ensure continued enhancement of flight safety. The Flight Standards Service must enhance operational safety through aggressive aviation education programs and seminars for industry and the flying public. In addition, Flight Standards Service must explore options for economic incentives and creative solutions for the improved safety compliance of operators. Specifically, the office is responsible for carrying out the following functions:

- Certification, operating methods, flight operations, and maintenance activities of U.S. air carriers and foreign air carriers operating in and over the United States
- Maintenance standards for U.S.-registered aircraft (including continued airworthiness)
- Certification and conduct of commercial, industrial, private, and general aviation operations
- Examination and certification (except medical) of airmen
- Examination and appointment of persons designated and authorized to act as representatives of the FAA administrator with respect to the certification of airmen and the maintenance of civil aircraft and products
- Use of air navigation facilities and appliances and systems used in civil aircraft; the minimum equipment capability of civil aircraft for operating in the National Airspace System (NAS) and other established environments; and the operational aspects of flight procedures, including en route and instrument approach procedures (except air traffic control)
- Approval of, and surveillance over, the aircraft maintenance programs of operators and pilot schools
- Assurance that appropriate operational considerations are accommodated with regard to aircraft maintenance policies, procedures, and practices
- Establishment of operating requirements and criteria for the use of aircraft systems
- Assurance that appropriate policies and practices and other operational considerations are accounted for in the operating limitations and information requirements in the development of airplane and rotorcraft flight manuals
- Recommendation of quantities, priorities, and locations for approach and landing navigation aids and visual aids for the National Airspace System Plan
- Issuance, amendment, and termination of rules and regulations promulgated under Title III, "Organization of Agency and Powers and Duties of Administrator," and Title VI, "Safety Regulation of Civil Aeronautics," of the FA Act that are within the purview of flight standards
- Issuance, amendment, and termination of standard instrument approach procedures, minimum en route altitudes, flight procedures, operational weather minimums, and minimum equipment requirements
- Granting or denial of exemptions from regulations and taking of final action on any request or petition for reconsideration

FUNCTIONAL ORGANIZATION OF THE FLIGHT STANDARDS SERVICE. The Flight Standards Service programs are carried out nationwide by a workforce of approximately 4500 aviation safety inspectors and support personnel. The functions of the Flight Standards Service are managed and executed through the Office of the Director (AFS-1), three staff elements, three policy divisions, one technical programs division, and one national field programs division at Washington headquarters. There are nine regional divisions with 85 flight standards district offices (FSDOs) and satellite offices located throughout the United States and its territories. The flight standards staff of the Europe, Africa, and Middle East Office (AEU) are located in Brussels, Belgium, with a district office in Frankfurt, Germany. The AEU office also reports to AFS-1.

The Office of the Director of Flight Standards Service consists of three subordinate staff organizations in addition to the director and deputy director's administrative staff. These staff organizations are the executive staff, the general aviation staff, and the Project Safe staff. They each serve as an extension of the director and assist the director in carrying out management functions for accomplishing the Flight Standards Service mission. The executive staff provides support to the director for servicewide management activities. It also is responsible for ensuring that Flight Standards Service resource requirements, such as people, automation, and facilities, are adequately identified, planned, and budgeted for through Flight Standards Service national systems. These resource requirements must be appropriately distributed to adequately meet the certification, surveillance, and enforcement workload demands generated by the industry nationwide. It is also responsible for administrative support services for Washington headquarters' managers and employees. The general aviation staff is the focal point for the aviation community at the national level concerning general-aviation affairs, accident prevention, air shows, and sport aviation.

There are three Flight Standards Service (AFS) policy divisions. These policy divisions are the air transportation division (AFS-200), the aircraft maintenance division (AFS-300), and the general-aviation and commercial division (AFS-800). These divisions are responsible for the development and interpretation of regulations, policies, and guidance for the certification, inspection, and surveillance of air operators, air agencies, and airmen. Each policy division is assigned an area of functional responsibility according to specific expertise and organizational alignment. These divisions are responsible for determining the standards to be used for the certification of air operators, air agencies, and airmen.

The technical programs division (AFS-400) provides coordination and leadership for research and development programs, all-weather programs, and human-factors programs. This division also sets national operational requirements for en route procedures and instrument approach procedures.

The field programs division (AFS-500) provides nationwide oversight and coordination for the implementation of operational programs. This division develops and publishes national program guidelines for the annual work program and executes the *National Aviation Safety Inspection Program (NASIP)*. This division also provides national standardization and guidance for the administration of large air carrier certificates. The AFS-500 division has responsibility for inspector training, including the annual call for training requirements. AFS-500 also provides national oversight for the maintenance of human resource management systems concerning inspector job performance (such as currency of job task analysis, position descriptions, and performance standards).

The regional flight standards divisions and the flight standards staff of the AEU office are responsible for managing and executing the daily operational programs of the Flight Standards Service through a system of district offices. The AFS division managers within the regions have the responsibility for all flight standards activities within their respective regions. The regional flight standards division staffs provide management support to the district offices for the execution of certification, surveillance, investigation, and enforcement functions. The AEU flight standards staff is the focal point for aviation safety activities in the European, African, and Middle Eastern areas. Most domestic regional flight standards divisions have international responsibilities for specific geographic areas outside the United States. For example, the Western Pacific Flight Standards division is responsible for Asia and the Pacific territories and countries. Divisions with international responsibilities also provide consultative and liaison services to other countries on flight safety, certification, surveillance, and enforcement.

AIR CARRIER RESPONSIBILITY FOR SAFETY.   Section 601(b) of the FA Act specifies, in part, that when prescribing standards and regulations and when issuing certificates, the FAA shall give full consideration "to the duty resting upon air carriers to perform their services with the highest possible degree of safety in the public interest." The FA Act charges the FAA with the responsibility for promulgating and enforcing adequate standards and regulations. At the same time, the FA Act recognizes that the holders of air carrier certificates have a direct responsibility for providing air transportation with the highest possible degree of safety. The meaning of Section 601(b) of the FA Act should be clearly understood. It means that this responsibility rests directly with the air carrier, irrespective of any action taken or not taken by an FAA inspector or the FAA.

Most of the day-to-day inspections, reviews, and sign-offs are performed by the manufacturers, airlines, and airports; the system depends on self-inspections, and it is simply not possible for the FAA to make every inspection on every airplane in every location around the world. This self-inspection, or "designee," concept is startling to many of the general public, but it has worked effectively for many decades. The airlines and the manufacturers have a great concern for the safety of their airplanes and operations; it is in their business interests to place a high priority on safety.

Before certification, the FAA's objective is to make a factual and legal determination that a prospective certificate holder is willing and able to fulfill its duties as set forth by the FA Act and complies with the minimum standards and regulations prescribed by the FAA. This objective continues after certification. If a certificate holder fails to comply with the minimum standards and regulations, Section 609 of the FA Act specifies that the FAA may reexamine any certificate holder or appliance. As a result of an inspection, a certificate may be amended, modified, suspended, or revoked, in whole or in part. Additionally, Section 605(b) generally provides that whenever an inspector finds that any aircraft, aircraft engine, propeller, or appliance used or intended to be used by any air carrier in air transportation is not in a condition for safe operation, the inspector shall notify the air carrier, and the product shall not be used in air transportation until the FAA finds the product has been returned to a safe condition.

The following conditions or situations could indicate that an air carrier's management is unable or unwilling to carry out its duties as set forth by the FA Act:

- Repetitive noncompliance with minimum standards and regulations
- Insufficient training programs and guidance
- Lack of concern or enthusiasm on the part of air carrier management for compliance with the FA Act and the FARs
- Lack of operational control of aircraft
- Lack of ensuring the airworthiness of aircraft
- Inaccurate record-keeping procedures

The FA Act and the FARs contain the principle that air carriers offering services to the public must be held to higher standards than the general-aviation community. Inspectors must also be aware of the private rights of citizens and air carriers. Since public safety and national security are among the FAA's highest priorities, FAA inspectors must be prepared to take action when any air carrier does not, or cannot, fulfill its duty to perform services with the highest possible degree of safety.

FAA SAFETY INSPECTION PROGRAM.   Aviation safety depends in part on the quality and thoroughness of the airlines' maintenance programs and on oversight and surveillance by safety inspectors of the FAA. Even though the frequency of maintenance-related accidents has not increased since 1978, airline deregulation has been accompanied by increasing concern that maintenance standards might have been lowered at some carriers and that pressures of the marketplace might lead to unsafe operating practices. At the same time, deregulation has increased the stress on FAA inspection programs. The existing regulatory inspection program, with its local and regional structure, does not have sufficient flexibility to adapt to a dynamic industry environment.

The 85 FSDOs nationwide handle the dual functions of safety inspection and advice for airlines. In addition to scheduled airline surveillance, the local offices are responsible for safety inspections of nonscheduled air taxis and other operations, such as flight schools, engine overhaul shops, and private pilots. An air carrier's operating certificate is held at a specific flight standards office, typically the one nearest the carrier's headquarters or primary operations or maintenance base. For each carrier, a principal inspector is assigned to operations (flights, training, and dispatch), airworthiness (maintenance), and avionics (navigation and communications equipment). For large airlines, each of the principal inspectors can have one or two assistants.

Local FSDOs conduct several types of inspections on each airline's operations and maintenance functions. Inspectors periodically conduct maintenance-base inspections, which focus on the records kept by an airline. For example, records demonstrating that an airline has complied with airworthiness directives might be inspected. Inspectors conduct shop inspections to observe maintenance procedures and carry out ramp inspections to observe the airworthiness of aircraft. A similar operations-base inspection focuses on records concerning the hours of training and checkrides given pilots and the rest periods between duty shifts given crews as required by regulations. En route inspections involve observations of actual flight operations, with the inspector riding in the jump seat in the cockpit.

Base inspections are preannounced. There is a tendency to focus on records rather than to probe deeply into the data underlying carrier records concerning maintenance, training, and flightcrew logs. Inspectors at the local level try to work with air carriers to achieve compliance when they find discrepancies. Violations and fines are viewed as a last resort.

Every major airline has a reliability program that monitors maintenance activities and looks for emerging problems. For example, most airlines monitor engine temperatures, oil consumption, and the metal content of oil. They then use these tests to determine when an individual engine needs to be overhauled or repaired. Some airlines also use statistical measures such as the number of engines requiring premature overhaul, engines that are shut down in flight, the number of mechanical discrepancies that are left outstanding at flight time, and the rate at which these discrepancies are cleared. In some companies, analysts search for adverse trends that might indicate, for example, a shop procedure that needs to be revised, both to ensure safety and to reduce maintenance expense. Some FSDOs have taken advantage of these statistical data to monitor the effectiveness of airline maintenance. In some cases, flight standards inspectors have encouraged airlines to set up or expand their statistical reliability programs.

INSPECTOR WORKLOAD.   The inspector workload has been affected greatly by airline deregulation. Every time a new airline has been formed, an airline has placed a new aircraft type into service, or two airlines have merged, flight standards have been obliged to devote resources to certificating the new or changed air carrier. Certification involves page-by-page approval of the airline's operations and maintenance manuals, checkrides for the airline's senior pilots, and final proving runs for the operation as a whole.

Certification is an activity that competes for inspectors' time with the entire safety inspection program. Because it is a potential bottleneck in the establishment of new and changing airlines, requests for certification divert resources from other surveillance activity and usually bring pressure on the FAA to address the certification on a priority basis. After all certification requests have been fulfilled, any remaining inspector time is devoted to surveillance. The surveillance program consequently experiences curtailment and fluctuations as a result of this operating philosophy.

The situation has improved in recent years. However, the large number of airline mergers during the mid-1980s has affected flight standards personnel in a unique way. When airlines merge, the acquired carrier often is kept intact as an operating entity. This merger is temporarily convenient for the merged company due to differences in aircraft types, crew training, and maintenance programs.

Consequently, the FAA inspection force assigned to the acquiring carrier might become responsible for the operations of a carrier that is much different and is located in another region.

Because principal inspectors can be far away from their newly assigned airlines, there is a far greater dependence on inspections by the local offices that happen to be nearby. Inspections performed by one local office on another office's carrier are called *geographic* inspections by the FAA. Although the FAA is taking steps to remedy the situation, at present the geographic inspections do not count in an office's work-force staffing standards. Offices that have fewer assigned certificates but greater geographic responsibilities tend to be understaffed.

A localized problem exists in parts of the country where the cost of living is high. For example, the Los Angeles FSDO has trouble attracting inspectors to move to or stay in the Los Angeles area because living there requires either a very high housing cost or a very long commute, which has resulted in a high turnover rate in some major metropolitan areas. Although this has been a continual problem over the years, a differential pay scale has been initiated to alleviate this situation.

REEXAMINATION OF AIR CARRIERS.   When one carrier is merged with another, Aviation Flight Standards Service local office acquires responsibility for the subsidiary airline, which is for that office a new air carrier certificate, operations program, and maintenance program. However, the airline and its certificates and programs are not new—they have been approved by and have been in operation under supervision of a different local office prior to the merger. In some cases, the new local office might be concerned that the merged operation for which it is responsible is not ready to operate safely in its new configuration. Yet the new local office might feel that it cannot reinspect or reopen the certification of business units that have already been approved but that might not meet its own standards in their present combination.

The ability of airlines to move their operations to different geographic areas is one of the prime contributors to airline competitiveness and efficiency. However, the mobility of airlines has left FAA surveillance behind in certain cases. For example, in one case an airline maintains business offices in one city, while its principal inspectors are relocated to a nearby flight standards local office. However, the airline no longer has flight operations within hundreds of miles of that city. This case can lead to reduced efficiency in the surveillance program and undue dependence on geographic surveillance by other local offices.

PUBLIC COMPLAINTS.   When the public complains to the FAA about an air carrier, every complaint must receive a complete investigation and response by a Flight Standards Service inspector. In many cases, the complaints are about small items that take an inordinate amount of time to research, yet pose no safety hazard. The FAA estimated that only 1 of 100 public complaints leads to discovery of a violation. The DOT Secretary's consumer hot line has led to increases in public complaints received by the FAA. The hot line performs an important public service, but many of the individual complaints that are generated can divert inspectors' attention from other problems.

FAA RULEMAKING

General rulemaking procedures followed by the FAA are explained in Part 11 of the FARs. These procedures require that a public docket be established and maintained as official FAA records of each rulemaking action. Certain rulemaking responsibilities have been delegated to FAA regional directors. For example, responsibility for processing aircraft and engine regulatory proposals and final rules is delegated to certification directorates. However, it is important to remember that the administrator is the final authority with respect to all aviation safety rulemaking actions.

To fulfill the FAA's regulatory responsibility, the administrator gives full consideration to the obligation of air carriers to perform their services with the highest degree of safety in the public interest. The administrator also considers any differences between air transportation and air commerce. Safety standards, rules, regulations, and certificates that recognize those differences are prescribed and revised from time to time by the FAA. For example, the regulatory requirements for issuance of a private pilot certificate are less stringent than the requirements established for the issuance of an airline transport pilot certificate. Accordingly, privileges of the private pilot certificate are limited compared to those granted to a holder of an airline transport pilot certificate.

Prior to deregulation, the FAA had considerable regulatory autonomy, overseeing an industry in which profits were protected through the extensive rate and entry rules of the CAB. Over the past two decades, vigorous industry economic competition has made rulemaking a distinctly adversarial process. Carriers, labor groups, aircraft manufacturers, and general-aviation supporters carefully scrutinize every proposed safety regulation and question its efficacy and impact on costs. Often such activities, in concert with administrative policies and bureaucratic labyrinths, have effectively blocked safety regulations for years.

RULEMAKING PROCESS.   The FAA is a very decentralized regulatory organization. Four regional offices, the FAA aeronautical center, and FAA headquarters all have distinct regulatory responsibilities. The responsibilities of each are as follows:

Northeast region	Engines and propellers
Central region	Small aircraft
Northwest mountain	Transport-category aircraft
Southwest region	Helicopters
Aeronautical center	Registration and markings
Headquarters	Maintenance and operations
	Designs not delegated
	Airports
	Air traffic
	Medical security
	Environment
	Registration not delegated

Although each FAA office responsible for rulemaking (aviation standards, air traffic, etc.) develops its own process, the basic steps for all are similar. This section discusses the Aviation Flight Standards headquarters' rulemaking process.

First, Aviation Flight Standards decides which projects to address. There are three kinds of projects: *P projects,* which are instigated by public petition; *I projects,* for which 80 hours of study is authorized to determine a need to pursue rulemaking; and *R projects,* which are approved for work as time permits.

Individual offices select their highest priorities and submit them to the Aviation Flight Standards regulatory review board. The associate administrator for Aviation Flight Standards chairs the board, which includes the chief counsel, the director of the Office of Aviation Policy and Plans, and the directors of Aviation Flight Standards. The board meets four times per year and selects priority projects that form a top-26 list. The board controls the regulatory work schedule, reviews the status of rules, and adds or removes projects. The office developing the rule devises a work schedule that the associate administrator for Aviation Flight Standards coordinates and approves.

The Secretary of Transportation reviews programs of interest semiannually. The secretary notifies the FAA of schedules for certain projects, expresses general departmental policy issues, and identifies specific areas of concern.

The FAA, through the secretary, submits to the *Office of Management and Budget (OMB)* an annual draft regulatory program. The program covers regulatory policies, goals, and objectives and also provides information concerning significant regulatory actions or actions that might lead to rulemaking. The OMB reviews the program to ensure that all proposed regulatory actions are consistent with administration principles.

After a rule is cleared to proceed, the appropriate program office organizes a team to manage each rule. The team consists of one technician, one writer/editor, one attorney, one economist, and other interested parties. The team agrees on a good first draft of the *Notice of Proposed Rulemaking* (NPRM), which is submitted to the Office of Aviation Policy and Plans for regulatory impact analysis when appropriate. An impact analysis includes potential costs and benefits imposed on society, lower-cost approaches that were not chosen and why, and an explanation of any legal preclusions from cost/benefit criteria. When the team receives the regulatory impact analysis, it develops a new draft and briefs the principals (associate administrators, etc.) of interest. Following the principals' briefing, the team coordinates the package at the branch and division levels. After the branches and divisions concur, the package is coordinated with the principals. Next, the Associate Administrator for Aviation Standards reviews the package and forwards it to the chief counsel, who prepares the package for Office of the Secretary of Transportation and OMB review.

After the FAA review process is complete, the FAA chief counsel submits the NPRM package to the general counsel of the Department of Transportation (DOT). The DOT general counsel coordinates the package within the department. The Assistant Secretary for Policy and International Affairs, the Assistant Secretary for Budget and Programs, the Assistant Secretary for Governmental Affairs, the Assistant Secretary for Public Affairs, the general counsel, and the Safety Review Task Force review most FAA NPRMs. Other offices within the secretary's office, such as the Office of Civil Rights, the Office of Small and Disadvantaged Business, the Office of

Commercial Space Transportation, and other Department of Transportation model offices, review particular issues as appropriate. For example, the National Highway Traffic Safety Administration reviewed the FAA regulation setting standards for child restraint seats in aircraft. The assistant secretaries often coordinate the review within their own offices. For example, the Assistant Secretary for Policy and International Affairs sometimes asks five or six offices within the policy office to review a single rule.

The standard scheduled review period for the DOT offices is 7 days for a significant rule and 24 hours for other rules. After the DOT offices submit comments or concurrences, the general counsel forwards the package to the deputy secretary and the Secretary of Transportation for review. After they sign off, the package is forwarded to the OMB.

The OMB coordinates the package with the appropriate offices, which usually include the Office of Information and Regulatory Affairs and the Office of Economics and Government. The OMB review period is 60 days for major NPRMs (30 days for major final rules), but the OMB may request an extension if not prohibited by law. For nonmajor NPRMs, the OMB review period is 10 days plus extensions. The NPRM is published in the *Federal Register* after it is cleared by the OMB.

If anyone at any stage in the process objects, the package is reworked until the problem is resolved. At that point, the FAA recoordinates the package with the appropriate people. This process continues until everyone concurs or at least until all comments are considered.

After the Notice of Proposed Rulemaking is published in the *Federal Register*, the public comments on the rule for a set period. At the close of the comment period, the FAA compiles the comments and considers appropriate changes in the rule. It develops a draft of the final rule and the whole process starts again. The final rule is reviewed by the FAA, the Office of the Secretary of Transportation (OST), and the OMB. After the rule clears the process again, the OMB releases it to the FAA. The administrator issues the final rule and has it published in the *Federal Register*. Once a regulation is completed and has been printed in the *Federal Register* as a final rule, it is "codified" by being published in the Code of Federal Regulations (CFR). The CFR is the official record of all regulations created by the federal government. These regulations are organized into 50 topics called titles or volumes. FAA regulations are found in Title 49 of the CFR.

From petition for rulemaking to publication of the final rule, the FAA Aviation Standards headquarters rulemaking process includes 217 steps. Fifty-six of the 217 steps (26 percent) involve the Department of Transportation or the Office of Management and Budget, although some rules skip many of these steps and emergency rules skip most of these steps. If anyone disagrees on aspects of the rule, some of the steps are repeated. FAA-initiated rules skip some steps unique to the petition process. These petition steps are all within the FAA.

Most analysts of the FAA rulemaking process agree that the area most in need of improvement is that of timeliness in identifying and responding to safety issues. Critics complain that excessive delays in the rulemaking process complicate and delay new or amended certification programs and make them more costly. They complain of excessive rewriting and divergent rules for different aircraft categories when such rules should be identical. Some critics blame the Office of the Secretary of Transportation and the Office of Management and Budget for much of the delay.

Some critics complain that the OST and the OMB lack the necessary technical background to understand and review FAA rules. Others suggest that rulemaking quality and timeliness would improve if one person were accountable for the whole process. On the other hand, some argue that the OST reduces the delay over the long term by helping the FAA deal with the OMB. Most disagreements between the FAA and either the OST or the OMB concern economic analysis. OST officials often question the quality of the cost/benefit analysis coming from the FAA. FAA officials maintain that lack of understanding hinders the quality of those evaluations.

The actual delay from involvement by three different parties in this process is difficult to determine. Sometimes the parties negotiate while the rule sits in a particular office. Sometimes the OST and the FAA work together to build consensus when they develop the rule. Other times, the OST or the OMB sends a rule back to the FAA for further work. The OST or the OMB might say the responsibility for the delay lies with the FAA because the proposal was incomplete or flawed. The FAA might place blame for the same delay on the OST or the OMB because the FAA thought the package was acceptable and needed no additional analysis.

The problem of rulemaking delay seems to lie with the whole process, not just the OST, the OMB, or the FAA. In any case, most analysts agree that the rulemaking process needs to be streamlined. The essential steps in developing sound public safety policy must be identified, and the rest eliminated.

OSHA ORGANIZATION STRUCTURE AND RULEMAKING

OSHA has 8 directorate offices and implements its enforcement and assistance activities through 10 regions across the country.

OSHA ORGANIZATION

ASSISTANT SECRETARY (**OSHA**).   This is the Office of the Assistant Secretary of Labor, which consists of the following offices:

- Office of Equal Employment Opportunity (EEOC)
- Public Affairs Office

DIRECTORATE OF ADMINISTRATIVE PROGRAMS.   This division houses the

- Office of Human Resources
- Office of Administrative Services

- Office of Management Systems and Organization
- Office of Program Budgeting, Planning, and Financial Management

DIRECTORATE OF CONSTRUCTION.   This directorate serves as OSHA's principal source for standards, regulations, policy, programs, and assistance to OSHA, other federal agencies, the construction industry, and the general public with respect to occupational safety and health. Offices within this directorate include

- Office of Construction Services
- Office of Construction Standards and Guidance
- Office of Engineering Services

DIRECTORATE OF ENFORCEMENT PROGRAMS.   This is the compliance enforcement directorate of OSHA. OSHA inspectors enforce employer compliance to the OSHA code of federal regulations and other safety regulations through site audits and inspections and assess citations and fines where appropriate. Offices within this directorate include

- Office of Investigative Assistance
- Office of Maritime Enforcement
- Office of Health Enforcement
- Office of General Industry Enforcement
- Office of Federal Agency Programs

DIRECTORATE OF COOPERATIVE AND STATE PROGRAMS.   This directorate carries out compliance assistance, training, education programs, and outreach services and alliances that are designed to reach out to the business communities. Offices within this directorate include

- Office of Small Business Assistance
- Office of Partnership and Recognition
- Office of Outreach Services and Alliances
- Office of State Programs
- Office of Training and Education

DIRECTORATE OF STANDARDS AND GUIDANCE.   This directorate develops and promulgates standards. In addition this office plans, develops, and manages nonregulatory approaches to safety and health guidelines which supplement the agency's rule-making efforts. Offices within this directorate include

- Office of Physical Hazards
- Office of Chemical Hazards (metal)
- Office of Chemical Hazard (nonmetal)
- Office of Biological Hazards
- Office of Maritime Enforcement
- Office of Safety Systems
- Office of Engineering Safety

DIRECTORATE OF INFORMATION TECHNOLOGY (DIT).   The only office of significance within this division is the Office of Management Data Systems.

DIRECTORATE OF EVALUATION AND ANALYSIS.   This directorate monitors and measures the agency's performance to track its progress against strategic objectives. In addition, customer service–related functions and on-site audits of regional and area office operations are also conducted by this office. Offices within this directorate include

- Office of Program Review
- Office of Evaluations and Audit Analysis
- Office of Statistical Analysis
- Office of Regulatory Analysis

DIRECTORATE OF SCIENCE, TECHNOLOGY, AND MEDICINE.   The mission of the directorate is to provide field and staff support to the rest of OSHA by providing specialized technical expertise and advice in occupational safety and health. Some of these areas include industrial hygiene, ergonomics, occupational health nursing and medicine, and safety engineering along with sample analysis and equipment calibration and repair. Offices within this directorate include

- Cincinnati Technical Center
- Office of Ergonomic Support
- Office of Occupational Health Nursing
- Office of Occupational Medicine
- Office of Science and Technology Assessment
- Salt Lake Technical Center

- Technical Data Center
- Office of Technical Programs and Coordination Activities

OSHA REGIONAL OFFICES

- OSHA region 1 (CT, ME, NH, RI, VT)
- OSHA region 2 (NJ, NY, PR, VI)
- OSHA region 3 (DE, DC, MD, PA, VA, WV)
- OSHA region 4 (AL, FL, GA, KY, MS, NC, SC, TN)
- OSHA region 5 (IL, IN, MI, MN, OH, WI)
- OSHA region 6 (AR, LA, NM, OK, TX)
- OSHA region 7 (IA, KS, MO, NE)
- OSHA region 8 (CO, MT, ND, SD, UT, WY)
- OSHA region 9 (AZ, CA, HI, NV)
- OSHA region 10 (AK, ID, OR, WA)

OSHA RULEMAKING

OSHA can initiate new standards on its own or if petitioned by other parties of relevance such as the Secretary of Health and Human Services (HHS), the *National Institute for Occupational Safety and Health (NIOSH)*, state and local governments, any nationally recognized standards-producing organization (e.g., National Fire Protection Association, or NFPA), employer or labor representatives, or any other interested individual or citizen groups.

ADVISORY COMMITTEES AND **NIOSH.** If it is determined by OSHA that a specific standard is warranted, one or more of several advisory committees may be asked to work on specific recommendations. OSHA has two standing committees at its disposal, and in addition to these an ad hoc committee may be appointed to examine special concerns. All advisory committees, standing or ad hoc, are required to have representation from management, labor, state agencies, and one or more designees of the Secretary of Health and Human Services (HHS). The two standing advisory committees are

- *National Advisory Committee on Occupational Safety and Health (NACOSH)*, which advises and makes recommendations to the Secretary of Labor on specifics pertaining to the administration of the Act and the Secretary of HHS
- *Advisory Committee on Construction Safety and Health (ACCSH)*, which advises the Secretary of Labor on the development of health and safety standards for the construction industry

NIOSH, the research arm of OSHA, may also make recommendations to modify and/or promulgate standards. This agency conducts research and provides technical assistance to OSHA and industry on various safety and health hazards. During the course of its research, NIOSH may conduct workplace investigations, gather testimony from employers and employees, and require that employers measure and report employee exposure to potentially hazardous materials. NIOSH also may require employers to provide medical examinations and tests to determine the incidence of occupational illness among employees.

STANDARDS ADOPTION. OSHA publishes its plans to recommend, modify, or reject standards in the *Federal Register* as a Notice of Proposed Rulemaking. Sometimes the earlier option of Advance Notice of Proposed Rulemaking is used when information is needed to draft a proposal. Before publishing any major regulation, OSHA is required to consult with the Office of Management and Budget as mandated by Executive Order. OSHA is also required by the Small Business Regulatory Enforcement and Fairness Act (SBREFA) to consult with small businesses on regulations that significantly affect them.

The Notice of Proposed Rulemaking will include the details of the new rule together with a time frame of at least 30 days (usually 60 days or more) from the date of publication for the public to respond. Concerned parties may submit written comments together with supporting documentation of their concerns. The citizenry also has a right to request a public hearing on the proposal if OSHA has not proposed one. If requested, OSHA must schedule a public hearing and publish the time and place of its occurrence, in advance, in the *Federal Register*. After the close of the comment and public hearing period, the full text of the regulation and the date it becomes effective must be published in the *Federal Register*. OSHA must also include the reasons for implementing the regulation together with the preamble that led up to this regulation and a cost/benefit analysis of the regulation. Once this is done, the regulation is coded and placed in the Code of Federal Regulation. Alternatively, OSHA may decide after due consideration that a standard is not warranted and publish the reasons for such a determination. OSHA Safety and Health Standards for General Industry are placed in 29 CFR Part 1910, and the Construction Standards are located in 29 CFR Part 1926.

EPA ORGANIZATION STRUCTURE AND RULEMAKING

EPA has 12 major offices and implements its regulatory authority through 10 regions across the country.

EPA ORGANIZATION

OFFICE OF THE ADMINISTRATOR. This office provides overall direction and supervision of the Agency and reports directly to the President of the United States. Divisions within the Office of the Administrator include Administrative Law Judges, Children's Health Protection, Civil Rights, Communications, Education and Media Relations, Congressional and Intergovernmental Relations, Cooperative Environmental Management, Regional Operations, Small and Disadvantaged Business Utilization, Science Advisory Board, Pollution Prevention Policy Staff, Environmental Appeals Board, and Office of Policy, Economics and Innovation.

OFFICE OF ADMINISTRATION AND RESOURCES MANAGEMENT.   This office is responsible for policy and procedures governing administrative services, environmental health, and safety. In addition, this office handles facilities services, information resources management, automated data processing systems, grants management, human resources management, organizational and management analyses, and procurement and contracts management.

OFFICE OF AIR AND RADIATION.   This office is responsible for the air and radiation protection agenda of the Agency, including establishing a technical vision and direction. Sample activities include developing national programs, technical policies, and regulations for air pollution control; developing national standards for air quality; setting emission standards for stationary and hazardous pollutant sources; and providing field training for air pollution control activities. The office also provides technical assistance to states and relevant agencies on surveillance, measurement, and inspection programs for radiation levels in the environment. In addition, this office conducts comprehensive research on acid rain, studies the impact of occupational exposure to pollutants, and analyzes the global effects of pollution.

OFFICE OF ENFORCEMENT AND COMPLIANCE ASSURANCE.   This office is the regulatory enforcement wing of the EPA and advises the EPA Administrator on matters concerning administrative, civil, and criminal enforcement; compliance assurance; environmental-equity efforts; and compliance monitoring and assurance activities. There is also a strong focus on encouraging pollution prevention practices within communities.

CHIEF FINANCIAL OFFICER.   This office develops, manages, and coordinates EPA's strategic planning and goals-based management system. In addition, it manages the budget function (including payroll and disbursements), provides resource and financial management services, and manages agencywide internal controls and audit resolution.

OFFICE OF ENVIRONMENTAL INFORMATION.   This office serves as a strategic resource and is responsible for advancing the creation, management, and use of environmental information for protecting the environment by improving decision making, documenting performance, and quantifying success. In consultation with all relevant stakeholders it recommends and monitors policies that address the concerns of local, state, and federal governments including tribes, regulated communities, and other interest groups.

OFFICE OF INSPECTOR GENERAL.   This office audits and investigates EPA programs and operations. It is also responsible for detecting and preventing fraud and coordinating and recommending policies and procedures for the Agency that are efficient and do not waste taxpayer money. This office also keeps senior EPA management and Congress appraised of any serious abuses and deficiencies encountered in EPA programs and operations.

OFFICE OF GENERAL COUNSEL.   This office provides legal services to all offices within the Agency with respect to the Agency's workings and programs. This office offers legal opinions, legal counsel, and litigation support and provides legal advice for the formulation and administration of the Agency's policies and programs.

OFFICE OF INTERNATIONAL AFFAIRS.   This office develops a direction for the Agency's involvement in international policies and programs. It positions the Agency to play a leading role in solving international environmental issues and to help influence policies that protect the global environment.

OFFICE OF PREVENTION, PESTICIDES AND TOXIC SUBSTANCES.   This office is responsible for implementing, integrating, and assessing the Agency's pollution prevention, pesticides, and toxic substances control programs, and developing programs for implementing the Pollution Prevention Act. It sets priorities for the Agency's research activities and monitors regulatory activities relating to implementing the Pollution Prevention Act and pesticides and toxic substances control programs. In addition, this office monitors and assesses EPA headquarters and regional offices' programs that address pollution prevention, and pesticides and toxic substances control programs.

OFFICE OF RESEARCH AND DEVELOPMENT.   This office is responsible for the research and development needs of the Agency and provides technical and scientific advice on agencywide technical programs. It is the principal source of scientific advice to the Agency and participates in the development of Agency policy and regulations. This office sets the direction and conduct of environmental research. Some of the topics studied include health risk assessment, engineering and technology advancements in pollution control, acid rain deposition, and monitoring systems, and quality assurance. The office is also responsible for disseminating state-of-the-art scientific and technical information relating to all aspects of pollution control including measurement, monitoring, and modeling methodologies.

OFFICE OF SOLID WASTE AND EMERGENCY RESPONSE.   This office is responsible for policy, guidance, and development of guidelines for land disposal of hazardous wastes and underground storage tanks. In addition, it provides guidance on solid waste management, analyses on the recovery of useful energy from solid waste, development and implementation of programs to respond to hazardous waste sites and spills (including oil spills), source reduction of wastes, and the Superfund program.

OFFICE OF WATER.   This office provides guidance and direction for the Agency's water quality activities including the development of national programs, technical policies, and regulations relating to drinking water, water quality, groundwater, pollution source standards, and the protection of wetlands, marine, and estuarine areas.

EPA REGIONAL OFFICES (STATES COVERED)

- Region 1 (CT, ME, MA, NH, RI, VT)
- Region 2 (NJ, NY, PR, and VI)
- Region 3 (DE, MD, PA, VA, WV, DC)
- Region 4 (AL, FL, GA, KY, MS, NC, SC, TN)
- Region 5 (IL, IN, MI, MN, OH, WI)
- Region 6 (AR, LA, NM, OK, TX)
- Region 7 (IA, KS, MO, NE)
- Region 8 (CO, MT, ND, SD, UT, WY)
- Region 9 (AZ, CA, HI, NV)
- Region 10 (AK, ID, OR, WA)

EPA RULEMAKING

The EPA rulemaking process is similar to the one described under OSHA rulemaking. Initially, an authorized agency such as the EPA decides that a regulation may be needed. The Agency then researches the need for the regulation and proposes it, if needed. The proposal is listed in the *Federal Register*, and comments are invited from members of the public. The Agency reviews all comments, makes changes where appropriate, and issues a final rule. During the standards development process, all information regarding the original proposal, requests for public comment, notices about meetings (time and place) for public discussions, and the text of the final regulation is published in the *Federal Register*. Biannually, the Agency publishes a report that documents its efforts on all the regulations it is working on or has recently finished. These are published in the *Federal Register*, usually in April and October, as the "Unified Agenda of Federal and Regulatory and Deregulatory Actions." Environmental regulations appear in Title 40 of the CFR. This title is revised every July, with one-fourth of the volumes within it updated every quarter.

KEY TERMS

Flight Standards Service (AFS)

National Aviation Safety Inspection Program (NASIP)

Supplemental inspection documents (SIDs)

P projects

I projects

R projects

Notice of Proposed Rulemaking (NPRM)

Advance Notice of Proposed Rulemaking (ANPRM)

National Institute for Occupational Safety and Health (NIOSH)

Occupational Safety and Health Review Commission (OSHRC)

Secretary of Health and Human Services (SHHS)

National Fire Protection Association (NFPA)

National Advisory Committee on Occupational Safety and Health (NACOSH)

Advisory Committee on Construction Safety and Health (ACCSH)

Office of Management and Budget (OMB)

Small Business Regulatory Enforcement and Fairness Act (SBREFA)

REVIEW QUESTIONS

1. Discuss the organizational structure of the FAA, and describe the functions of the six major functional offices.

2. What is the primary mission of the Flight Standards Service? List 10 specific functions for which the Flight Standards Service is responsible.

3. Briefly describe the organizational structure of the Flight Standards Service together with the function of the Policy and Technical Programs division, and explain the duties of inspectors located at the regional offices.

4. What does Section 601(b) of the FA Act say about an air carrier's responsibility for safety?

5. Why are most day-to-day inspections, reviews, and sign-offs performed by the airlines, not the FAA? Give examples of how an air carrier might demonstrate its inability or unwillingness to carry out its duties as set forth by the FA Act.

6. How has inspector workload been affected since airline deregulation? Why has it been difficult attracting inspectors to major metropolitan areas?

7. Explain the tendency for inspectors to focus on records during maintenance-base inspections, and discuss how public complaints are handled.

8. What is the purpose of the Air Transportation Oversight System (ATOS)? How does it differ from the way in which the FAA oversees the airlines?

9. How will the new Safety Performance Analysis System (SPAS) enhance the dissemination of safety data? What are P, I, and R projects? What is the purpose of an impact analysis after an NPRM has been submitted to the Office of Aviation Policy and Plans?

10. Discuss the FAA rulemaking process, and explain the basis for the NPRM being published in the *Federal Register*.

11. Explain the rulemaking process of OSHA and EPA.

12. Discuss the organizational structure of OSHA, and describe the functions of its eight directorate offices.

13. Discuss the organizational structure of the EPA, and describe the functions of its 12 major offices.

REFERENCES

SELECTED FAA HANDBOOKS

Federal Aviation Administration. Various dates. Selected handbooks. Washington, D.C.: Supt. of Documents, U.S. Government Printing Office.

Serial Number	Title
2100.13	*FAA Rulemaking Policies*
2150.3	*Compliance and Enforcement Program*
8040.1A	*Airworthiness Directives*
8130.2B	*Airworthiness Certification of Aircraft and Related Approvals*
8300.9	*Airworthiness Inspector's Handbook*
8410.D	*Air Carrier Inspector's Handbook Part 135*
8430.6C	*Air Carrier Operations Inspector's Handbook*
8430.17	*Air Carrier Operations Bulletins*

FAA/DOT REGULATIONS

FAA/DOT. 2000. *Flight Standards 2000 Business Plan*. Washington, D.C.: U.S. Government Printing Office, February.

Part Number	Title
Part 11	General Rulemaking Procedures
Part 13	Investigation and Enforcement Procedures
Part 21	Certification Procedures for Products and Parts
Part 23	Airworthiness Standards: Normal, Utility, Acrobatic, and Commuter Category Airplanes
Part 25	Airworthiness Standards: Transport Category Airplanes
Part 27	Airworthiness Standards: Normal Category Rotorcraft
Part 29	Airworthiness Standards: Transport Category Rotorcraft
Part 33	Airworthiness Standards: Aircraft Engines
Part 36	Noise Standards: Aircraft Type and Airworthiness Certification
Part 39	Airworthiness Directives
Part 43	Maintenance, Preventive Maintenance, Rebuilding, and Alterations
Part 121	Certification and Operations: Domestic, Flag, and Supplemental Air Carriers and Commercial Operators of Large Aircraft

Part 125	Certification and Operations: Airplanes Having a Seating Capacity of 20 or More Passengers or a Maximum Payload of 6000 Pounds or More
Part 127	Certification and Operations of Scheduled Air Carriers with Helicopters
Part 129	Operations of Foreign Air Carriers
Part 135	Air Taxi Operators and Commercial Operators

Federal Aviation regulations. Various dates. Washington, D.C.: Supt. of Documents, U.S. Government Printing Office.

National Research Council, Assembly of Engineering, Committee on Federal Aviation Administration Airworthiness Certification Procedures. 1980. *Improving Aircraft Safety: FAA Certification of Commercial Passenger Aircraft.* Washington, D.C.: National Academy of Sciences.

WEB REFERENCES

Federal Aviation Administration, http://www.faa.gov.
http://www2faa.gov/index.cfm/1038
Flight standards regions and field offices, http://www.faa.gov/avr/afs/fsdo/index.cfm.
http://www.osha.gov
http://www.osha.gov/html/RAmap.html
http://www.epa.gov
http://www.epa.gov/epahome/postal.htm

THE NATIONAL TRANSPORTATION SAFETY BOARD

LEARNING OBJECTIVES

After completing this chapter, you should be able to

- Describe the purpose of the National Transportation Safety Board (NTSB) and its organizational structure.
- List the types of aviation accidents investigated by the NTSB.
- Explain the steps involved in investigating a major commercial aviation accident.
- Discuss the composition, function, and working of the go-team, the party system, and the board of inquiry as they relate to accident investigation.

- Summarize the responsibilities of the FAA during an investigation.
- Discuss some of the other functions of the NTSB.
- Recognize the reason for the renewed national focus on air safety.

NATIONAL TRANSPORTATION SAFETY BOARD

The *National Transportation Safety Board (NTSB)* is an independent agency that determines the probable cause of transportation accidents and promotes transportation safety through the recommendation process. The NTSB also conducts safety studies, evaluates the effectiveness of other government agencies' transportation safety programs, and reviews appeals of adverse actions by the U.S. Department of Transportation (DOT) involving pilot and mariner certificates and licenses.

To help prevent accidents, the NTSB develops and issues safety recommendations to other government agencies, industry, and organizations that are in a position to improve transportation safety. These recommendations are always based on the NTSB's investigations and studies and are the focal point of its efforts to improve safety in U.S. transportation systems.

The NTSB's origins can be found in the Air Commerce Act of 1926, in which Congress charged the Department of Commerce with investigating the causes of aircraft accidents. Later that responsibility was given to the Civil Aeronautics Board's Bureau of Aviation Safety. In 1966, Congress consolidated all transportation agencies into a new Department of Transportation and established the National Transportation Safety Board as an independent agency within the department. The new board was also charged with determining the probable cause of

- Highway accidents selected in cooperation with the states
- All passenger train accidents, fatal railroad accidents, and any railroad accident involving substantial damage
- Major marine accidents, including any marine accident involving a public vessel and a nonpublic vessel
- Pipeline accidents involving a fatality or substantial property damage
- Fatalities or serious injuries caused by the release of hazardous materials

In creating the NTSB, Congress envisioned that a single agency could develop a higher level of safety than the individual modal agencies working separately. Unlike the Bureau of Safety, the NTSB was to make its recommendations for safety reforms publicly. In summary, the NTSB's mission is to determine the "probable cause" of transportation accidents and to formulate safety recommendations to improve transportation safety.

With the passage of the *Independent Safety Board Act of 1974*, Congress made the NTSB completely independent outside the DOT, because "no Federal agency can properly perform such functions unless it is totally separate and independent from any other...agency of the United States." Because the DOT is charged with both the regulation and the promotion of transportation in the United States, and accidents may suggest deficiencies in the system, the NTSB's independence is necessary for objective oversight.

The NTSB has no authority to regulate, fund, or be directly involved in the operation of any mode of transportation. Therefore, it has the ability to oversee the transportation system, conduct investigations, make recommendations from a totally objective viewpoint, and make recommendations for needed safety improvements. Its effectiveness depends on an ability to make timely and accurate determinations of the cause of accidents, along with comprehensive and well-considered safety recommendations.

The most visible portion of the NTSB involves major accident investigations. Under its accident selection criteria, the NTSB's investigative response depends primarily on

- The need for independent investigative oversight to ensure public confidence in the transportation system
- The need to concentrate on the most significant and life-threatening safety issues
- The need to maintain a database so that trends can be identified and projected

NTSB investigations include the participation of modal agencies and other parties (such as manufacturers, operators, and employee unions). Within the transportation network, each government organization has been established to fulfill a unique role. Each modal agency investigates accidents to varying degrees of depth and with different objectives. As the only federal agency whose sole purpose is promoting transportation safety, the NTSB conducts detailed, open, and thorough accident investigations that often uncover significant systemwide problems that need to be corrected to prevent future similar accidents.

Aviation is the largest of the NTSB's divisions. Under the Independent Safety Board Act of 1974, the NTSB investigates hundreds of accidents annually, including

- All accidents involving 49 Code of Federal Regulations (CFR) Parts 121 and 135 air carriers
- Accidents involving public (i.e., government) aircraft
- Foreign aircraft accidents involving U.S. airlines and/or U.S.-manufactured transport aircraft or major components
- Accidents involving air traffic control, training, midair collisions, newly certified aircraft/engines, and in-flight fire or breakup
- General aviation accidents, some of which are delegated to the Federal Aviation Administration (FAA) for fact finding (probable-cause determinations are not delegated)

In addition, based on the agency's mandate under Annex 13 to the Convention and International Civil Aviation (known as the Chicago Convention) and related international agreements, the NTSB participates to a greater or lesser degree in the investigation of commercial aviation accidents throughout the world. The NTSB enjoys a worldwide reputation. The major share of the NTSB's air safety recommendations are directed to the FAA. These recommendations have resulted in a wide range of safety improvements in areas such as pilot training, aircraft maintenance and design, air traffic control procedures, and survival equipment requirements. The NTSB is also empowered to conduct special studies of transportation problems. A special study allows the NTSB to break away from the mold of the single accident investigation to examine a safety problem from a broader perspective. In the past, for example, the NTSB has conducted special studies in weather, crashworthiness, in-flight collisions, and commuter airlines.

NTSB ORGANIZATION

The NTSB is composed of five members appointed by the President and confirmed by the Senate, two of whom are designated by the President for 2-year terms to serve as chair and vice chair. The full term of a member is 5 years. The NTSB's headquarters are in Washington, D.C. Regional offices are located in Parsippany, N.J., Miami, Chicago, Dallas/Ft. Worth, Seattle, and Los Angeles, with field offices in Washington, D.C., Atlanta, Denver, and Anchorage. An organization chart for the NTSB is shown in Fig. 3-1.

Policy at the NTSB is established by the chair, vice chair, and members of the board, and is carried out by the offices of the Managing Director, Government, Public, and Family Affairs; Safety Recommendations and Accomplishments; General Counsel; Finance; Administrative Law Judges; and Aviation Safety.

To carry out the responsibilities of the NTSB as prescribed in the Independent Safety Board Act of 1974, *board members* establish policy on transportation safety issues and problems and on NTSB goals, objectives, and operations. Board members review and approve major accident reports, as well as all safety recommendations, and decide appeals of FAA and Coast Guard certificate actions. Individual members preside over hearings and testify before congressional committees.

The *Office of the Managing Director* implements the NTSB's programs by coordinating the day-to-day operations of the staff. The office schedules and manages the NTSB's review of major reports and provides executive secretarial services to the NTSB. The *Office of Finance* works very closely with the managing director in managing NTSB funds so that they are properly controlled and spent. To accomplish this, the financial management staff prepares annual budget requests to the Office of Management and Budget (OMB) and Congress. It also evaluates program operations and conducts reviews to ensure that appropriated funds are expended in accordance with approved programs.

This office maintains an accounting system that provides accountability for expenditures and furnishes timely external and internal financial management

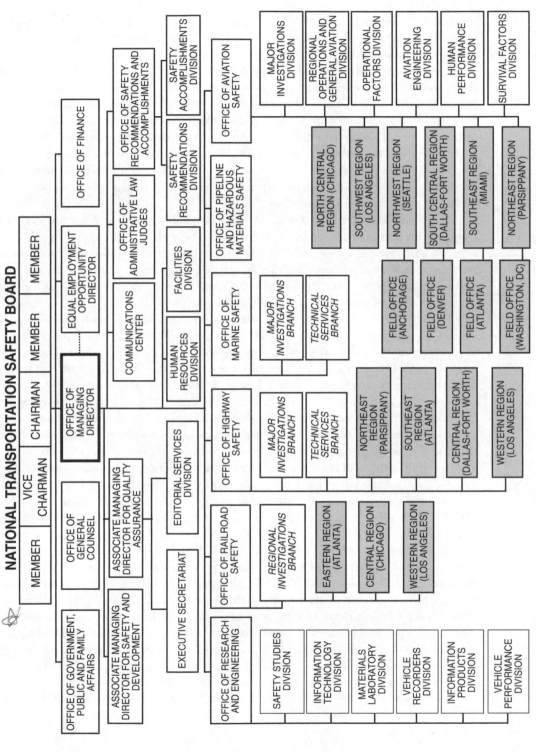

FIGURE 3-1 National Transportation Safety Board organizational chart.

reports. It audits and certifies bills and vouchers for payment and conducts audits of NTSB functions. The procurement and the contracting for needed goods and services also are handled by this office.

The *Office of Government, Public, and Family Affairs* keeps Congress and federal, state, and local government agencies informed of the NTSB's efforts to improve transportation safety. This office responds to oral and written inquiries and addresses problems and concerns raised by Congress and other government entities. It prepares testimony for NTSB participation in congressional hearings and provides information on legislation at the federal, state, and local government levels.

The NTSB's state liaison program serves as an advocate for NTSB recommendations to state and local governments and provides information and insights to the NTSB on state policies and activities.

The office also answers questions from the public, the news media, and the transportation industry. In addition, staff members work with the media at accident sites, NTSB meetings, and hearings, and they disseminate safety information to increase public awareness of the NTSB's activities in transportation safety. At major accidents, members of the NTSB conduct regular media briefings with assistance from the Office of Government, Public, and Family Affairs.

The *Office of Safety Recommendations and Accomplishments* helps to ensure that the NTSB issues appropriate and effective recommendations for enhancing safety in all transportation modes. The office develops other programs to increase the acceptance of NTSB recommendations and also coordinates the "Most Wanted" safety recommendations program, alcohol and drug policy, and international accident prevention activities. In the latter regard, the office has assisted in the establishment of the International Transportation Safety Association (ITSA), a new global organization.

The *Office of General Counsel* provides legal advice on policy, legislation, NTSB rules, and other legal matters. The office helps to ensure that the NTSB's review of airman and seaman certificate and license appeals is timely and objective and assists the Department of Justice in representing the NTSB in court proceedings. The general counsel's office also provides legal assistance and guidance to the NTSB's other offices regarding hearings, appearances as witnesses, and the taking of depositions.

By necessity, much of the NTSB's work deals with the inanimate—aircraft structures, railroad tracks, pipelines, operating rules and procedures, and so forth. However, there is one unit of the NTSB, called the *Office of Administrative Law Judges*, that most often deals directly with the individual. Basically, the role of the Office of the Administrative Law Judge is to act as an initial appeals court for persons who might have had licenses or certificates suspended, revoked, or modified by the Department of Transportation. The license holders range from pilots and aircraft mechanics to merchant seamen and flight dispatchers. But the authority of the law judge also extends beyond the individual to include the hearing appeals that might involve the loss or suspension of operating certificates issued for individual aircraft models or to airline firms.

The law judges function as trial judges, administering oaths, receiving evidence, ruling on motions, issuing subpoenas, and regulating the course of the hearing. Ninety percent of all hearings are held outside the Washington, D.C., area. Pursuant to authority in the Equal Access to Justice Act of 1980, the judges also review applications from airmen who prevail over the FAA in appeals brought under Section 609 of the Federal Aviation Act of 1958. The review of the applications for attorney fees and expenses for the most part is a determination of whether an award will be granted based on the written record of the earlier proceeding. However, the law judge assigned to the application might set the matter for informal conference or an evidentiary hearing when necessary for full and fair resolution of the issues arising from the application.

The law judge's initial decisions and orders are appealable to the full NTSB. Either party to the proceeding, the airman or the FAA, may appeal the judge's decision to the NTSB. After the NTSB has issued its opinion and order, either party may petition the NTSB for reconsideration. If a petition for reconsideration is not filed, then the NTSB's order becomes final if not appealed to the U.S. Court of Appeals. Only the airman or seaman can take an appeal to the U.S. Court of Appeals. The FAA and the U.S. Coast Guard, in the case of seamen, do not have the right of appeal to the court. On review, the court has the power to affirm, modify, or set aside the full NTSB's opinion and order, in whole or in part, and if need is found, to order further proceedings by the NTSB.

The *Human Resources Division* manages personnel and the training program to improve productivity and morale, comply with federal laws and regulations, and maintain a highly skilled and efficient workforce. The division recruits applicants for all vacant positions; processes personnel actions for all hires, promotions, and salary adjustments; and provides employees with employee services and benefit information. This division develops personnel and training programs and oversees the performance appraisal systems and incentive award program.

The *Facilities Division* manages building facilities, telephones, printing, and mail and messenger services. The division maintains the working environment and the NTSB's accountable property and coordinates printing, graphics, and photographic services. It also provides mail and messenger services and ensures adequate physical security.

The *Executive Secretarial Division* provides services to NTSB members and staff involving the review, distribution, control, processing, and record keeping for major agency documents.

The *Office of Aviation Safety* is responsible for fulfilling a number of functions. It has the primary responsibility for investigating aviation accidents and incidents and proposing probable causes for NTSB approval. Working with other NTSB offices, the office also formulates aviation safety recommendations. The staff is located in 10 regional and field offices in major metropolitan areas throughout the United States. This office is composed of six divisions: Major Investigations, Field Operations and General Aviation, Operational Factors, Human Performance, Aviation Engineering, and Survival Factors.

ACCIDENT INVESTIGATION PROCESS

When a major commercial aviation accident occurs, an NTSB go-team, led by an *investigator-in-charge (IIC)*, is dispatched from the agency's Washington, D.C., headquarters to the accident site, usually within a couple of hours of notification of the event. The IIC, a senior air safety investigator with the NTSB's Office of Aviation Safety (OAS), organizes, conducts, and manages the field phase of the investigation, regardless of whether a board member is also present on the scene. This activity includes investigating the factual circumstances of the crash (on site and afterward), preparing final reports for submission to the board members, initiating safety recommendations to prevent future accidents, and participating in foreign accident investigations. OAS also encompasses the six regional offices and four field offices that are responsible for investigating general-aviation accidents. The IIC has the responsibility and authority to supervise and coordinate all resources and activities of the field investigators. The NTSB go-team will form as many as 10 investigative groups. Discipline teams will be formed around subject matter areas, such as power plants, systems, structures, operations, air traffic control, human factors, weather, and survivability. Cockpit voice recorder and flight data recorder groups are formed at the NTSB laboratory in Washington. All NTSB staff assigned to a particular investigation are under the direction of the IIC.

PARTY PROCESS

Increasingly, the NTSB has no choice but to conduct its investigations in the glare of intense media attention and public scrutiny. As commercial air travel has become routine for millions of passengers, major accidents have come to be viewed as nothing short of national catastrophes. At the same time, an NTSB statement of cause may be nothing short of catastrophic for the airline, aircraft manufacturer, or other entity that may be deemed responsible for a mishap. A very real, albeit unintended, consequence of the NTSB's safety investigation is the assignment of fault or blame for the accident by both the courts and the media. Hundreds of millions of dollars in liability payments, as well as the international competitiveness of some of the most influential U.S. corporations, rest on the NTSB's conclusions about the cause of a major accident. This was not the system that was intended by those who supported the creation of an independent investigative authority more than 30 years ago, but it is the environment in which the investigative work of the agency is performed today.

The NTSB relies on teamwork to resolve accidents, naming "parties" to participate in the investigation that include manufacturers, operators, and, by law, the FAA. The *party system* enables the NTSB to leverage its limited resources and personnel by bringing into an investigation the technical expertise of the companies, entities (such as the pilots' union), and individuals who were involved in the accident or who might be able to provide specialized knowledge to assist in determining the probable cause. Except for the FAA, party status is a privilege, not a right. The IIC has the discretion to designate the parties that are allowed to participate in an investigation, and each party representative must work under the direction of the IIC or senior NTSB investigators at all times. No members of the news media, lawyers, or insurance personnel are permitted to participate in any phase of the investigation. Claimants or litigants (victims or family members) are also specifically prohibited from serving as party members.

The specialists that any party assigns to an investigation must be employees of the party and must possess expertise to assist the NTSB in its investigation. Providing the safety board with technical assistance gives parties many opportunities to learn what happened and to formulate theories as to the cause of the accident. Party representatives are not permitted to relay information to corporate headquarters without the consent of the IIC, and then only when necessary for accident prevention purposes. Information is not to be used for litigation preparation or for public relations. Sanctions for failing to abide by the NTSB party rules and procedures include the dismissal of individuals or even the party from the investigation team. Party representatives must sign a party pledge, a written statement agreeing to abide by the NTSB rules governing the party process.

The first 2 days following an accident are critical because the evidence is fresh and undisturbed. After people start going through the wreckage, the clues begin to disappear. An airspeed indicator's needle might be moved, or a fuel line might drain. Subtle clues are lost that could reveal possible causes of the accident. Consequently, crash sites are protected from the untrained until the go-team arrives on the scene.

THE GO-TEAM

On 24-hour alert, *go-team* personnel possess a wide range of accident investigation skills. In aviation, a go-team roster could include one of the five members of the NTSB, an air traffic control specialist, a meteorologist, a human-performance expert, an expert trained in witness interrogation, an engine specialist, as well as experts in hydraulics, electrical systems, and maintenance records. Some go-team members are completely intermodal in that their area of expertise is applicable to each mode. Human-factors experts fall into this category, as do the NTSB's metallurgists, meteorologists, and hazardous-materials experts.

Go-team duty is rotated. Immediately after one team has been dispatched, a new list is posted. Like firefighters, go-team members spend many hours doing office work and working on special studies until the inevitable call comes. The FAA usually gets the first word of an accident, then the director of the NTSB's regional or field office. This office notifies the go-team, the board member on duty, the NTSB chair, and the public affairs office. The team is normally on its way within 2 hours. Until it arrives, an investigator from the nearest NTSB field office secures the crash site with the help of local authorities. Representatives from the aircraft manufacturer, the airline, the engine manufacturer, and the FAA also arrive. If the accident is major, a member of the NTSB accompanies the team. The investigator-in-charge calls a meeting and assigns each of these individuals to a section of the go-team.

ACCIDENT SITE

The length of time a go-team remains on the accident site varies with need, but generally a team completes its work in 10 to 14 days. However, accident investigations often can require off-site engineering studies or laboratory tests that might extend the fact-finding stage. In cases of crew fatalities, a local coroner usually performs autopsies on the flightcrew to determine at the outset whether pilot incapacitation might have been a factor. An autopsy can also reveal who was sitting where in the cockpit and who was flying the aircraft.

After the preliminary steps are completed, the detailed work begins. The go-team is organized into groups of experts, each of which focuses on specific aspects of the investigation. Each group, headed by a group chairperson, concentrates on a specific portion of the investigation. Coordination is effected among group chairpersons to ensure investigative coverage in areas where more than one group may have a responsibility. Using their combined knowledge of flying in general and of this aircraft in particular, they compare what they know with what they find in the wreckage. Simple cameras are an important tool of the trade. Before the team members touch any of the wreckage, they take pictures from various angles and distances and make verbal notes into tape recorders.

Operational factors experts in three disciplines (air traffic control, operations, and weather) support major investigations with intensive work in their specialties. Air traffic control (ATC) specialists examine ATC facilities, procedures, and flight handling, including ground-to-air voice transmissions, and develop flight histories from Air Route Traffic Control Center (ARTCC) and terminal facility radar records. Other specialists examine factors involved in the flight operations of the carrier and the airport and in the flight training and experience of the flightcrew. Weather specialists examine meteorological and environmental conditions that may have caused or contributed to an accident.

Human-performance specialists examine the background and performance of persons associated with the circumstances surrounding an accident, including the person's knowledge, experience, training, physical abilities, decisions, actions, and work habits. Also examined are company policies and procedures, management relationships, equipment design and ergonomics, and the work environment.

Aviation engineering experts in four areas provide strong technical investigative skills. Power plant specialists examine the airworthiness of aircraft engines, while structures experts examine the integrity of aircraft structures and flight controls as well as the adequacy of design and certification. Systems specialists examine the airworthiness of aircraft flight controls and electrical, hydraulic, and avionic systems. And maintenance specialists examine the service history and maintenance of aircraft systems, structures, and power plants.

Survival-factors experts investigate factors that affect the survival of persons involved in accidents, including the causes of injuries and fatalities. These investigators also examine cabin safety and emergency procedures, crashworthiness, equipment design, emergency responsiveness, and airport certification.

LABORATORY

While the investigators work on site, the NTSB's materials laboratory in Washington, D.C., performs detailed analyses on items found at the site. One of the finest of its kind in the world, the laboratory is designed to support investigators in the field. For example, the laboratory has the capability to "read out" aircraft cockpit voice recorders (CVRs) and decipher flight data recorders (FDRs), which provide investigators with such key factors as airspeed, altitude, vertical acceleration, and elapsed time. These two *black boxes* provide investigators with a profile of an aircraft during the often crucial last minutes of flight.

Metallurgy is another of the laboratory's skills. NTSB metallurgists perform postaccident analysis of wreckage parts. The laboratory is capable of determining whether failures resulted from inadequate design strength, excessive loading, or deterioration in static strength through fatigue or corrosion.

The investigation of the American Airlines DC-10 that lost its left engine after takeoff from Chicago's O'Hare Airport in May 1979 probably could not have been concluded without the help of the materials lab. Preliminary investigations led metallurgists to focus on the aft bulkhead of the left engine pylon—the vertical member of the wing from which the engine is suspended. They found the overstressed area where the engine broke off. As suspected, a trail of fatigue marks also was found leading up to the overstressed area. But the real mystery turned up when the metallurgists then followed the fatigue marks to their point of origin, only to discover another overstressed area, and nothing else. The first overstress had caused the fatigue, and the fatigue had caused the final break. But what had caused the initial overstress?

The metallurgists and specialists reviewed the aircraft's maintenance records and found that when removing the engines, a maintenance crew had used a forklift to help lower the entire engine-pylon assembly. Although the crew didn't realize it at the time, the method was causing hidden damage at the points where the engine and pylon were fastened to the wing. As a result of the findings, the engine removal procedure was changed.

ACCIDENT REPORT PREPARATION

Following completion of the on-scene phase of the investigation (which may last for several days or weeks), each NTSB group chair (the senior investigator overseeing a specific area of the investigation) completes a factual report on his or her area of responsibility. The reports are likely to include proposed safety recommendations to correct deficiencies and prevent future similar accidents. All factual material is placed in the public docket that is open and available for public review. Thereafter, the investigators involved in the case begin an often lengthy period of further fact gathering, usually involving one or more public hearings, and final analysis of the factual information collected.

There is no time limit on NTSB investigative activity. Safety board procedures have a target date for completion of the final accident report within 1 year of the date of the accident, but recent major commercial aviation accident investigations have taken as little as 4 months and as much as more than 4 years.

A key milestone in the report preparation process is the group chairs' preparation of analytical reports in their respective areas of expertise. The parties may contribute to the analytical reports through their continued contact with the NTSB group chairs and the IIC, but parties are not allowed to review, edit, or comment on the analytical reports themselves. The parties also contribute to the safety board's analytical process through written submissions, which are sometimes extensive and become part of the public docket.

SAFETY RECOMMENDATION

The *safety recommendation* made to the FAA is the NTSB's end product. Nothing takes a higher priority, and nothing is more carefully evaluated. In effect, the recommendation is vital to the NTSB's basic role of accident prevention because it is the lever used to bring changes and improvements in safety to the nation's transportation system. Close to 90 percent of the recommendations made to the FAA are acted upon favorably. With human lives involved, timeliness also is an essential part of the recommendation process. As a result, the NTSB issues a safety recommendation as soon as a problem is identified without necessarily waiting until an investigation is completed and the probable cause of an accident determined. In its mandate to the NTSB, Congress clearly emphasized the importance of the safety recommendation, saying the NTSB shall "advocate meaningful responses to reduce the likelihood of recurrence of transportation accidents." Each recommendation issued by the NTSB designates the person, or the party, expected to take action, describes the action that the NTSB expects, and clearly states the safety need to be satisfied.

Recommendations are based on findings of the investigation and may address deficiencies that do not pertain directly to what is ultimately determined to be the cause of the accident. For example, in the course of its investigation of the crash landing of a DC-10 in Sioux City, Iowa, in 1989, the NTSB issued recommendations on four separate occasions before issuing its final report. In the case of the crash of an ATR-72 in Roselawn, Ind., in 1994, the NTSB issued urgent safety recommendations within 1 week of the accident. In the TWA Flight 800 investigation, once it was determined that an explosion in the center fuel tank caused the breakup of the aircraft, the NTSB issued urgent safety recommendations aimed at eliminating explosive fuel/air vapors in airliner fuel tanks.

To emphasize the importance of the safety recommendation, Congress has required the DOT to respond to each NTSB recommendation within 90 days.

PUBLIC HEARING

Following an accident, the NTSB might decide to hold a public hearing to collect added information and to discuss at a public forum the issues involved in an accident. Every effort is made to hold the hearing promptly and close to the accident site.

A hearing involves NTSB investigators, other parties to the investigation, and expert witnesses called to testify. At each hearing, a *board of inquiry* is established that is made up of senior safety board staff, chaired by the presiding NTSB member. The Board of Inquiry is assisted by a technical panel. Some of the NTSB investigators who have participated in the investigation serve on the technical panel. Depending on the topics to be addressed at the hearing, the panel often includes specialists in the areas of aircraft performance, power plants, systems, structures, operations, air traffic control, weather, survival factors, and human factors. Those involved in reading out the cockpit voice recorder and flight data recorder and in reviewing witness and maintenance records also might participate in the hearing.

Parties to the hearing are designated by the NTSB member who is the presiding officer of the hearing. They include those persons, government agencies, companies, and associations whose participation in the hearing is deemed necessary in the public interest and whose special knowledge will contribute to the development of pertinent evidence. Typically, they include the FAA, operator, airframe manufacturer, engine manufacturer, pilots' union, and any other organization that can assist the safety board in completing its record of the investigation. Except for the FAA, party status is a privilege, not a right. Parties are asked to appoint a single spokesperson for the hearing.

Expert witnesses are called to testify under oath about selected topics to assist the safety board in its investigation. The testimony is intended to expand the public record and to demonstrate to the public that a complete, open, and objective investigation is being conducted. The witnesses who are called to testify are selected because of their ability to provide the best available information on the issues related to the accident.

News media, family members, lawyers, and insurance personnel are not parties to the investigation and are not permitted to participate in the public hearings.

Following the hearing, investigators will gather additional needed information and conduct further tests identified as necessary during the hearing. After the investigation is complete and all parties have had an opportunity to review the factual record, from both the hearing and other investigative activities, a technical review meeting of all parties is convened. That meeting is held to ensure that no errors exist in the investigation and that there is agreement that all that is necessary has been done.

On rare occasions, the hearing may be reopened when significant new additional information becomes available or follow-up investigation reveals additional issues that call for an airing in a public forum such as a hearing. This was most recently done in the NTSB investigation of the September 8, 1994, accident involving USAir Flight 427 at Aliquippa, Pa., near Pittsburgh.

FINAL ACCIDENT REPORT

With the completion of the fact-finding phase, the accident investigation process enters its final stage—analysis of the factual findings. The analysis is conducted at

[Handwritten margin notes:]

other functions
- *Carry out study*
- *special investigation*
- *Assessment on aviation related issue*
- *Fulfills U.S. obligation (ICAO treaties)*
- *Noncatastrophic accident*
- *G.A accident*

the NTSB's Washington, D.C., headquarters. The final accident report includes a list of factual findings concerning the accident, analysis of those findings, recommendations to prevent a repetition of the accident, and a probable-cause statement.

The IIC and the NTSB senior staff create a final draft report, called the *notation draft,* for presentation to the board members. This draft includes safety recommendations and a finding of probable cause. Following a period for review of the draft report, a public meeting (referred to as the "sunshine meeting") of the board members is held in Washington. The NTSB staff will present and comment on the draft report; party representatives are permitted to attend but may not make any kind of presentation or comment. At this meeting, the board members may vote to adopt this draft, in its entirety, as the final accident report; may require further investigation or revisions; or may adopt the final accident report with changes that are discussed during the meeting.

Safety recommendations resulting from major investigations generally are included in the final accident report; however, in the interest of safety, they may be issued at any time during the course of an investigation if the NTSB deems it necessary.

Technically, NTSB investigations are never closed. Parties to the investigation may petition the board to reconsider and modify the findings and/or probable-cause statement if the findings are believed to be erroneous or if the party discovers new evidence. Petitions from nonparties will not be considered.

INVESTIGATING A GENERAL-AVIATION ACCIDENT

The investigation of general-aviation accidents is a simpler process requiring fewer staff members per accident. Inasmuch as the NTSB investigates many general-aviation accidents per year, abbreviated investigations are generally necessary, given the agency's limited staff and budgetary resources. Most general-aviation accident investigations are conducted by one of the NTSB's regional or field offices. In a *field investigation,* at least one investigator goes to the crash site; a *limited investigation* is carried out by correspondence or telephone. Some, but by no means all, general-aviation accidents generate safety recommendations approved by the NTSB members.

INTERNATIONAL ACCIDENT INVESTIGATIONS

The NTSB is the government agency charged with the responsibility for ensuring compliance with U.S. obligations under Annex 13 to the Chicago Convention, the international treaty that provides the structure for the governance of civil aviation throughout the world. The NTSB's international responsibilities represent a significant portion of the agency's overall aviation workload and are mounting. In the event of a civil aviation accident outside of U.S. territory, the NTSB appoints the accredited U.S. representatives to the investigation and oversees advisers from the U.S. aviation industry. The NTSB provides an objective representative to assist the authorities charged with the management of an investigation in foreign countries whether the accident involved a U.S. airline or U.S.-manufactured aircraft or components.

In many instances, the NTSB provides direct assistance to the state conducting the investigation. Depending on the sophistication of its own investigative capabilities, the state where the accident occurred might delegate all or part of its responsibilities to the NTSB. In addition, NTSB involvement enables U.S. authorities to take necessary accident prevention measures based on the findings of the investigation. The NTSB also provides needed technical support, such as the readout of cockpit voice recorders, to foreign investigators.

FAMILY ASSISTANCE AND THE OFFICE OF FAMILY AFFAIRS

Following the enactment of the *Aviation Disaster Family Assistance Act of 1996*, the President designated the NTSB as the lead federal agency for the coordination of federal government assets at the scene of a major aviation accident and as the liaison between the airline and the families. The role of the NTSB includes integrating the resources of the federal government and other organizations to support the efforts of state and local governments and the airlines to aid aviation disaster victims and their families. The NTSB's Office of Family Affairs assists in making federal resources available to local authorities and the airlines, for example, to aid in rescue and salvage operations and to coordinate the provision of family counseling, victim identification, and forensic services. The safety board has sought to maintain a distinct separation between family assistance activities and the NTSB's technical investigative staff.

FAA RESPONSIBILITIES DURING AN INVESTIGATION

Accident investigation is largely the responsibility of each FAA Flight Standards District Office (FSDO), which maintains a preaccident plan that is tailored to that office's specific requirements (e.g., geographic location, climate, staffing, and resources). The FAA works very closely with the NTSB, and the formal agreement between agencies can be found in *Order 8020.11, Aircraft Accident and Incident Notification, Investigation, and Reporting*. FAA accident investigation responsibilities include the following:

Ensuring that
- All facts and circumstances leading to the accident are recorded and evaluated.
- Actions are taken to prevent similar accidents in the future.

Determining if
- There was a violation of the Federal Aviation Regulations.
- The performance of FAA facilities or functions was a factor.
- The airworthiness of U.S.-certificated aircraft was a factor.
- The competency of U.S.-certificated aircrew, air agencies, commercial operators, air carriers, or airports was a factor.
- The Federal Aviation Regulations were adequate.

- The airport certification safety standards or operations were involved.
- The air carrier/airport security standards or operations were involved.
- Aircrew medical qualifications were involved.

The FAA conducts investigations and submits factual reports of the investigations to the NTSB on accidents delegated to the FAA by the NTSB. This delegation of certain NTSB accident investigation responsibilities is exercised under Section 304(a)(1) of the Independent Safety Board Act of 1974.

The FAA's principal investigator at an accident is called the *investigator-in-charge*. This individual directs and controls all FAA participation in the accident until the investigation is complete. Included is the authority to procure and use the services of all needed FAA personnel, facilities, equipment, and records.

The FAA investigator-in-charge is under the control and direction of the NTSB investigator-in-charge in an NTSB-conducted investigation. When accident investigations are delegated to the FAA by the NTSB, the FAA investigator-in-charge becomes an authorized representative of the NTSB. All the investigative authority prescribed in the applicable NTSB regulations falls to this person. All other FAA personnel report to the investigator-in-charge and are responsible to that person for all reports they have prepared or received during the investigation.

NTSB ACCIDENT DATABASES

The NTSB aviation accident database and synopses, which go back as far as 1962, are available online at http://www.ntsb.gov/ntsb/query.asp. These databases contain information on civil aviation accidents and selected incidents that occur within the shores of the United States, within its territories, and in international waters. Within a few days of an accident a preliminary report is normally available online. This is followed by a factual preliminary report that gets replaced with a final report outlining probable cause, when the investigation is completed. Complete information may not be available for cases under revision or where the NTSB did not have primary responsibility for investigating an accident. A summary of information available on this site includes

- Interactive search capability for the NTSB database that is updated daily. Search information is possible by

  Day, month, and year since 1962
  City and state in which the accident occurred
  Severity—incident, accident, fatal, nonfatal
  Aircraft category (airplane, helicopter, etc.)
  Type of assembly—amateur or commercial
  Aircraft make, model, and registration
  Operation (Part 121, Part 135, etc.)
  Name of airline

NTSB accident number
Word strings

- Monthly listing of accidents sorted by date that is updated daily.

- List of completed investigations scheduled for release in the next 2 weeks.

- Complete downloadable data sets for each year beginning from 1982 and updated monthly in Microsoft Access 95 MDB format.

- Complete description of Government Information Locator System (GILS) accident databases, including definition of *accident* and *incident* as well as instructions for requesting information from pre-1982 databases.

- Complete FAA incident database information about incidents, including those not investigated by NTSB.

- Lists of data, information products, and other sources of information about aviation accidents, including publications, dockets, and press releases.

OTHER NTSB FUNCTIONS

The NTSB is charged with carrying out studies, special investigations, evaluations, and assessments on issues that are aviation-related. For example, in 1993, these duties included a special report on a commercial space-launch-procedure anomaly involving the Pegasus/SCD-1. It was conducted under the agreement with the U.S. Department of Transportation, prompted by concern over safety, given the growth in commercial space launch and recovery activities.

Because of the international nature of the industry and the United States' leading role in aviation technologies, the NTSB's investigation of domestic accidents and participation in international aviation investigations are essential to the enhancement of worldwide airline safety. The NTSB fulfills the U.S. obligations for international aviation accident investigations established by an International Civil Aviation Organization (ICAO) treaty by sending an accredited representative to the investigation of major accidents where U.S. interests exist. The board is also an active participant in the International Transportation Safety Association (ITSA).

Foreign governments often request special assistance and expertise from the NTSB. The NTSB's major aviation accident reports, safety recommendations, and accident statistics are disseminated worldwide and have a direct influence on the safety policy of foreign airlines.

A less visible segment of the NTSB is the investigation of noncatastrophic and general-aviation accidents. The NTSB also investigates accidents involving property damage only, in which data are collected in a relatively limited but highly focused investigation. The investigation of selected incidents not meeting the definition of an accident can often provide preventive solutions that may help preclude similar accidents from recurring.

Although the NTSB is considered the primary source for aviation accident data, its approach goes beyond statistics or establishing probable cause. Examination of all factors that led to an accident or incident, for example, ensures that regulatory agencies are provided with a thorough and objective analysis of actual as well as potential deficiencies in the transportation system. Only then can solutions be proposed to correct the deficiencies that may have caused the accident.

KEY TERMS

National Transportation Safety Board (NTSB)

Independent Safety Board Act of 1974

Board members

Office of the Managing Director

Office of Finance

Office of Government, Public, and Family Affairs

Office of Safety Recommendations and Accomplishments

Office of General Counsel

Office of Administrative Law Judges

Office of Aviation Safety

Investigator-in-charge

Go-team

Party system

Probable cause

Black boxes

Safety recommendation

Board of Inquiry

Final accident report

Notation draft

Field investigation

Limited investigation

Aviation Disaster Family Assistance Act of 1996

Order 8020.11, Aircraft Accident and Incident Notification, Investigation, and Reporting

REVIEW QUESTIONS

1. What are the primary responsibilities of the National Transportation Safety Board (NTSB)? How did passage of the Independent Safety Board Act of 1974 affect the NTSB? Describe the types of accidents investigated by the NTSB. Describe the

organizational structure of the NTSB. What are the functions of the following offices: Safety Recommendations and Accomplishments; Administrative Law Judges; Government, Public, and Family Affairs; General Council; and Aviation Safety?

2. Explain the role of the investigator-in-charge (IIC) and the go-team. What is the so-called party system that enables the NTSB to leverage its limited resources? Identify the steps taken in a major accident investigation. What types of activities are performed at the NTSB's laboratory in Washington, D.C.? When are safety recommendations made?

3. What is the purpose of a public hearing? Are hearings ever reopened? What information is included in the final accident report? Distinguish between a field investigation and a limited investigation of a general-aviation accident.

4. Discuss the role of the NTSB in international aviation accident investigations. What is the role of the NTSB under the Aviation Disaster Family Assistance Act of 1996?

5. Describe the responsibilities of the FAA during a major accident investigation. Describe some of the functions of the NTSB besides accident investigation. What were some of the contributing factors that led to a renewed national focus on air safety?

REFERENCES

Chandler, Jerome G. 1986. *Fire & Rain*. Austin, Tex.: Texas Monthly Press.

Collins, Richard L. 1986. *Air Crashes*. New York: Macmillan Publishing Co.

Ellis, Glenn. 1984. *Air Crash Investigation of General Aviation Aircraft*. Greybull, Wyo.: Capstan Publishing Co.

Lebow, Cynthia C., et al. 1999. *Safety in the Skies*. Washington, D.C.: Institute for Civil Justice, RAND.

NTSB aircraft accident reports. Various dates. Springfield, Va.: National Technical Information Service.

NTSB Annual Reports

Ramsden, J. M. 1976. *The Safe Airline*. London, England: MacDonald and Jane's Publishers, Inc.

NTSB WEB SITES

http://www.ntsb.gov

http://www.ntsb.gov/ntsb/query.asp

RECORDING AND REPORTING OF SAFETY DATA

LEARNING OBJECTIVES

After completing this chapter, you should be able to

- Recognize the importance of studying accidents and incidents for the purpose of developing insights, information, and recommendations leading to accident prevention.

- Explain and give examples of an accident, incident, and injury as defined by the NTSB.
- List and briefly describe the characteristics of effective incident reporting systems.
- Distinguish between mandatory and voluntary reporting systems.
- Identify organizations within the FAA that are primarily responsible for collecting and managing most of the safety-related data.
- Identify and describe some of the accident and incident reporting systems that are maintained by organizations within the FAA.
- Describe the Aviation Accident Data System managed by the NTSB.
- Discuss the role of NASA in managing the Aviation Safety Reporting System.
- List and briefly describe the data required by the air carriers to submit to the Research and Special Programs Administration under Economic Regulations, Parts 217, 234, 241, 291, and 298.
- Describe the information obtained from flight data recorders and cockpit voice recorders.
- Identify several reports provided by ICAO based on ADREP data.
- Explain OSHA's definition of an injury and illness.
- Highlight some of OSHA's recording requirements.
- Recognize what constitutes an environmental spill and what are its reporting requirements.

ACCIDENTS

Before any discussion on accident reporting, investigation, and prevention can occur, there should be an understanding of the definitions used to classify accidents and incidents. Some NTSB definitions used in measuring aviation safety in commercial passenger transportation include these:

- *Accident*—an occurrence associated with the operation of an aircraft that takes place between the time any person boards the aircraft with the intention of flight and the time all such persons have disembarked, and in which any person (occupant or nonoccupant) suffers a fatal or serious injury or the aircraft receives substantial damage.
- *Fatal injury*—any injury that results in death within 30 days of the accident.
- *Serious injury*—any injury that requires hospitalization for more than 48 hours, results in a bone fracture, or involves internal organs or burns.
- *Substantial damage*—damage or failure that adversely affects the structural strength, performance, or flight characteristics of the aircraft and that would normally require major repair or replacement of the affected component.

- *Incident*—an occurrence other than an accident associated with the operation of an aircraft that affects or could affect the safety of operations.

- *Major accident*—an accident in which a Part 121 aircraft was destroyed, or there were multiple fatalities, or there was 1 fatality and a Part 121 aircraft was substantially damaged.

- *Serious accident*—an accident in which there was 1 fatality without substantial damage to a Part 121 aircraft, or there was at least 1 serious injury and a Part 121 aircraft was substantially damaged.

- *Injury*—a nonfatal accident with at least 1 serious injury without substantial damage to a Part 121 aircraft.

- *Damage*—an accident in which no person was killed or seriously injured, but in which any aircraft was substantially damaged.

The accident and its investigation remain the most conspicuous source of insights and information leading to accident prevention. Accidents provide compelling and incontrovertible evidence of the severity of hazards. The often catastrophic and very expensive nature of accidents provides the incentive for allocating resources to accident prevention to an extent otherwise unlikely.

In an accident investigation, it is essential that a clear and accurate analysis of the relevant factors be developed without delay. Further, the focus of the investigation should be directed toward effective preventive action. This focus applies particularly to government authorities and operators. With the investigation directed away from "pursuit of the guilty party" and toward effective preventive action, cooperation is fostered among those involved in the accident, facilitating the discovery of the true causes of the accident. It is emphasized that the short-term expediency of finding someone to blame for an accident is detrimental to the long-term goal of preventing accidents.

Since, by definition, an accident involves at least serious injury or substantial aircraft damage, there is a likelihood that a legal process will result from an accident. As the official authority on the accident, the investigator is often seen as a ready source of information with which to establish culpability in the courts. Consequently, witnesses and other persons involved in an accident may be inclined to withhold information from the investigator, thereby preventing a full understanding of what occurred, particularly with respect to the human-factor elements involved.

An accident investigation includes an analysis of the evidence to determine all the causes that induced the accident—a process leading to the formulation of safety recommendations. Safety recommendations regarding serious hazards should be made as soon as the hazards have been positively identified, rather than waiting until the investigation is completed. These safety recommendations should be included in the final report on the investigation. This publicity of safety recommendations fulfills several functions:

- It helps ensure that the recommendations are reasonable and realistic in the circumstances.

- It enables other countries, organizations, and individuals to see what action was recommended. Although the recommendation was not specifically addressed to them, it may enable them to take actions that avoid similar hazards.

- It can provide pressure for a prompt and reasonable response.

Recommendations must cover all hazards revealed during the investigation, not just those directly concerned with the causes. In this way, accident investigation forms the basis of an effective accident prevention program.

INCIDENTS

Incidents are events that can be defined loosely as *near-accidents.* Causal factors leading to accidents also lead to incidents, and all accidents begin as near-accidents. The various combinations of possibly unsafe acts and conditions that occur each day usually end as incidents rather than accidents, and the larger number of incidents offers wider opportunities for safety trend analyses and for suggesting potential accident prevention measures. However, for an aviation incident to be widely known, it must be reported by at least one of the people involved. Yet, the definition of an incident is subject to the interpretation of the observer, and what appears to be an incident to one person might not appear so to another. Thus some information might be lost and measurement error might occur. Similar errors result from incidents that are recognized but not reported. Various sampling techniques can be employed for testing database consistency, and valid trend analyses are possible if errors in the data can be estimated. Incident types include these:

- *Near-midair collision.* This is an incident associated with the operation of an aircraft in which the possibility of collision occurs as a result of proximity of less than 500 feet to another aircraft or an official report received from an aircrew member stating that a collision hazard existed between two or more aircraft.

- *Runway incursion.* This is an occurrence at an airport involving an aircraft, vehicle, person, or object on the ground that creates a collision hazard or results in loss of separation with an aircraft taking off, intending to take off, landing, or intending to land.

- *In-flight fire.* A fire that occurs aboard an aircraft, whether or not damage occurs, is an in-flight fire. Fire is extremely dangerous to aircraft and passengers because of the confined nature of cockpits and cabins, the amount and flammability of fuel, and the time involved in landing and evacuating an aircraft. Flightcrews are required to report occurrences of in-flight fires to the NTSB.

- *Flight-critical equipment failure.* The term *flight-critical* is subject to various interpretations. Some examples are control system malfunctions and engine failures.

Accidents provide little more than after-the-fact evidence that safety in the aviation industry was inadequate. Some have suggested that incidents provide insight into changing probabilities of aviation accidents. The implication is that there is a correlation between incidents and accidents and that careful tracking of changes in incidents might point to problems before accidents occur. There is, however, no evidence of such a correlation. This lack of correlation might suggest that corrective measures were taken before accidents occurred, that incidents are not good indicators of impending accidents, or that there has not yet been sufficient analysis of the possible correlation either because of inadequate data or because attention has been focused elsewhere.

There are at least two difficult problems that must be confronted in searching for a correlation between incidents and accidents. First, if there is a time lag between incidents and accidents, what is the length of this lag, and is the length of the lag constant over time? Second, there is little reason to believe the relationship between incidents and accidents is static. With strides in the development of technology, increased knowledge, and greater experience, the correlation between accidents and some incidents should weaken over time. For example, despite numerous engine failures on Boeing 727s over the years, there have been no accidents attributed to this cause. The reason might be due to efforts over time to institute redundancies and develop procedures to avoid accidents from engine failures.

The reporting, investigation, and analysis of incidents are a highly effective means of accident prevention. The most important characteristics of incidents are that

- They are similar to accidents, except that they lack the terminal event that causes the injury or damage in an accident. Incidents can, therefore, reveal the same hazards as accidents, without the associated injury or damage.

- They are far more numerous than accidents (estimates range from 10 to 100 times more numerous). Thus, they are a plentiful source of hazard information.

- The people involved in incidents are available to provide additional information on the hazards that caused them.

It stands to reason that the introduction of a comprehensive incident reporting and investigation system requires money and labor hours. However, experience has shown that such systems are cost-effective, as incident investigation offers true "before the accident" prevention.

CHARACTERISTICS OF INCIDENT REPORTING

Although many incidents occur in aviation, they are not always made known to those responsible for safety. Often, reporting systems are lacking, or people are not sufficiently motivated to report incidents. Experience indicates that successful incident reporting systems employ most of the following characteristics:

- *Trust.* Persons reporting incidents must be able to trust the recipient organization and be confident that any information they provide will not be used against them. Without such confidence, people are reluctant to report their mistakes, and they may also be reluctant to report other hazards they are aware of. For an incident reporting system to be successful, it needs to be perceived as being non-punitive with regard to unintentional errors or mistakes. On the other hand, most people do not expect an incident reporting system to exempt criminal acts or deliberate violations from prosecution or disciplinary action.

- *Independence.* Ideally, an incident reporting system should be run by an organization divorced from the federal agency that is also responsible for the enforcement of aviation regulations. Accordingly, some countries, including the United States, use a *third party* for the management of so-called voluntary reporting systems. The third party receives, processes, and analyzes the submitted incident reports and feeds the results back to the federal agency and the aviation community. With so-called mandatory reporting systems, it may not be possible to employ a third party. Nevertheless, it is desirable that the federal agency give a clear understanding that any information received will be used for accident prevention purposes only. This principle also applies to an airline or any other aircraft operator that uses incident reporting as part of its accident prevention program.

- *Ease of reporting.* The task of submitting incident reports should be as easy as possible for the reporter. Reporting forms should be readily available so that anyone wishing to file a report can do so easily. They should be simple to compile, with adequate space for a descriptive narrative, and they should also encourage suggestions on how to improve the situation or prevent a recurrence. Classifying information such as type of operation, light conditions, type of flight plan, weather, and so forth can be presented in a "check" format. The forms should ideally be self-addressed and postage-free.

- *Acknowledgment.* The reporting of incidents requires considerable time and effort by the user and should be appropriately acknowledged.

- *Motivation and promotion.* The information received from an incident reporting system should be made available to the aviation community as soon as possible, as this may help to motivate people to report further incidents. Such promotion activities may take the form of monthly newsletters or periodic summaries. Ideally, all such methods would be used with a view to achieving maximum effort.

- *Feedback.* Procedures or channels for forwarding hazard information to the regulatory authorities or management are needed, as it is they who are ultimately responsible for hazard elimination.

INCIDENT REPORTING SYSTEMS

Effective incident reporting systems can be organized in different ways. However, there are two main types that characterize the majority of systems used by federal authorities: mandatory and voluntary.

MANDATORY INCIDENT REPORTING SYSTEMS

In a *mandatory reporting system,* people are required to report certain types of incidents, which necessitates detailed regulations outlining who shall report and what shall be reported. Otherwise, the mandatory system could not be enforced. To achieve this goal and avoid unnecessary duplication, those items requiring an incident report must be segregated from the day-to-day problems, defects, and so forth for which adequate control systems and procedures should already exist. In effect, this means establishing a *base level,* in terms of hazards, below which an incident report is not necessary. Unless this is done, the mandatory system may be flooded with reports, possibly obscuring important items. It is important to concentrate what are usually limited resources where they will be most effective.

The number of variables in aircraft operations is so great that it is very difficult to provide a complete list of items or conditions that should be reported. For example, loss of a single hydraulic system on an aircraft with only one such system is critical; on a type with three or four systems, it may not be. A relatively minor problem in one set of circumstances can, when the circumstances change, result in a hazardous situation. The rule should be: *If in doubt, report.*

Because mandatory systems deal mainly with specific and concrete matters, they tend to collect more information on technical failures than on the human-factor aspects. To help overcome this problem, some countries with a well-developed mandatory reporting system also have a voluntary incident reporting system aimed specifically at acquiring more information on the human-factor aspects.

VOLUNTARY INCIDENT REPORTING SYSTEMS

In a *voluntary reporting system,* pilots, controllers, and others involved in aviation are invited (rather than required) to report hazards, discrepancies, or deficiencies in which they were involved or which they observed. Experience in a number of countries, including the United States, has shown that a voluntary system requires a trusted third party to manage the system. The reason is simply that people are reluctant to report their mistakes to the carrier that employs them or the government agency that licenses them.

In voluntary systems, confidentiality is usually achieved by deidentification, or not recording any identifying information. Because of this confidentiality, voluntary systems tend to be more successful than mandatory systems in collecting human factor–related information.

The next sections discuss the salient features of accident/incident recording and reporting systems that are related to aviation, general industry, and the environment.

AVIATION RECORDING AND REPORTING SYSTEMS

The U.S. federal government collects vast amounts of aviation data to support its responsibility for overseeing aviation safety. The four major organizations that collect and analyze aviation safety and safety-related data are the FAA, the Research and Special Programs Administration (RSPA), the NTSB, and the National Aeronautics and Space Administration (NASA).

FEDERAL AVIATION ADMINISTRATION REPORTING SYSTEMS

The FAA collects a wide range of aviation information and operates more than 280 automated data systems. Three organizations within the FAA collect and manage most of the safety-related data: the *Associate Administrator for Aviation Standards,* the *Associate Administrator for Air Traffic,* and the *Office of Aviation Safety.*

ASSOCIATE ADMINISTRATOR FOR AVIATION STANDARDS.   Aviation Standards personnel, working out of regional and field offices across the United States, collect and review large quantities of data, as well as certificate aircraft, aircrew, and airlines; oversee and enforce Federal Aviation Regulations; and investigate aircraft accidents and incidents. Many of these data are entered into the numerous databases maintained in Oklahoma City at the Mike Monroney Aeronautical Center and the *Aviation Standards National Field Office.* Some of the systems, required for the daily operation of Aviation Standards, are less important for analyzing system safety. Examples include databases containing aircrew and airline certification records, medical records, aircraft registry and airworthiness information, and regulatory history. The Aviation Standards National Field Office does maintain four data systems that are used, or can be used, for safety analyses. These databases, containing information on aviation accidents and incidents, mechanical difficulties, regulation violations, and aircraft utilization and reliability, are discussed in this section.

The FAA has developed the *Aviation Safety Analysis System* to integrate and standardize current and future databases and maintain them on a central host computer linked via a telecommunication network to workstations located at all Aviation Standards facilities. An overview of the Aviation Safety Analysis System is presented later in this section.

FAA ACCIDENT INCIDENT DATA SYSTEM. Accident data provide the key means of measuring aviation safety. An understanding of underlying accident causes and trends leads to preventive measures. Responsibility for investigating all civil aircraft accidents in the United States rests with the NTSB, although authority is delegated to DOT and the FAA for certain accidents. Both FAA and NTSB officials collect accident data, but the NTSB alone determines probable causes. The FAA is responsible for ensuring aviation safety and investigates accidents primarily to assess whether corrective action is required in the aviation system. In January 1984, both agencies began using common forms, the NTSB series 6120, for the reporting of accident data. While efforts are underway to develop a joint NTSB/FAA accident database, both agencies currently maintain separate data systems. There is considerable, but not complete, overlap between the two systems. The NTSB's *Aviation Accident Data System* contains all U.S. civil aircraft accidents and selected incidents, while the FAA's *Accident Incident Data System* has fewer accident records but substantially more incident data than the NTSB system. The FAA Incident Data System contains incident data that do not meet the aircraft damage or human injury thresholds established in the NTSB's definition of an accident. For example, a bird that collides with an aircraft during takeoff may not cause sufficient damage to reach the threshold for an NTSB accident. This incident, however, will make it into the FAA's Incident Data System. The FAA's Incident Data System data are categorized as follows: location information, aircraft information, operator information, narrative, findings, weather/environmental information, and pilot information.

The Accident Incident Data System contains general-aviation and air carrier incidents dating from 1978 and general-aviation accidents from 1973. In 1982, as a step toward the common NTSB/FAA accident database, air carrier accident information was introduced to the system. Although the NTSB database is considered the definitive source for aircraft accident data, the Accident Incident Data System is more accessible to FAA personnel on a daily basis. Copies of completed accident reports are forwarded from the NTSB to the Aviation Standards National Field Office, where the data are entered into the Data General MV-15000 minicomputer.

Although NTSB investigators also use the common series 6120 forms for reporting incidents, Aviation Standards personnel use the less detailed FAA Form 8020-5. By regulation, aircraft operators must notify the NTSB of five types of incidents, which may be investigated depending on the circumstances and NTSB workload. This regulation results in approximately 50 air carrier reports per year from the NTSB, compared with more than 1500 reports by the FAA investigators.

The completed FAA reports are sent to Oklahoma for processing and review, where personnel classify the incidents and assign probable-cause factors. Other Aviation Standards National Field Office employees encode and enter the incident information into the data system. Aviation Incident Data System data are available to FAA regional offices and headquarters either by the commercial computer timeshare system operated by Boeing Computer Services or by printouts from the Aviation Standards National Field Office.

Although the NTSB and NASA provide detailed analyses of the accident and incident data they maintain, the FAA examines air traffic incident data only. In 1984, the Safety Analysis Division of Aviation Standards was moved to the newly formed Office of Aviation Safety. Consequently, Aviation Standards does not have the resources to analyze air carrier incidents or other data maintained in Oklahoma City. While sufficient information, such as causes and factors, is collected, it is not used in measuring and monitoring aviation system safety or to assist in setting regulations.

ENFORCEMENT INFORMATION SYSTEM.   The Enforcement Information System, which is managed by the Aviation Standards National Field Office on the MV-15000 minicomputer in Oklahoma City, was designed and is used primarily for administrative purposes. In support of Aviation Standards and General Counsel personnel, the Enforcement Information System tracks the complete history of each enforcement case and keeps copies of all documentation. Electronic records are available from 1963 to the present. Because of the sensitivity of the data, only closed cases are available to the public.

The Enforcement Information System is the only Aviation Standards National Field Office system that allows input directly from the field offices; the others require that the field personnel send paper copies of the data to Oklahoma City for processing by Aviation Standards National Field Office personnel.

SERVICE DIFFICULTY REPORTING SYSTEM.   The mechanical reliability of aircraft and components is monitored by Aviation Standards National Field Office analysts through the *Service Difficulty Reporting System*. Reports, required by regulation, are filed by air carriers, repair stations, manufacturers, FAA inspectors, and others concerning specific types of aircraft failures or malfunctions. These reports arrive at the Aviation Standards National Field Office in paper form, where the data are encoded and entered into the MV-15000 minicomputer.

Although it contains data for more than 10 years, the Service Difficulty Reporting System is most useful for detecting short-term safety problems. The Service Difficulty Reporting System program automatically tracks trends in reports according to aircraft and component type. If the monthly or annual trend in reports exceeds a preset value, then the system automatically alerts Aviation Standards National Field Office analysts. An airworthiness directive, warning, or alert is issued to the public if, after review, the trend alert proves serious.

Service Difficulty Reporting System data are rarely used for long-term analyses. Due to the nature of the system, long-term adverse trends avoid detection because they have such shallow slopes and do not set off the alerting system. Also, because mechanical difficulties are often discovered during maintenance inspections, the frequency and depth of these inspections, along with the willingness of the airlines to file reports, affect the Service Difficulty Reporting System database.

AIR OPERATOR DATA SYSTEM.   Aviation Standards personnel must frequently refer to information about air carriers and other commercial operators and the structure of their organizations, fleets, and facilities. Although such information is available in fragments from many sources within the DOT, the *Air Operator Data System* attempts to consolidate the vital data available from within the FAA. Of interest for safety analysis are data involving aircraft operations, such as utilization and engine reliability.

Unlike other Aviation Standards data-gathering efforts previously discussed (accident/incident, enforcement, and service difficulties), there is no regulatory requirement for air carrier reporting or FAA collection of *air operator data* as such. Air carriers must report organizational, operational, and financial data to the Research and Special Programs Administration's Office of Aviation Information Management (and previously to the Civil Aeronautics Board). Certain engine problems must also be submitted.

Air carrier inspectors follow general guidelines for collecting the data monthly. They send air operator data to Oklahoma City by mail for processing. Although Aviation Standards National Field Office employees ensure accurate transcription of data, there are no procedures in effect for ensuring accuracy at the source.

Air operator data provide the opportunity for analyzing certain air carrier operating practices, by individual company or industrywide. When used in conjunction with other system information, daily utilization data give one view of the amount of schedule pressure placed on aircraft fleets. Engine reliability data, which are the basis for over-water flight certification, indicate the final product of equipment design and airline maintenance and operating procedures. The Aviation Safety Analysis System, mentioned earlier, will consolidate and standardize new and existing safety databases. In contrast to the present system, FAA personnel without extensive training in computer programming will have access to a wide range of safety data via desktop workstations.

AVIATION SAFETY ANALYSIS SYSTEM.   The Aviation Safety Analysis System was conceived in 1979 to build on the general office automation program for regional and field offices then in development at the FAA. New office equipment, proposed as part of the automation program, was to have sufficient processing and network capabilities for an integrated safety data system. The numerous compatibility and communication difficulties created by the data systems then in use (for the most part, still in use) at the FAA were to be addressed by the Aviation Safety Analysis System. An Aviation Safety Analysis System program office was established in 1982, and a long-term phased development plan was proposed. The initial phase integrated and standardized current data systems. Subsequent phases implemented and developed new databases. The types of Aviation Safety Analysis System databases fall into four categories:

1. Airworthiness data
2. Regulatory data
3. Operational data
4. Organizational information

Airworthiness data are mainly historical information on aircraft, such as mandatory modifications specified by the FAA. Regulatory data consist of background information, such as Notices of Proposed Rulemaking, legal opinions, and previous regulations. Data describing the aviation environment are included in the operational category. These databases track aircrew, aircraft, and operators along with accidents, incidents, mechanical reliability reports, and enforcement actions. The work management subsystems to monitor Aviation Standards tasks, such as airline inspections, fall into the category of organizational information.

The Aviation Safety Analysis System altered many of the tasks previously performed by Aviation Standards personnel. Data are entered and validated where they are collected and generated, at the field office level.

ASSOCIATE ADMINISTRATOR FOR AIR TRAFFIC.   In managing the national airspace system, Air Traffic personnel control traffic, operate facilities, and develop procedures and standards for airways, airspace, and flight operations. On a daily basis, information is collected and reviewed concerning air traffic levels, national airspace system status, system errors, controller errors, pilot deviations, and delays, although most of the data are entered into automated systems only after reaching specific offices within FAA headquarters. Other offices, regions, or field facilities within Air Traffic do not have ready access to many of these systems. However, the *Office of Air Traffic Evaluations and Analysis* specialists monitor every report on operational errors, midair near-collisions, and pilot deviations and communicate findings to the field facilities.

Although Air Traffic tracks and analyzes air traffic safety data, it does not manage the data systems dealing with incidents or systemwide operational information of interest to this study. The Office of Aviation Safety (discussed in the next section) handles the incident data while the air traffic activity data are processed by the FAA Office of Management Systems. The Office of Air Traffic Evaluations and Analysis developed its own data system, the *Operational Error Reporting System,* to receive and track operational error reports in a timely fashion. The system has been online, linking a number of regional offices with headquarters, since June 1987.

NEAR MIDAIR COLLISIONS SYSTEM (NMACS) DATABASE. An NMACS is an incident in which a possibility of a collision occurs as a result of aircraft being less than 500 feet from each other, or a report is received from a pilot or flightcrew member stating that a collision hazard existed between two or more aircraft. A report does not necessarily involve the violation of regulations or error by the air traffic control system, nor does it necessarily represent an unsafe condition. Historically, an NMACS was classified as an incident, and reports of NMACS events were maintained in the FAA's Incident Data System, together with reports of a variety of other events. By the mid-1980s, NMACS were considered to be sufficiently important to warrant the establishment of a separate reporting system and investigative procedures. All NMACS reports are thoroughly investigated by FAA inspectors in coordination with air traffic controllers. Data obtained from NMACS reports are used to develop programs, policies, and procedures to reduce NMACS occurrences and thereby enhance the safety and efficiency of the air transportation system. This voluntary reporting system, which has no legal or regulatory requirement for reporting, can get subjective and hence has similar shortcomings as the Aviation Safety Reporting System (ASRS).

AIR TRAFFIC ACTIVITY DATABASE. An essential exposure measure for air safety analysis is the level of traffic. One parameter, departures, is the best exposure reference for general safety comparisons. Although departure data are available for specific carriers from Civil Aeronautics Board records and research and the Special Programs Administration, systemwide traffic data, including departures, are available from the Air Traffic Activity Database.

Air traffic control personnel keep track of the daily activity at ATC facilities. Monthly summaries of various operations, including the number of takeoffs and landings at airports with control towers and the number of aircraft handled by radar control facilities, are submitted to the Office of Management Systems at FAA headquarters. There the data are encoded for entry into the Boeing Computer Services System, where they are processed and cross-checked. Due to the large volume of monthly data, the Boeing system is used not for analysis or storage, but as a tool for preparing summary reports. Annual Air Traffic Activity Reports are published and are available to the public.

Facility, region, or system-total data are available with tables categorizing information by aircraft operator (air carrier, air taxi, general aviation, and military). This study uses historical tower activity data to illustrate the growth of hubs and is the exposure reference for air traffic incidents. The number of aircraft handled by en route radar controllers is an alternate measure of traffic trends.

OFFICE OF AVIATION SAFETY.   Reporting directly to the FAA administrator, the Office of Aviation Safety conducts accident investigations, safety analyses, and special programs. In this role, it monitors or manages several databases. The Office of Aviation Safety operates the *National Airspace Incident Monitoring System*, an automated system containing near-midair collisions, operational error, and pilot deviation databases. The FAA maintains contact with the NASA-administered, but FAA-funded, Aviation Safety Reporting System through the safety analysis division with the Office of Aviation Safety.

NEAR-MIDAIR COLLISION DATABASE.   The FAA learns about near-midair collisions primarily from pilot reports, though air traffic controllers, passengers, and ground observers also serve as notifiers. In each case, a preliminary report is filed and must be investigated by the FAA within 90 days.

Although the Aviation Standards Accident Incident Data System tracks near-midair collisions, they are not included in its database. All incident reports involving air traffic operations, including near-midair collisions, end up in the Office of Aviation Safety. There, the data are encoded and entered into an IBM/AT personal computer system located at FAA headquarters. Near-midair collision information from 1980 to the present is available in the system.

OPERATIONAL ERROR DATABASE.   The loss of legal flight separation around an aircraft that is attributed to the ATC system is an operational error. For example, during en route operations, controllers are required to keep aircraft apart by 5 miles horizontally and 1000 feet vertically for flights below 29,000 feet and 2,000 feet vertically for flights above 29,000 feet. Operational deviations, generally less serious than operational errors, do not involve loss of separation between two aircraft, but result from an aircraft passing too close to a restricted airspace or landing zone.

From 1983 to 1985, the FAA instituted two changes. First, the en route ATC computers were reconfigured with the Operational Error Detection Program that automatically records and reports any loss of proper separation for aircraft in the system. Second, the responsibility for maintaining an operational error report database was shifted to the Office of Aviation Safety. Preliminary reports of operational errors and deviations are filed from the ATC facility within 48 hours after the event. All reported operational errors and deviations are investigated, and depending on the outcome, a final report is submitted. Personnel from the Office of Aviation Safety encode and enter preliminary and final report data into the IBM/AT.

PILOT DEVIATION DATABASE.   An ATC facility that observes a pilot deviation is responsible for reporting it to the appropriate Flight Standards office for investigation. Prior to 1985, incidents involving pilot deviations were entered into Aviation Standards National Field Office Accident Incident Data System, although they were not specifically categorized as pilot deviations. Presently, the results of pilot deviation investigations are sent directly to the Office of Aviation Safety where the data are entered into an IBM PC. The Office of Aviation Safety is responsible for tracking and reporting trends in pilot deviations. The office published its first statistical report of pilot deviations in October 1987. Similar to the operational error data, pilot deviation information stored electronically extends back only to 1985.

MISCELLANEOUS PROGRAM REPORTING INITIATIVES

AVIATION SAFETY ACTION PROGRAM (ASAP).   The goal of ASAP is to prevent accidents and incidents by identifying unsafe practices and correcting them. Voluntary reporting of safety issues by employees is encouraged. The program's strategy is to create a nonpunitive environment through enforcement-related incentives for employees to report safety issues, even though these issues may involve an alleged violation of aviation regulations. In ASAP an airline enters into a safety partnership with the FAA and could include a third party such as the employee's labor union. Since the FAA does take enforcement action (even though this action is limited) against ASAP, its policy on ASAP remains controversial. Also, information contained in ASAP reports may be considered sensitive by some airlines. The FAA has tried to overcome this problem by proposing regulations that would protect ASAP reports and other sensitive safety and security information from being disseminated to the public.

FLIGHT OPERATIONAL QUALITY ASSURANCE (FOQA).   Flight operational quality assurance programs involve collecting and analyzing data recorded during flight to improve the safety of flight operations (for more on this see automatic recording systems later). In excess of 200 flight parameters can be accessed from digital flight data recorders downloaded frequently through quick-access recorders. While FOQA programs are quite popular with international airlines, only a few U.S. airlines use FOQA because of data confidentiality issues. In 2001 the FAA took steps to address the confidentiality issue by codifying carrier data and ensuring confidential enforcement protection for any operator who operates aircraft under an FOQA program approved by the FAA. The regulations would ban disclosure of sensitive safety information and security information that was voluntarily submitted to the FAA.

AIR TRANSPORTATION OVERSIGHT SYSTEM (ATOS).   The Air Transportation Oversight System was implemented in 1998 as a new approach to FAA certification and surveillance oversight, using system safety principles and systematic processes to ensure that air carriers are in compliance with FAA regulations and have safety built into their operating systems. Unlike the traditional oversight methods, ATOS incorporates the structured application of new inspection tasks, analytical processes, and data collection techniques into the oversight of individual air carriers. This approach enables Flight Standards inspectors to be more effective in the oversight of air carriers by focusing on the most critical safety aspects of an air carrier's operation. As currently applied, ATOS provides a systematic process for conducting surveillance, identifying and dealing with risks, and providing data and analysis to guide the oversight of each carrier. ATOS is now being applied to 9 air carriers that handle 95 percent of U.S. passengers. These airlines include Alaska Airlines, American Airlines, America West Airways, Continental Airlines, Delta Airlines, Northwest Airlines, Southwest Airlines, United Airlines, and US Airways. Ultimately it is envisioned that all U.S. airlines will participate in this program.

Under ATOS, an air carrier's operations have been separated into 7 systems, 14 subsystems, and 88 underlying component *elements,* which provide the structure for conducting surveillance, collecting data, and identifying risks or areas of concern. Surveillance is effectively implemented through two distinct types of inspection, the *safety attribute inspection (SAI)* and the *element performance inspection (EPI).* An SAI is planned at the subsystem level and conducted at the element level by a team of inspectors to determine if the air carrier has the safety attributes of responsibility, authority, procedures, controls, process measurement, and interfaces adequately designed into its system element processes. EPIs are also conducted at the element level but are accomplished by individual inspectors to determine if the carrier's system element processes meet established performance requirements, if the air carrier's procedures and controls are adhered to, and if proper records are maintained. In addition, more than 2200 specific regulatory requirements have been incorporated into the SAIs and EPIs to ensure that air carriers are in full compliance with all applicable Codes of Federal Regulations (CFRs). More information on ATOS is available at www.faa.gov/avr/afs/atos/overview/ATOS_Model.htm.

By collecting and analyzing data on the many airline systems, FAA inspectors are better able to target areas for improvement. ATOS is a significant shift in the way in which the FAA oversees airlines and how its inspectors operate. It should lead to a more collaborative partnership toward system evaluation and may open doors for additional information sharing between air carriers and the FAA. With the cooperation of the air carriers, ATOS will foster a more proactive relationship with the FAA on safety-related issues as emerging trends and concerns are identified. This will aid in achieving the AFS mission, which is to provide the public with the safest aircraft operations in the world.

AFS is in the process of developing a centralized analysis and information management system to accumulate, analyze, and disseminate safety data and information within Flight Standards and to assist airlines in the interpretation of data. Information will be disseminated from a computer-based decision support tool called the *Safety Performance Analysis System (SPAS).* It will include selected air carrier data and data summaries from ATOS and other sources including Aviation Safety Reporting System (ASRS) reports and Office of System Safety information. This information will assist inspectors and air carriers in decision making with respect to targeting surveillance resources and taking corrective actions to mitigate safety risks. Planned enhancements will continue to be validated for effectiveness and added to the system as appropriate.

NTSB ACCIDENT/INCIDENT REPORTING SYSTEM

Since its inception in 1967, the NTSB has kept records of civil aircraft accidents and its Aviation Incident/Accident Database contains information on every known civil aviation accident in the United States. Accidents involving only military or public-use aircraft are not usually investigated by the National Transportation Safety Board. While this incident/accident reporting system was primarily designed for administrative purposes, the system does have analytical capabilities. The NTSB Aviation Accident/Incident database is the official repository of aviation accident data and causal factors. The NTSB publishes annual reviews of aircraft accident data and occasional special studies, which are supported by statistical analyses accomplished with the data system. As discussed earlier, the NTSB is responsible for investigating all aircraft accidents and certain incidents, determining their probable causes, and making recommendations to the FAA. The NTSB uses Form 6120.19A to generate a *preliminary report* that is completed within 5 working days. It then develops the more informative *factual report* using Form 6120.4. The *final report* may take months and even years to complete. Preliminary reports contain a limited amount of information (date, location, aircraft operator, type of aircraft, etc.). The NTSB Aviation Accident/Incident database is presented in the following categories: location information, aircraft information, operator information, narrative, sequence of events, findings, injury summary, weather/environmental information, and pilot information.

The NTSB's data collection methods evolved over the years into a single format which was used until 1982, when the procedure and report form were revised. The documentation was again changed in 1983, when NTSB accident investigators began submitting data in the format that was eventually adopted as NTSB series 6120.4. Data from the reports are entered into the computer, along with the findings of probable cause and contributing factors. Computer searches are possible with any data block or group of blocks as selection criteria (see Chap. 3).

Differences in data formats impose some restrictions on possible computer-assisted analyses. For example, in 1982 the NTSB changed its method of classifying accidents. Accidents are now categorized by the first occurrence in the sequence of events that led to the accident. Earlier, groupings were made by the accident type. The NTSB has developed a matrix for comparing occurrences and types. For broad safety studies, the effect of the format changes is small. Although the collection of data has essentially remained the same, the latter format allows a more detailed analysis of accident circumstances.

DEPARTMENT OF TRANSPORTATION REPORTING SYSTEM

The *Bureau of Transportation Statistics* (*BTS*), a unit of the U.S. Department of Transportation, compiles and analyzes data on the nation's transportation system. By law, the airlines are required to provide financial, activity, and other data to BTS. They are used to develop and provide information to the flying public on the condition and performance of the nation's air transportation system. Some of the measures cataloged include the number of departures, flight hours, and miles performed in domestic commercial service during the most recent 5 calendar years, by airline. These are also the same measures tracked by government and industry to calculate accident or incident rates for the air transportation system. These data only apply to airlines that operate aircraft with 30 or more seats.

NATIONAL AERONAUTICS AND SPACE ADMINISTRATION REPORTING SYSTEM

The National Aeronautics and Space Administration provides and supports aviation research and development and administers the confidential and voluntary *Aviation Safety Reporting System* (*ASRS*). The Aviation Safety Reporting System is designed to encourage reports by pilots and air traffic controllers concerning errors and operational problems in the aviation system by guaranteeing anonymity and immunity from prosecution for all reporters. About 400,000 incident reports have been submitted to date without any reporter's identity ever being revealed. System data can provide an alternate federal insight into the nature and trends of aviation incidents.

NASA AVIATION SAFETY REPORTING SYSTEM.   The Aviation Safety Reporting System is a joint effort by FAA, NASA, and the Battelle Memorial Institute to provide a voluntary reporting system where pilots, controllers, and others can submit accounts of safety-related aviation incidents. The system is funded mainly by the FAA, administered by NASA, and maintained by Battelle. Reports are sent to the Aviation Safety Reporting System office at NASA Ames Research Center, where the data are analyzed and entered into a computer by employees of Battelle. The database is maintained at Battelle Laboratories in Columbus, Ohio.

Prior to the establishment of the system in 1976, attempts at providing voluntary incident reporting programs met with little success. Potential reporters feared liability and disciplinary consequences. Even after the FAA introduced its Aviation Safety Reporting Program, which offered limited immunity and anonymity to participants, few reports were submitted. The aviation community feared that the FAA, responsible for setting and enforcing regulations, would misuse the data. The FAA acknowledged these concerns and transferred control of the Aviation Safety Reporting Program to a neutral third party, NASA. A Memorandum of Agreement was executed between the FAA and NASA in August 1975, establishing the Aviation Safety Reporting System. The agreement provided for a limited waiver of disciplinary action, confidentiality of reporting sources, and an advisory committee comprised of representatives of the aviation community. The Aviation Safety Reporting System became operational on April 15, 1976.

The Aviation Safety Reporting System report form was designed to gather the maximum amount of information without discouraging the reporter. Structured information blocks and key words are provided, not only to guide the reporter, but also to aid subsequent data retrieval and research. Narrative descriptions are encouraged. Space is provided for the reporter's name, address, and telephone number, which permits NASA to acknowledge the report's receipt by return mail and also allows the Battelle analyst to contact the reporter for follow-up data. Information that identifies the reporter is deleted before data are entered into the computer.

Under the guidance of NASA, Battelle receives the incident reports, processes and analyzes the data, and publishes reports of the findings. Human factors in aviation safety, a continuing concern at the NASA Ames Research Center, were a major consideration in the development of the Aviation Safety Reporting System. The data analysts, primarily experts in aircraft operations and air traffic control, provide insight into the nature of the human error or other underlying factors in the incidents. Although the reports are encoded in detail, the complete narrative text of each report is retained for later reevaluation.

Because the Aviation Safety Reporting System is voluntary and reporters are deidentified, a concerted effort among a number of individuals can distort the database. For example, air traffic controllers at certain facilities increased their reporting of incidents associated with a display system that they wanted upgraded. This reporting campaign ended with the air traffic controllers strike in August 1981.

RESEARCH AND SPECIAL PROGRAMS ADMINISTRATION REPORTING SYSTEMS

The Office of Aviation Information Management of the Research and Special Programs Administration assumed the former Civil Aeronautics Board's responsibility for collecting data on airline operations, traffic, and finances beginning in 1985. Airlines submit data periodically in accordance with Economic Regulations Parts 217, 234, 241, 291, and 298. Although these data do not directly indicate safety, they do provide measures of exposure such as departures, hours, and miles. However, the airline categories for exposure data reporting do not correspond to the operating categories used by the NTSB for classifying accidents, which results in some gaps and inaccuracies in statistics. Financial statistics also have potential uses in analyses, because many people in industry and government believe that economics influences safety to some degree.

AIR CARRIER STATISTICS DATABASE.   *Part 217*, "Reporting Data Pertaining to Civil Aircraft Charters Performed by U.S. and Foreign Air Carriers," requires U.S. and foreign air carriers to file traffic data on any civilian international charter flight flown to or from the United States in large aircraft (more than 60 seats or 24,000 pounds of payload). The information reported quarterly shows the charter passengers or tons of cargo flown between the origin and the destination point of the charter. The information is reported by aircraft type by month.

Part 234, "Airline Service Quality Reports," requires 14 certificated U.S. air carriers (a carrier with more than 1 percent of total domestic scheduled passenger revenues) to file monthly flight performance information for every domestic nonstop scheduled passenger operation to or from the 29 largest U.S. airports (airports with more than 1 percent of domestic scheduled passenger enplanements). Carriers are voluntarily reporting data for each domestic scheduled flight, instead of limiting their reporting to the 29 airports. For the origin airport of each nonstop segment, the carrier reports published departure times versus actual departure times; for the destination airport, the published arrival times versus the actual arrival times are reported. This information is reported by date and day. Flights delayed because of mechanical reasons, as defined by the FAA, are not reported.

Part 241, "Uniform System of Accounts and Reports for Large Certificated Air Carriers," prescribes the accounting and reporting regulations for large U.S. certificated air carriers (Section 401 certificate). A large carrier is defined as a carrier operating aircraft that are designed to accommodate more than 60 seats or a cargo payload of more than 18,000 pounds. All large carriers, according to the level of their operations as measured by annual operating revenues, are placed into one of four groups: group I, small ($10 million and under); group II, large ($10,000,001 to $99.9 million); group III ($100 million to $1 billion); and group IV (more than $1 billion). The amount and detail of reporting increase with carrier size. Data are submitted on individual schedules of the DOT Form 41 Report or by electronic media. In general, carriers report exposure data such as aircraft departures, hours, miles, and passenger enplanements in total and by aircraft types. A broad range of financial data, including categories of revenues and expenses, are also reported, with those related to operations being indexed by aircraft type.

Part 291, "Domestic Cargo Transportation," prescribes the reporting required of carriers providing domestic all-cargo operations exclusively under Section 418 certificates. These carriers are required to file Form 291-A, a one-page annual report, that contains seven profit-and-loss items and seven traffic and capacity items. The data are not reported by aircraft type.

Part 298, "Exemptions for Air Taxi Operations," prescribes the reporting for small certificated air carriers (Section 401 certificate) and commuter air carriers. Both classes of carriers operate aircraft that are designed for 60 seats or fewer or for 18,000 pounds of cargo capacity or less. A commuter air carrier is defined as a special classification of air taxi operator that provides passenger service consisting of at least five round trips per week between two or more points. Commuters report only traffic exposure data totals with no indexing by aircraft type. Small, certificated air carriers submit the same information as commuters plus revenue and expense data. The direct expense data and three operational items (block hours, departures, and gallons of fuel issued) are indexed by aircraft type on small certificated air carrier reports. Air taxi operators that are not commuters have no reporting requirements.

Various reports, including electronic submissions, are sent monthly, quarterly, semiannually, and annually to the Office of Aviation Information Management. There the data are entered into the Amdahl computer located in the DOT head-quarters building in Washington, D.C. Most of these data are published or loaded on magnetic tapes and are available to the general public by subscription.

AUTOMATIC RECORDING SYSTEMS

Many modern air transport aircraft have automatic recording devices installed. The *flight data recorder* (FDR), which monitors selected parameters of the flight, and the *cockpit voice recorder* (CVR), which records voices and cockpit sounds, are installed to assist with the investigation of accidents and, in some cases, incidents. An engineering recorder may also be installed to monitor aircraft systems. The data from this recorder can be used to detect impending failures and to verify the adequacy of component life and overhaul schedules.

Automatic recorders that are installed in air traffic control and communication systems primarily for accident investigation purposes may also be used as a check on correct operating procedures.

FDRs and CVRs were initially installed to assist accident investigators in determining accident causes, particularly for catastrophic accidents to large aircraft. In some countries, professional groups such as pilots and air traffic controllers accepted the philosophy that the installation of these recorders would be of significant benefit to the aviation industry in helping to determine accident causes. Accordingly, they agreed to their use, provided that guarantees by operational and administrative agencies were negotiated and honored, preventing disciplinary action from being taken on the basis of information determined from the recorders unless willful negligence or dereliction of duty could be proved (see FOQA earlier).

Many countries routinely use flight recorder information for accident prevention. They regard this information as an invaluable source of safety insights and information on the operation of their aircraft. Standard flight profiles are usually programmed into a computer along with acceptable deviations. Recorded data are then compared with these standard profiles. Significant deviations are then examined to see if hazards could be present. If so, corrective action can then be taken. This method need not require the identification of individuals, since it is often the number and type of deviations that reveal hazards.

Some carriers that routinely examine FDR records for indications of hazards or deviations from standard operating procedures have the findings reviewed by a committee consisting of retired captains or flightcrews. This group has the respect of both management and pilots and thus avoids direct employer/employee contact. The fear of job loss or punishment is avoided, and the accident prevention insights are more readily obtained.

INTERNATIONAL REPORTING SYSTEMS

The international exchange of accident and incident data provides a broad range of experience on which to base safety guidance. Such information can be of particular value to smaller countries or carriers that are not in a position to maintain an accident or incident reporting system, or whose database is too limited to permit the identification of potential hazards. Safety data interchange is, therefore, encouraged among civil aviation authorities and safety organizations. For maximum effectiveness, compatibility of the basic coding of the data in these systems is needed. This applies equally to simple manual systems or more complex *electronic data processing (EDP)* systems.

For aircraft components and systems, the ATA 100 specifications—an internationally accepted code—provide a great deal of compatibility. These standard codes are used in the *International Civil Aviation Organization's (ICAO's) Accident/Incident Reporting System (ADREP)* and some national systems.

A number of countries use EDP systems for the storage, processing, and dissemination of accident and incident data. Most of these countries use compatible EDP formats and codes, making it possible to exchange data tapes and thus benefit from each other's experience.

In addition to the exchange of data from reporting systems, many of the larger countries and organizations publish material dealing with many aspects of aviation safety. These publications include such things as films, magazines, and summaries of accidents and incidents.

ICAO ADREP SYSTEM.   The ICAO ADREP system is a databank of worldwide accident and incident information for large commercial aircraft. Thus, ICAO can provide countries with accident prevention information based on wide international experience. The ICAO Accident/Incident Reporting Manual (ADREP) (Document 9156) contains detailed information on this system. ICAO provides the following information based on ADREP data:

- *ADREP Summary* is a computer-generated publication containing the ADREP preliminary reports and data reports received by ICAO during a 2-month period. It is issued six times per year. The last issue of each year contains a comprehensive index of the accident and incident reports reported to ICAO during that year.

- *ADREP Annual Statistics* is an ICAO circular containing annual statistics from the databank. These statistics may be useful for safety studies and accident prevention programs.

- *ADREP Requests* are computer printouts provided by ICAO in response to specific requests from countries. Guidance for the formulation of ADREP requests is contained in the ADREP manual.

The ADREP computer programs are available to countries wishing to utilize their computer systems for accident or incident recording. These programs, in addition to promoting standardized coding, offer significant financial savings to countries.

OTHER ICAO SAFETY INFORMATION.   The ICAO also publishes the following aircraft accident and incident information:

- *Aircraft Accident Digest* is a publication that contains narrative-type accident or incident final reports, selected for their contribution to accident prevention or the use of new or effective investigative techniques.

- *List of Final Reports Available from Countries* is a listing of narrative-type aircraft accident final reports available on request from the reporting countries. The list is updated every 6 months on the basis of information supplied by countries.

No single measurement or statistic provides a complete picture of commercial aviation safety. Although accident and fatality statistics are the best measures of long-term past risk in commercial aviation, they are of limited value over short periods and are not suitable monitors of short-term effects of policy decisions. For example, the consequences of rulings requiring collision avoidance systems on commercial transports and transponders on many general-aviation aircraft might not be apparent in the accident data for 10 years or more.

Nonaccident safety data, while not substitutes for accident and fatality data, are valuable supplements. If properly collected and maintained, nonaccident data can help identify and estimate the magnitude of safety problems and permit the monitoring of safety programs.

GENERAL INDUSTRY (OSHA) RECORDING AND REPORTING SYSTEMS

APPLICABILITY

OSHA regulations listed in 29 CFR 1904 require every employer covered by the OSHA Act with 11 or more employees to record and report all employee occupational (work-related) deaths, injuries, and illnesses on Forms 300 and 301. Employers with 10 or fewer employees are required to keep injury and illness records only if OSHA or the Bureau of Labor Statistics (BLS) specifically notifies them in writing that they must keep these records. However, all employers covered by the OSHA Act must report all deaths and events that cause in-patient hospitalization of 3 or more employees orally to OSHA (via phone or in person to the OSHA area office) within 8 hours of being aware of the events. Injuries and illnesses that require only first aid are exempt from being recorded. *Injuries* include, but are not limited to, cuts, fractures, sprains, or amputations. *Illnesses* include both acute and chronic illnesses such as skin disease, respiratory disorder, or poisoning.

Injury and illness records are used for several purposes. OSHA collects data through the OSHA Data Initiative (ODI) to help direct its programs and measure its own performance. Inspectors also use the data during inspections to help direct their efforts to the hazards that are hurting workers. Records are used by employers and employees to implement safety and health programs at individual workplaces. Analysis of the data is a widely recognized method for discovering workplace safety and health problems and for tracking progress in solving those problems. Records also provide the base data for the BLS Annual Survey of Occupational Injuries and Illnesses, the nation's primary source of occupational injury and illness data. Each year, the BLS uses a stratified random data collection survey process to collect injury and operational data (e.g., hours worked) from companies across the United States. It then analyzes, categorizes, and publishes the data by industry groupings called *standard industrial classification* (SIC) codes. Companies can use these data to compare themselves to their peer industries. More on SIC codes and injury statistics will be covered in Chap. 5 on injury statistics.

OSHA FORM 300

This form requires the employer to record all occupationally related injuries and illnesses other than first aid and requires the following information:

- Employee's name, employee's job title, and the place (department) where the injury/illness occurred

- A very short description of the injury/illness including the body part that was affected and what caused it
- The number of lost or restricted workdays that the injury/illness caused
- The type or classification of the injury/illness

OSHA FORM 301

This incident report Form 301 contains more detailed information on the injuries/illnesses than that listed in Form 300. Every injury or illness listed in Form 300 must have an accompanying Form 301. Information on Form 301 describes in greater detail how the injury or illness occurred, what time of day it happened, what the employee was doing when he or she was injured, and the extent of medical treatment the injury or illness required. Actual copies of OSHA Forms 300 and 301 can be downloaded from the OSHA website (www.osha.gov). Note that the FAA uses the Mishap Report Form 3900-6 (or succeeding form) for its own internal facilities and operations.

RECORD RETENTION

OSHA Form 300, the annual summary of illnesses and injuries form (Form 300A), and the OSHA Form 301 incident report must be retained on file by an employer for 5 years following the end of the calendar year that these records cover (29 CFR 1904.33).

FORM POSTING

An employer is required to post a summary (Form 300A) of the total number of injuries and illnesses for the (calendar) year in a conspicuous place (preferably on the employee bulletin boards) at each establishment. Totals for the previous year must be displayed continuously from February 1 through April 30 of the year following the year covered by the form.

ENVIRONMENTAL (EPA) RECORDING AND REPORTING

Under federal environmental laws, any facility or vessel (e.g., oil spills from ships) must report to government authorities any hazardous substance that is released into the environment in quantities that exceed a threshold amount. These threshold amounts are referred to as *reportable quantities* (RQs) and must be reported in a timely manner. In addition to release reporting required under CWA, CERCLA, EPCRA (SARA Title III), RCRA, and TSCA, state statutes may require additional release reporting. The environment can be air, water, soil, or a combination of the three media. A release includes any contact of a hazardous substance with the environment through either intentional or unintentional discharge and by any method (e.g., spilling, leaking, dumping, leaching, and abandonment of barrels or other closed containers containing hazardous substances). Notification of releases is usually required immediately or within 24 hours of knowledge of the release. EPA regulations cover a variety of reporting requirement that include

- Media (environments) into which the release is applicable
- Listing of hazardous substances together with their quantities
- Persons responsible for reporting
- Time line for reporting
- The agency to contact in case of a release
- Fines for not reporting

In some cases there is also a requirement for follow-up reporting which must include the original notification, response actions taken, acute or chronic health risks associated with the release, and advice on medical care to exposed victims.

All these reporting release requirements help ensure accountability for cleanup, provide data for regulatory and public policy purposes, and enable federal, state, and local authorities to effectively prepare for and respond to emergencies that could harm humans and the environment. The EPA does not necessarily take enforcement action against releasers, but is quite strict against those violators who fail to report releases.

CERCLA REPORTING REQUIREMENTS

Under this act, if CERCLA-defined hazardous substances greater than or equal to their RQs are released into the environment, then the person in charge of the vessel or facility from where the release occurred must immediately notify the *National Response Center* (NRC). CERCLA hazardous substances are listed in 40 CFR, Table 302.4. The NRC (housed within the U.S. Coast Guard), acts as the single federal point of contact for all release reporting and is continuously staffed for handling emergency responses.

CERCLA-defined hazardous substances include substances listed under certain sections of the CAA, CWA, RCRA, and TSCA. As and when substances get regulated under the CAA, CWA, RCRA, or TSCA, they automatically become CERCLA hazardous substances. When a substance is delisted from the CAA, CWA, RCRA, or TSCA lists, it may still be regulated under CERCLA if the EPA administrator independently determines the substance to be hazardous. The CWA contains two sets of chemicals that are reportable under both CERCLA and EPCRA.

EPCRA REPORTING REQUIREMENTS

EPCRA (SARA Title III) was enacted in 1986, with the primary objective of informing communities of chemical hazards in their areas. All regulations pertaining to this section are found in 40 CFR Subchapter J, Superfund, Emergency Planning, and Community Right-to-Know Programs (Parts 300 through 399). EPCRA reporting requirements are the same as CERCLA requirements, but in addition, EPCRA regulations list *extremely hazardous substances* (*EHSs*) and requires their reporting in case of a release in excess of the RQs [EPCRA Section 304(a)]. This regulation applies only to facilities (vessels are excluded) and does not require reporting if releases and human exposures occur solely within the confines of a facility. EPCRA requires facilities to notify *state emergency response commissions* (*SERCs*) and *local emergency planning committees* (*LEPCs*). SERCs and LEPCs are dedicated emergency response groups that exist at the state and local levels and were established under EPCRA. The governor of each state must designate a SERC. Many SERCs include state department agencies involved with the environment, natural resources, emergency services, public health, occupational safety, and transportation. The SERC designates local emergency planning districts and appoints LEPCs for each district. The LEPC has representation from elected state and local officials, police, fire, civil defense, public health professionals, environmental, hospital, emergency planning, and transportation officials.

If an EPCRA release also qualifies as a CERCLA release, then in addition to the SERC and the LEPC, the NRC must be notified. Since different criteria were used to compile the list of chemicals, there are chemicals common to both CERCLA and EPCRA while there are some chemicals that appear on one list, but not the other.

Also 40 CFR Parts 311 and 312 of EPCRA require businesses to report the locations and quantities of chemicals stored on site to state and local governments, so that communities can be prepared to respond to chemical releases and similar emergencies.

EPCRA's emergency planning requirements (40 CFR Parts 302 to 303 and 355.30) requires SERC and LEPC notification if an EHS equal to or in excess of its *threshold planning quantity* (*TPQ*) is stored or used at a facility. EHSs and their TPQs are identified in 40 CFR Part 355. For example, the FAA triggers this reporting requirement as it has sulfuric acid (used in battery electrolyte) present at the Air Route Traffic Control Centers (ARTCCs) and the Air Traffic Control Towers (ATCTs) in excess of its 1,000-pound TPQ. In addition, the FAA's Aeronautical Center and Technical Center has chlorine (used in water and wastewater treatment) in excess of 1,00 pounds, calcium hypochlorite (used in water and wastewater treatment) in excess of 10 pounds, and methyl chloroform (used as a solvent cleaner) in excess of 100 pounds.

EPCRA Section 313 requires EPA and the states to collect data on chemical releases and make the data available to the public through the *Toxics Release Inventory* (*TRI*) program. And 40 CFR Part 370 (Hazardous Chemical Reporting: Community Right-to-Know) establishes reporting requirements of hazardous chemicals to enhance community awareness and to facilitate the development of state and local emergency response plans. And 40 CFR Part 372 (Toxic Chemical Release Reporting: Community Right-to-Know) requires timely reporting of chemical releases to the EPA. There are several different types of reports that cover diverse topics ranging from oil spills and chemical accidents to releases of dangerous substances into the water or air. The EPA compiles and organizes all this reported information annually, and it makes this information available to the public through printed documents and online databases such as the Toxics Release Inventory. The goal of TRI is to educate the public by disseminating this information so that communities may be empowered to hold companies and local governments accountable for how toxic chemicals are managed. Currently, there are about 650 chemicals included in the TRI. Extended information on TRI including chemical lists, toxicity, and other reporting and program information can be found at http://www.epa.gov/tri/.

RCRA SUBTITLE I (HAZARDOUS AND SOLID WASTES AMENDMENTS OF 1984)

The sections within this part (40 CFR Parts 280 and 281) set technical standards and corrective action requirements for owners and operators of *underground storage tanks* (USTs). Underground fuel tanks are of primary concern in fueling handling and storage at airports and fuel supplier work sites.

Reporting requirements. Facilities must submit the following information:
- Notification for all UST systems which includes certification of installation for new UST systems
- Reports of all releases including suspected releases, spills and overfills, and confirmed releases
- Corrective actions planned or taken including initial abatement measures, initial site characterization, free product removal, and investigation of soil and groundwater cleanup
- A notification before permanent closure or change in service

Recordkeeping requirements. Facilities must maintain the following information:
- A corrosion expert's analysis of site corrosion potential if corrosion protection equipment is not used
- Documentation of operation of corrosion protection equipment
- Documentation of UST system repairs
- Recent compliance with release detection requirements
- Results of the site investigation conducted at permanent closure

OTHER REPORTING REQUIREMENTS

Other regulations (in addition to CERCLA and EPCRA) may also require reporting of releases of hazardous materials. Some of these are as follows:

- The Hazardous Materials Transportation Act (HMTA) requires reporting of hazardous material releases during shipping and handling. These regulations are administered by the Department of Transportation (DOT).
- The Toxic Substances Control Act (TSCA) requires reporting of releases of polychlorinated biphenyls.

- The Clean Water Act (CWA) requires reporting of hazardous substances listed under 40 CFR 116.4 and oil releases from vessels and facilities into navigable waters.
- RCRA *listed* hazardous wastes (F, K, P, U) under 40 CFR Part 261, Subpart D, and *characteristic* wastes (I, C, R, TC) under 40 CFR Part 261, Subpart C, are reportable if the release equals or exceeds the designated RQ.

CONTROL OF AIR POLLUTION FROM AIRCRAFT AND AIRCRAFT ENGINES

The sections within this part (40 CFR Part 87) deal with setting limits on exhaust emissions of smoke from aircraft engines, ban venting of fuel emissions into the atmosphere from certain types of aircraft engines, and require the elimination of intentional discharge to the atmosphere of fuel drained from fuel nozzle manifolds after engines are shut down. This does not apply to normal fuel seepage from shaft seals, joints, and fittings.

KEY TERMS

Accident

Fatal injury

Serious injury

Substantial damage

Incident

NTSB accident classifications: major, serious, injury, damage

Runway incursion

Near-midair collision

Mandatory reporting system

Voluntary reporting system

Aviation Standards National Field Office

Aviation Accident Data System

Accident Incident Data System

Service Difficulty Reporting System

Air Operator Data System

Aviation Safety Analysis System

Office of Air Traffic Evaluations and Analysis

Operational Error Reporting System

Air Traffic Activity Database

National Airspace Incident Monitoring System

Aviation Safety Reporting System

Aviation Safety Action Program (ASAP)

Flight Operational Quality Assurance (FOQA)

Air Transportation Oversight System (ATOS)

Safety attribute inspection (SAI)

Element performance inspection (EPI)

Safety Performance Analysis System (SPAS)

Part 217

Part 234

Part 241

Part 291

Part 298

Flight data recorder (FDR)

Cockpit voice recorder (CVR)

ICAO Accident/Incident Reporting System (ADREP)

OSHA Forms 300, 300A, and 301

OSHA injury and illness

Standard industrial classification (SIC) codes

CWA, CERCLA, EPCRA (SARA Title III), RCRA, RCRA Subtitle I, TSCA, and HMTA reporting

State emergency response commissions (SERCs)

Extremely hazardous substances (EHS)

Local emergency planning committees (LEPCs)

Threshold planning quanity (TPQ

Toxic Release Inventory (TRI)

Underground storage tanks (USTs)

Pollution from aircraft and aircraft engines—40 CFR Part 87

REVIEW QUESTIONS

1. What is the difference between accidents and incidents? Identify several types of incidents.

2. List and define NTSB's classification of accidents.

3. Why do accident investigations still provide the best insight and information leading to accident prevention? What is the purpose of publicizing safety recommendations following an accident?

4. List and briefly describe the characteristics of a successful incident reporting system. What is the difference between a mandatory and voluntary reporting system? What characteristic of voluntary systems is particularly important?

5. Describe the primary responsibilities of the Associate Administrator for Aviation Standards. What is included in the FAA Accident Incident Data System? What is the purpose of the Service Difficulty Reporting System? How will the new Aviation Safety Analysis System work?

6. What is the function of the Office of Air Traffic Evaluations and Analysis? Describe the three databases managed by the Office of Aviation Safety under the National Airspace Incident Monitoring System.

7. Discuss the salient features of ASAP, FOQA, and ATOS.

8. Describe the database system managed by the NTSB. Why was NASA chosen to manage the Aviation Safety Reporting System? What is its function? The Office of Aviation Information Management of the Research and Special Programs Administration under the DOT collects a great deal of data on airline operations, traffic, and finances. Describe some of these databases and their relationship to safety.

9. What is the purpose of flight data recorders and cockpit voice recorders? How might they be used for accident prevention purposes? What is the ICAO ADREP? Give several examples of information provided by ICAO and based on ADREP data.

10. Discuss OSHA's recording requirements for injuries and illnesses and discuss the reasons for collecting these data.

11. Discuss the role that BLS plays in recording and analyzing occupational injury/illness data.

12. Discuss some of the salient features of EPA's hazardous substance reporting requirements, giving reasons for the same.

13. Discuss the differences in reporting requirements under CERCLA and EPCRA.

14. List one salient feature of reporting under CWA, RCRA, RCRA Subtitle I, and TSCA.

15. What does 40 CFR Part 87 address?

REFERENCES

International Civil Aviation Organization. 1987. *Accident/Incident Reporting Manual (ADREP Manual)*. Montreal, Canada: ICAO.

———. 1984. *Accident Prevention Manual* (Doc. 9422-AN/923). Montreal, Canada: ICAO.

———. 1994. *Aircraft Accident and Incident Investigation (Annex 13)*. Montreal, Canada: ICAO.

Monan, Capt. W., and Capt. H. Orlady. 1995. ASRS Report Update. *Air Line Pilot*. Washington, D.C.: ALPA, March.

U.S. Congress, General Accounting Office. 1980. *How to Improve the Federal Aviation Administration's Ability to Deal with Safety Hazards*. Washington, D.C.: GAO, February.

U.S. Department of Transportation, Federal Aviation Administration. 1985. *Information Resource Management Plan*, vol. 1: *Strategic Overview*. Washington, D.C.: Supt. of Documents, U.S. Government Printing Office, October.

———. 1986. *Information Resource Management Plan*, vol. 2: *Systems Plan FY 87–FY 89*. Washington, D.C.: Supt. of Documents, U.S. Government Printing Office, December.

WEB REFERENCES

EPA Laws and regulations, Chapter 1, Environmental Protection Agency, http://www.epa.gov/docs/epacfr40/chapt-I.info/

EPA Reporting Requirements

http://www.epa.gov/ebtpages/indureporting.html

http://www.aee.faa.gov/aee-200/emer_plan/EPCRAG3.htm

http://www.asy.faa.gov/safety_data/

http://www.epa.gov/ceppo/pubs/cerep.txt

http://www.epa.gov/tri/

http://www.faa.gov/avr/afs/atos/overview/ATOS_Model.htm

http://www.flightsafety.org/foqa.html

http://www.osha-slc.gov/recordkeeping/entryfaq.html

http://www2.faa.gov/avr/afs/afs200/afs230/asap/

REVIEW OF SAFETY STATISTICS

LEARNING OBJECTIVES

After completing this chapter, you should be able to

- Explain why the accident rate is a better measure of safety than accident counts.
- Discuss some of the issues to be aware of when analyzing and comparing commercial aviation accident statistics.
- Discuss the results of Boeing's *Summary of Commercial Jet Transport Aircraft Accidents.*
- Summarize the trend in aviation accident statistics as reported by the NTSB for Part 121 and Part 135 operators during the 1980s, 1990s, and 2000s.
- Define an OSHA illness and injury.
- Explain OSHA injury and illness rate calculations.
- Explain what SIC codes are and identify SIC codes relevant to aviation.

INTRODUCTION

In passenger transportation, *safety factors* (also referred to as *causal factors*) are events that are associated with or influence fatality rates. A *safety indicator* is a measurable safety factor. The probability of death (or injury) as a result of traveling on a given mode, when quantified, is the primary benchmark of passenger transportation safety. Vehicle accident rates also are commonly used as safety indicators since most passenger fatalities occur as a result of vehicle accidents.

If one were to look at risk as the probability of death, then the risk due to traveling can be established from fatality rates. However, commercial aviation fatality rates are poor indicators of short-term risk changes, as a single large jet can result in the deaths of a few hundred people. Hence, a single accident can greatly influence fatality rates. Therefore the trending of fatality rates requires data from extended time periods (5 years or more). As an alternative, accident rates can be used instead of fatality rates as indicators of safety levels. The number of fatalities, even for a specific type of accident, varies greatly with each crash. However, while the number of accidents may have a narrower range of annual variance than the number of fatalities, it has similar analysis issues. While accidents exceed fatalities in numbers, they are still quite small. Also, accidents can vary significantly from one year to the next, and hence accident rates are also poor indicators of short-term estimates of risk trends.

Before one can use accident statistics for safety analysis, it is necessary to understand the denominators of *exposure data* used in transportation accident or fatality rates. Exposure data are information that indicates the amount of opportunity for an event to occur. Cycles, distance, and time for passengers and vehicles are the principal exposure types. They are used in the denominator of rates, such as fatalities per passenger departure or electrical system failures per aircraft-hour. The choice of type of exposure data will affect how rates are compared between and within the transportation modes. Passenger-miles (the number of passengers multiplied by the miles traveled) are most appropriate when one is comparing air transportation with other modes of transportation, as they allow generalized broad-based system comparisons. Using trips between the same city pairs to compare risks for different modes of travel would be very appealing. However, the number of city pairs in the United States is extremely large, and some modes of transportation do not have passenger data in the format required for this type of comparison. In addition, risk per passenger-mile is not uniform over a trip and can vary by the type of route taken or the time of day.

The probability of an accident is significantly higher during takeoff or landing than in any other phase of flight. Hence, the bulk of commercial aviation passenger-miles occur under conditions when the overall average risk is much lower. Vehicle-miles (as opposed to passenger numbers) are preferred in the denominator for comparison of exposure and accident rates, as the number of passengers per vehicle usually vary. Finally, vehicle-miles traveled do not reflect the risks associated with the size of the vehicle and/or its speed.

Aviation accidents can vary greatly in severity, ranging from a flight attendant back injury gotten from pushing carts during in-flight service or handling overhead luggage, to several hundred deaths due to an aircraft crash. To render a more equitable perspective on things, some databases categorize accidents as fatal and nonfatal. However, even here, a single death on a ramp during an aircraft pushback would still be classified as an aviation accident. To address this issue, the NTSB has recently begun classifying accidents as *major, serious, injury,* or *damage* (see Chap. 4 for definitions). Another classification (popular with the insurance and air-craft manufacturing industries) of accidents includes (aircraft) *hull loss* and *non-hull loss.* A hull loss is airplane damage that is beyond economic repair.

Accident counts by themselves cannot be reliably used to measure relative safety among organizations and their products. All things being equal, an airline that has a larger fleet of aircraft could be expected (statistically speaking) to have a larger number of accidents than an airline with far fewer planes. Similarly, aircraft models that are flown more often would be expected to be involved in more acci-dents than less frequently used models. For this reason the *accident rate,* which is the number of accidents divided by some common base variable (e.g., flight hours, departures, and miles flown), is a more valid indicator of relative safety than just accident counts. In this context, time (flight hours) is a popular measure of expo-sure in many types of risk analysis. Since flight-hour data are needed for economic, operational, and maintenance reasons, airlines keep accurate records of such data which are readily available for safety rate comparisons. Also, rate data are used by the NTSB to compare safety performances among airlines and other sectors of the air transportation industry.

Takeoffs and landings are the riskiest phases of flight. There are more acci-dents during takeoffs and landings than in any other phase of flight. So, if takeoffs and/or landings are used as a base for rate calculations, then the airline that pre-dominately flies smaller aircraft more frequently on shorter routes would be favored more than the airline that predominately flies larger aircraft on long routes (and would consequently have fewer takeoffs and landings). The bias reverses in the above example if hours flown are used as the base for accident rate calculation. Then again, since most of the risk involved with air transportation is associated with takeoffs and landings, a 1000-mile trip is similar to a 100-mile trip when both are compared for safety. Overall, there is a stronger correlation between accidents and departures than there is between accidents and flight hours or between acci-dents and passenger-miles. Therefore, aircraft departures (or landings) are a valid exposure parameter for air transportation.

The air transportation industry is extremely dynamic with technological advances, operational strategies (point-to-point versus hub and spoke), and mergers, consolida-tions, and bankruptcies rapidly changing the aviation landscape. In addition, airline activity has nearly doubled over the past 20 years. The aviation industry of today is very different from what it was 20 years ago. Hence, one should exercise care when comparing safety statistics over long periods. While there is no clear answer as to what time domain is appropriate for analyzing and comparing aviation safety data, the most recent 5- to 10-year span appears to be the accepted norm among federal regulators.

Aviation accidents are rare occurrences, and the risk of death or serious injury by air travel is minuscule. The domestic airlines as a group average just under four cata-strophic accidents per year (even less if one were to exclude sabotage and terrorism).

With such small numbers, making statistical inferences from the data to evaluate safety performance of the industry is very difficult. Regulatory agencies and the aviation industry maintain a wide variety of safety-related information. However, one must take into consideration the accuracy and completeness of these data when developing safety trends and recommending control measures. Safety factors other than fatalities or accidents that include the nature and causes of accidents (covered in Chap. 6) should be studied and used for preemptive strikes against safety exposures and to provide timely feedback to management and policymakers.

In summary, no single measurement provides the complete safety picture. Passenger-exposure data are used when passenger risk is to be described. Passenger-miles are used when the influence of vehicle size or speed on the data is nonexistent. Departure-exposure data account for nonuniform risk over a trip. Time is the most widely used exposure measure, and time-based data are often readily available.

AVIATION ACCIDENT STATISTICS

Chapter 4 reviewed various data reporting systems, both voluntary and mandatory, that are used in aviation. While ensuring compliance or consistency in safety reporting is difficult in general, voluntary systems have added shortcomings in that they are less reliable. Reasons for this were also discussed. There are several reliable sources of accident data. One of the most easily accessible accident databases is maintained by Boeing, which publishes an annual *Statistical Summary of Commercial Jet Airplane Accidents*. Another good source document is the International Air Transport Association (IATA) *Safety Record (Jet)*, also published annually. Finally, the definitive and mandatory source of aircraft accident data is the NTSB database.

MANUFACTURERS' INVOLVEMENT WITH SAFETY DATA

Safety professionals with the airframe and engine manufacturers can be found at accident sites participating in the investigation. They collect and analyze data. They recommend improvements in the way aircraft are designed and built and in the way they are operated and maintained. The mandate is to learn to prevent future mishaps.

While manufacturers may undertake similar functions, the organizational framework for carrying out those functions can differ markedly between them. For example, accident investigation and safety data analysis are separate units at Boeing, while they are combined at other manufacturers and often linked to customer support in a single organization.

The closest thing to a constant among manufacturers' safety departments is their responsibility for accident investigation. When a company's product is involved in an accident, that company has a duty to help find the cause. In the parlance of the NTSB, the company becomes a party to the investigation. That means the manufacturer's representatives work alongside NTSB investigators in examining evidence at the accident site. The parties conduct subsequent tests and suggest findings, but the final analysis and published report are the NTSB's exclusively.

Safety departments scrutinize incidents as well as accidents. Committees are formed to represent various departments.

Based on the events reviewed, the committees recommend design changes or revisions in maintenance or operating procedures. Accident reports produce a lot of data. Still more data come from incident reports and other files compiled by airlines, manufacturers, and government agencies. The FAA, for example, records Service Difficulty Reports. Manufacturers' safety departments have come to view these data as a resource to be developed and cultivated. A company such as Boeing may be known as an airframe manufacturer, but it also is in the business of producing and analyzing data.

Boeing maintains several databases. One is the Jet Transport Accident File. It contains details on accidents worldwide involving not only its aircraft but all makes and models above 60,000 pounds. Another, called Jet Transport Safety Events, tracks incidents as well as accidents.

A separate file tracks jet-engine problems, again for all makes and models. Other equipment failures are tracked in a component reliability database, this time limited to Boeing aircraft. The data come from a variety of sources. Manufacturers' service representatives around the world report regularly on anomalous events both large and small.

Airlines often report directly to manufacturers. Other data sources include civil authorities and international organizations such as the International Air Transport Association, the U.K. CAA, and U.S. FAA, as well as insurance underwriters and publications. Even with all those collection efforts, safety staff don't claim that a record of every event unfailingly finds its way to the appropriate database. When little or no damage is involved, airlines sometimes make no external reports.

For example, Boeing had worldwide records of six 737 takeoff tail strikes for 1 year. But when a training team was visiting Brazil, conversation with pilots revealed that Transbrasil alone had 17 such strikes for that period. Boeing then asked USAir to survey its 737 fleet, which revealed a much higher strike rate than indicated originally.

While manufacturers of large airframes maintain what probably are the most extensive databases, others also are collecting such information. Makers of engines and other components are also in the business.

Because Sundstrand produces ground-proximity warning systems (GPWSs), for example, it focuses on controlled-flight-into-terrain (CFIT) accidents. Safety data are one area in which airframe makers usually cooperate.

Unfortunately, the situation is different on the propulsion side. Contact between GE, Pratt & Whitney, and Rolls-Royce is not encouraged because of the appearance of collusion among competitors. Airlines can obtain safety data from manufacturers but without direct access to the files. In our litigious society, none of the manufacturers has agreed to download its database directly to the carriers.

The purpose of gathering the data is trend analysis. In one study, Boeing staff looked for patterns in cases of flightcrew entering incorrect data into flight management computers. Boeing trend analysis is continually looking at interplay among component failures. Does one failure frequently serve as the precursor for another?

Computers may be able to discern patterns that a human analyst would miss. Boeing has taken a proactive stance in the area of safety. It no longer simply makes sure the product was designed correctly and leaves the rest to the carriers. A team assembled by Boeing uses accident records and other data to identify current safety issues. For example, Boeing was a leading proponent in stressing the importance of installing GPWS and the need for proper training in its use.

Similarly, to prevent approach and landing accidents, Boeing has urged that every runway used by commercial transports be equipped with an instrument landing system (ILS). And several training measures have been advocated for pilots and controllers to alleviate the dangers associated with nonstabilized approaches.

Boeing research also has addressed "crew-caused accidents." Its data and other sources identify flightcrew error as the primary cause in close to 70 percent of commercial jet hull loss accidents (see next section). However, accidents rarely have single causes. Manufacturers' prevention strategies address all the factors, not just the primary one. Removing even one link in the chain of events can prevent many accidents.

Airlines maintain databases and conduct trend analyses of their own. These analyses can benefit from work done by the manufacturers. For example, American studied a series of 757 tail strikes. After assessing its own records, the carrier checked with Boeing about the experiences of other 757 operators. Once the problem was identified, American and Boeing worked to solve it by revising training methods.

Manufacturers can alert carriers to problems they didn't know they had. While reviewing operating data from Metro customers in 1993, Fairchild safety staff noticed one carrier with what appeared to be an abnormal rate of transponder problems. Consultation with the transponder producer confirmed that the rate was higher than that for others operating the equipment. The carrier, it turned out, was using an improper repair procedure, which was corrected.

Fairchild not only builds aircraft but also operates and maintains them, and these functions help bring safety problems to light. A subsidiary, Fairchild Aircraft Services, provides repairs, modifications, and other support services in San Antonio. One high-time Metro arrived at that base with several cracks in flap ribs and skins. The cracks were repaired; and based on those findings, skin thicknesses were changed on the production line, doublers were added, and some fasteners were relocated.

Data flow steadily between manufacturers and their customers. An example is the Boeing magazine *Boeing Airliner*.

Metro operators receive a monthly trend monitoring report with statistics on factors such as dispatch reliability, in-flight engine shutdowns, and unscheduled component removals. The reports are based on operating data from more than 25 operators of close to 250 Metros.

The large airframe manufacturers also publish annual reports on their compilations of industrywide accident data, although they differ in the way the data are analyzed and presented. For day-to-day problem solving, airlines have more contact with a manufacturer's accident and incident investigators than its data analysts.

In the discussions that follow, databases from Boeing and the NTSB will be explored to analyze aviation accident statistics.

BOEING'S ACCIDENT STATISTICAL SUMMARY

According to the Boeing statistical summary, in the entire history (worldwide) of scheduled commercial jet operations (1959 to 2001) there have been 1307 accidents resulting in 24,700 onboard fatalities. Of the 1307 accidents, 498 accidents caused fatal injuries of which 421 were hull loss accidents, 19 were attributed to substantial damage accidents, and 58 were a result of less than substantial damage accidents. If one were to consider accidents by type of operation, 79 percent (1033) of the 1307 accidents were passenger operations, 13 percent (169) were cargo operations, and the remaining 105 accidents occurred during testing, training, demonstration, or ferrying. U.S. and Canadian operations collectively had about 34 percent (445) of the 1307 worldwide accidents, contributing to about 25 percent (6077) of the 24,700 global onboard fatalities. Over this period, there have been in excess of 395 million cumulative departures and more than 644 million cumulative flight hours. The industry is now operating 16,144 jet aircraft, with little over 17 million departures each year.

If we look at the plot of all accidents for the worldwide commercial jet fleet for the period 1959 to 2001, we can see that the rates for all accidents and those involving hull losses has been fairly stable for the past 27 years (Fig. 5-1). If we apply control limits to these data, generally 3 standard deviations on either side of the number of hull-loss accidents each year, we can clearly see that the jet airplane transportation process has been under statistical control for a number of years. However, even if this low accident rate remains constant over the next few years, we can expect to see an increase in the actual number of hull-loss accidents each year as the fleet increases in number of departures.

Hull losses and fatal accidents were also analyzed according to the phase of flight in which they occurred (Fig. 5-2). The combined final approach-and-landing phases accounted for 54 percent of the hull-loss and fatal accidents, followed by the combined phases from loading through initial climb (32 percent). Cruise, which accounts for about 57 percent of flight time in a 1.5-hour flight, occasioned only 6 percent of hull-loss accidents.

The summary also considered primary cause factors for commercial operation hull-loss accidents for the period 1992 to 2001 (Fig. 5-3). For accidents with

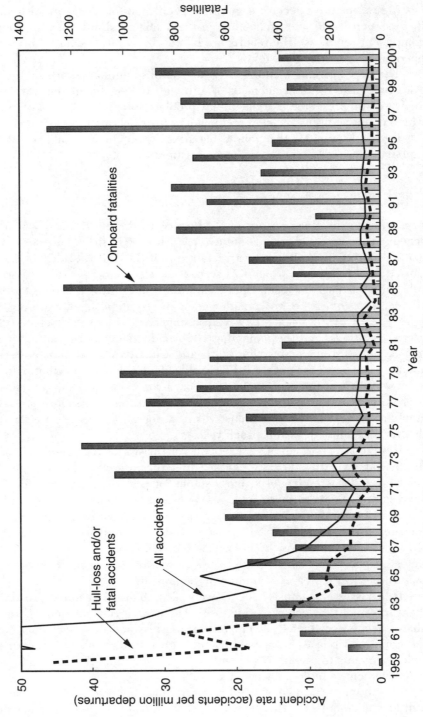

FIGURE 5-1 Accident rates and fatalities by year; all accidents, worldwide commercial jet fleet, 1959 through 2001. (*Source: Boeing 2001 Statistical Summary, June 2002.*)

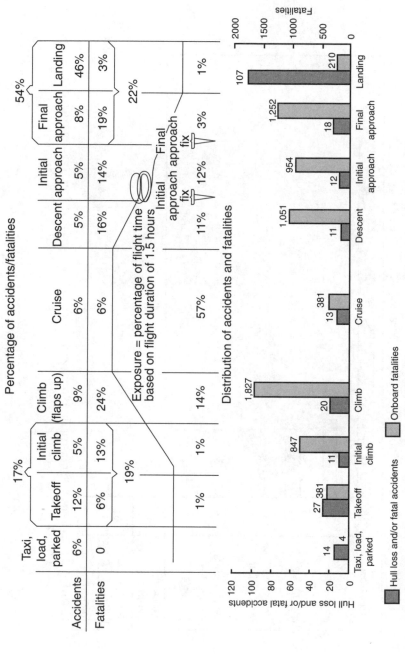

FIGURE 5-2 Accidents and onboard fatalities by phase of flight; hull-loss and/or fatal accidents, worldwide commercial jet fleet, 1992 to 2001. (*Source: Boeing 2001 Statistical Summary, June 2002.*)

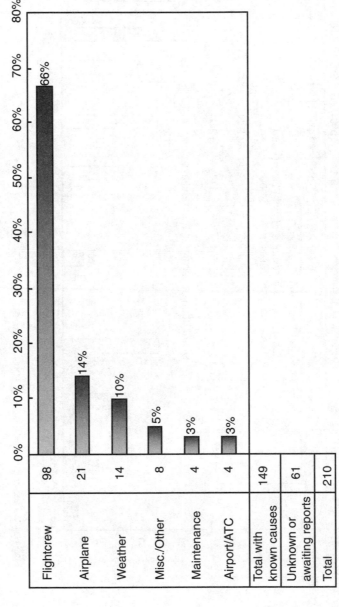

FIGURE 5-3 Accidents by primary cause,* hull loss, worldwide commercial jet fleet, 1992 through 2001. (*As determined by the investigating authority; *Source: Boeing 2001 Statistical Summary, June 2002.*)

known causes, flightcrews were considered the primary cause in most accidents—66 percent over the 10-year period.

Finally, fatalities by accident categories were covered for the period 1992 to 2001 (Fig. 5-4). Loss of control in flight accounted for the largest number, followed by controlled flight into terrain (CFIT).

Boeing's accident data exclude turboprop aircraft as well as those with maximum gross weight of 60,000 pounds (27,216 kilograms) or less; Soviet Union and Commonwealth of Independent States accidents; and accidents resulting from sabotage, hijacking, suicide, and military action.

Several important lessons can be learned from the Boeing data. The most obvious of these is the issue of human error. The human factor is clearly and consistently the most frequent cause of incidents and accidents in the airline industry. It must be kept in mind that Fig. 5-3 only shows the tip of the human-error problem; flight crewmembers are not the only source of error in the aviation system. Hidden in each of the other categories of primary cause factors is a significant, although sometimes hard-to-determine, human-error component. In the maintenance category, for example, recent unpublished data suggest that more than one-half of these involve human factors, and it seems likely that similar rates of human-error involvement can be found in the other categories as well. This is why estimates of the total contribution of human error to aviation incidents and accidents can range as high as 80 to 90 percent.

Based largely on data such as these, it has become increasingly apparent that the management and control of human error are the largest single challenge facing airline safety management. This recognition is not new. In 1974, the Flight Safety Foundation held its 28th Annual International Aviation Safety Seminar in Williamsburg, Va. The theme was human factors in flight operations, and it is believed that this seminar was the first major airline industry safety conference devoted to human error. In 1975, IATA sponsored its 20th Technical Conference: Safety in Flight Operations. The conference proceedings almost exclusively focused on human factors and human-error issues, which were termed "the last frontier of aviation safety." These early industry efforts were at the vanguard of present-day approaches to the problem of management and control of human error in aviation operations. The subject of human error is thoroughly explored in Chap. 7.

NTSB's Accident Statistical Summary

For the United States, the primary source of safety data is the NTSB. Tables 5-1 through 5-5 present a year-by-year summary of accidents, fatalities, and accident rates for U.S. carriers operating in scheduled and nonscheduled service under FAR Part 121 (major air carriers) and Part 135 (commuter air carriers and on-demand air taxis), respectively. Note that these data reflect the NTSB's official definition of an accident, of which fatal accidents are a subset. Because of the NTSB's very broad definition of an accident (any incident that involves, e.g., a broken bone is classified as an accident), the United States experiences an average of close to 30

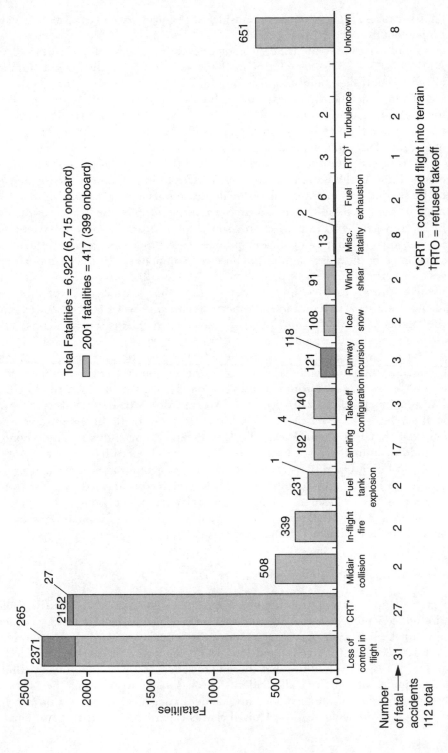

FIGURE 5-4 Fatalities by accident categories, fatal accidents, worldwide commerical jet fleet, 1992 through 2001. (Note: Accidents involving multiple, non-onboard fatalities are included; accidents involving single, non-onboard fatalities are excluded. *Source: Boeing 2001 Statistical Summary, June 2002.*)

TABLE 5-1 Accidents, Fatalities, and Rates, 1982 through 2001, for U.S. Air Carriers Operating Under 14 CFR 121, Scheduled and Nonscheduled Service (Airlines)

YEAR	ACCIDENTS		FATALITIES		FLIGHT HOURS	MILES FLOWN	DEPARTURES	ACCIDENTS PER 100,000 FLIGHT HOURS		ACCIDENTS PER 1,000,000 MILES FLOWN		ACCIDENTS PER 100,000 DEPARTURES	
	ALL	FATAL	TOTAL	ABOARD				ALL	FATAL	ALL	FATAL	ALL	FATAL
*1982	18	5	235	223	7,040,325	2,938,513,000	5,351,133	0.241	0.057	0.0058	0.0014	0.318	0.075
1983	23	4	15	14	7,298,799	3,069,318,000	5,444,374	0.315	0.055	0.0075	0.0013	0.422	0.073
1984	16	1	4	4	8,165,124	3,428,063,000	5,898,852	0.196	0.012	0.0047	0.0003	0.271	0.017
1985	21	7	526	525	8,709,894	3,631,017,000	6,306,759	0.241	0.080	0.0058	0.0019	0.333	0.111
*1986	24	3	8	7	9,976,104	4,017,626,000	7,202,027	0.231	0.020	0.0057	0.0005	0.319	0.028
*1987	34	5	232	230	10,645,192	4,360,521,000	7,601,373	0.310	0.038	0.0076	0.0009	0.434	0.053
*1988	30	3	285	274	11,140,548	4,503,426,000	7,716,061	0.260	0.018	0.0064	0.0004	0.376	0.026
1989	28	11	278	276	11,274,543	4,605,083,000	7,645,494	0.248	0.098	0.0061	0.0024	0.366	0.144
1990	24	6	39	12	12,150,116	4,947,832,000	8,092,306	0.198	0.049	0.0049	0.0012	0.297	0.074
1991	26	4	62	49	11,780,610	4,824,824,000	7,814,875	0.221	0.034	0.0054	0.0008	0.333	0.051
1992	18	4	33	31	12,359,715	5,039,435,000	7,880,707	0.146	0.032	0.0036	0.0008	0.228	0.051
1993	23	1	1	0	12,706,206	5,249,469,000	8,073,173	0.181	0.008	0.0044	0.0002	0.285	0.012
*1994	23	4	239	237	13,124,315	5,478,118,000	8,238,306	0.168	0.030	0.0040	0.0007	0.267	0.049
1995	36	3	168	162	13,505,257	5,654,069,000	8,457,465	0.267	0.022	0.0064	0.0005	0.426	0.035
1996	37	5	380	350	13,746,112	5,873,108,000	8,228,810	0.269	0.036	0.0063	0.0009	0.450	0.061
1997	49	4	8	6	15,838,109	6,696,638,000	10,318,383	0.309	0.025	0.0073	0.0006	0.475	0.039
1998	50	1	1	0	16,813,435	6,736,543,000	10,979,762	0.297	0.006	0.0074	0.0001	0.455	0.009
1999	52	2	12	11	17,555,208	7,101,521,000	11,308,762	0.296	0.011	0.0073	0.0003	0.460	0.018
2000	57	3	92	92	18,295,143	7,520,588,000	11,436,663	0.312	0.016	0.0076	0.0004	0.498	0.026
*2001	40	6	531	525	16,730,700	6,808,918,000	10,509,840	0.215	0.012	0.0053	0.0003	0.343	0.019

NOTES:
- 2001 data are preliminary. Flight hours, miles, and departures are compiled by the Federal Aviation Administration.
- Since March 20, 1997, aircraft with 10 or more seats used in scheduled passenger service have been operated under 14 CFR 121.
- Years marked with * are those in which an illegal act was responsible for an occurrence in this category. These acts, such as suicide and sabotage, are included in the totals for accidents and fatalities but are excluded for the purpose of accident rate computation. Fatalities resulting from the September 11 terrorist acts are excluded from this table.

SOURCE: NTSB.

TABLE 5-2 Accidents, Fatalities, and Rates, 1982 through 2001, for U.S. Air Carriers Operating Under 14 CFR 121, Scheduled Service (Airlines)

Year	Accidents		Fatalities		Flight Hours	Miles Flown	Departures	Accidents per 100,000 Flight Hours		Accidents per 1,000,000 Miles Flown		Accidents per 100,000 Departures	
	All	Fatal	Total	Aboard				All	Fatal	All	Fatal	All	Fatal
*1982	16	4	234	222	6,697,770	2,806,885,000	5,162,346	0.224	0.045	0.0053	0.0011	0.291	0.058
1983	22	4	15	14	6,914,969	2,920,909,000	5,235,262	0.318	0.058	0.0075	0.0014	0.420	0.076
1984	13	1	4	4	7,736,037	3,258,910,000	5,666,076	0.168	0.013	0.0040	0.0003	0.229	0.018
1985	17	4	197	196	8,265,332	3,452,753,000	6,068,893	0.206	0.048	0.0049	0.0012	0.280	0.066
*1986	21	2	5	4	9,495,158	3,829,129,000	6,928,103	0.211	0.011	0.0052	0.0003	0.289	0.014
*1987	32	4	231	229	10,115,407	4,125,874,000	7,293,025	0.306	0.03	0.0075	0.0007	0.425	0.041
*1988	29	3	285	274	10,521,052	4,260,785,000	7,347,575	0.266	0.019	0.0066	0.0005	0.381	0.027
1989	24	8	131	130	10,597,922	4,337,234,000	7,267,341	0.226	0.075	0.0055	0.0018	0.330	0.110
1990	22	6	39	12	11,524,726	4,689,287,000	7,795,761	0.191	0.052	0.0047	0.0013	0.282	0.077
1991	25	4	62	49	11,139,166	4,558,537,000	7,503,873	0.224	0.036	0.0055	0.0009	0.333	0.053
1992	16	4	33	31	11,732,026	4,767,344,000	7,515,373	0.136	0.034	0.0034	0.0008	0.213	0.053
1993	22	1	1	0	11,981,347	4,936,067,000	7,721,870	0.184	0.008	0.0045	0.0002	0.285	0.013
*1994	19	4	239	237	12,292,356	5,112,633,000	7,824,802	0.146	0.033	0.0035	0.0008	0.230	0.051
1995	34	2	166	160	12,776,679	5,328,969,000	8,105,570	0.266	0.016	0.0064	0.0004	0.419	0.025
1996	32	3	342	342	12,971,676	5,449,997,000	7,851,298	0.247	0.023	0.0059	0.0006	0.408	0.038
1997	44	3	3	2	15,061,662	6,339,432,000	9,925,058	0.292	0.02	0.0069	0.0005	0.443	0.030
1998	43	1	1	0	15,921,102	6,343,690,000	10,535,196	0.270	0.006	0.0068	0.0002	0.408	0.009
1999	47	2	12	11	16,693,365	6,689,327,000	10,860,692	0.282	0.012	0.0070	0.0003	0.433	0.018
2000	51	3	92	92	17,474,405	7,148,928,000	11,022,759	0.292	0.017	0.0071	0.0004	0.463	0.027
*2001	36	6	531	525	15,998,000	6,437,708,000	10,099,710	0.200	0.013	0.0050	0.0003	0.317	0.020

NOTES:
- Flight hours, miles, and departures are compiled by the Federal Aviation Administration.
- Since March 20, 1997, aircraft with 10 or more seats used in scheduled passenger service have been operated under 14 CFR 121.
- Years marked with * are those in which an illegal act was responsible for an occurrence in this category. These acts, such as suicide and sabotage are included in the totals for accidents and fatalities but are excluded for the purpose of accident rate computation. Fatalities resulting from the September 11 terrorist acts are excluded from this table.

SOURCE: NTSB.

TABLE 5-3 Accidents, Fatalities, and Rates, 1982 through 2001, for U.S. Air Carriers Operating Under 14 CFR 121, Nonscheduled Service (Airlines)

Year	Accidents		Fatalities		Flight Hours	Miles Flown	Departures	Accidents per 100,000 Flight Hours		Accidents per 1,000,000 Miles Flown		Accidents per 100,000 Departures	
	All	Fatal	Total	Aboard				All	Fatal	All	Fatal	All	Fatal
1982	2	1	1	1	342,555	131,628,000	188,787	0.584	0.292	0.0152	0.0076	1.059	0.53
1983	1	0	0	0	383,830	148,409,000	209,112	0.261	—	0.0067	—	0.478	—
1984	3	0	0	0	429,087	169,153,000	232,776	0.699	—	0.0177	—	1.289	—
1985	4	3	329	329	444,562	178,264,000	237,866	0.900	0.675	0.0224	0.0168	1.682	1.261
1986	3	1	3	3	480,946	188,497,000	273,924	0.624	0.208	0.0159	0.0053	1.095	0.365
1987	2	1	1	1	529,785	234,647,000	308,348	0.378	0.189	0.0085	0.0043	0.649	0.324
1988	1	0	0	0	619,496	242,641,000	368,486	0.161	—	0.0041	—	0.271	—
1989	4	3	147	146	676,621	267,849,000	378,153	0.591	0.443	0.0149	0.0112	1.058	0.793
1990	2	0	0	0	625,390	258,545,000	296,545	0.320	—	0.0077	—	0.674	—
1991	1	0	0	0	641,444	266,287,000	311,002	0.156	—	0.0038	—	0.322	—
1992	2	0	0	0	627,689	272,091,000	365,334	0.319	—	0.0074	—	0.547	—
1993	1	0	0	0	724,859	313,402,000	351,303	0.138	—	0.0032	—	0.285	—
1994	4	0	0	0	831,959	365,485,000	413,504	0.481	—	0.0109	—	0.967	—
1995	2	1	2	2	728,578	325,100,000	351,895	0.275	0.137	0.0062	0.0031	0.568	0.284
1996	5	2	38	8	774,436	423,111,000	377,512	0.646	0.258	0.0118	0.0047	1.324	0.53
1997	5	1	5	4	776,447	357,206,000	393,325	0.644	0.129	0.0140	0.0028	1.271	0.254
1998	7	0	0	0	892,333	392,853,000	444,566	0.784	—	0.0178	—	1.575	—
1999	5	0	0	0	861,843	412,194,000	448,070	0.580	—	0.0121	—	1.116	—
2000	6	0	0	0	820,738	371,660,000	413,904	0.731	—	0.0161	—	1.45	—
2001	4	0	0	0	732,700	371,210,000	410,130	0.546	—	0.0108	—	0.975	—

NOTES: 2001 data are preliminary. Flight hours, miles, and departures are compiled by the Federal Aviation Administration.
SOURCE: NTSB.

TABLE 5-4 Accidents, Fatalities, and Rates, 1982 through 2001, for U.S. Air Carriers Operating Under 14 CFR 135, Scheduled Service

Year	Accidents		Fatalities		Flight Hours	Miles Flown	Departures	Accidents per 100,000 Flight Hours		Accidents per 1,000,000 Miles Flown		Accidents per 100,000 Departures	
	All	Fatal	Total	Aboard				All	Fatal	All	Fatal	All	Fatal
1982	26	5	14	14	1,299,748	222,355,000	2,026,691	2.000	0.385	0.1169	0.0225	1.283	0.247
1983	16	2	11	10	1,510,908	253,572,000	2,328,430	1.059	0.132	0.0631	0.0079	0.687	0.086
1984	22	7	48	46	1,745,762	291,460,000	2,676,590	1.260	0.401	0.0755	0.0240	0.822	0.262
1985	18	7	37	36	1,737,106	300,817,000	2,561,463	1.036	0.403	0.0598	0.0233	0.703	0.273
1986	14	2	4	4	1,724,586	307,393,000	2,798,811	0.812	0.116	0.0455	0.0065	0.500	0.071
1987	33	10	59	57	1,946,349	350,879,000	2,809,918	1.695	0.514	0.0940	0.0285	1.174	0.356
1988	18	2	21	21	2,092,689	380,237,000	2,909,005	0.860	0.096	0.0473	0.0053	0.619	0.069
1989	19	5	31	31	2,240,555	393,619,000	2,818,520	0.848	0.223	0.0483	0.0127	0.674	0.177
1990	15	3	6	4	2,341,760	450,133,000	3,160,089	0.641	0.128	0.0333	0.0067	0.475	0.095
1991	23	8	99	77	2,291,581	433,900,000	2,820,440	1.004	0.349	0.0530	0.0184	0.815	0.284
*1992	23	7	21	21	2,335,349	507,985,000	3,114,932	0.942	0.300	0.0433	0.0138	0.706	0.225
1993	16	4	24	23	2,638,347	554,549,000	3,601,902	0.606	0.152	0.0289	0.0072	0.444	0.111
1994	10	3	25	25	2,784,129	594,134,000	3,581,189	0.359	0.108	0.0168	0.0050	0.279	0.084
1995	12	2	9	9	2,627,866	550,377,000	3,220,262	0.457	0.076	0.0218	0.0036	0.373	0.062
1996	11	1	14	12	2,756,755	590,727,000	3,515,040	0.399	0.036	0.0186	0.0017	0.313	0.028
1997	16	5	46	46	982,764	246,029,000	1,394,096	1.628	0.509	0.0650	0.0203	1.148	0.359
1998	8	0	0	0	353,670	50,773,000	707,071	2.262	—	0.1576	—	1.131	—
1999	13	5	12	12	342,731	52,403,000	672,278	3.793	1.459	0.2481	0.0954	1.934	0.744
2000	12	1	5	5	373,649	45,520,000	610,661	3.212	0.268	0.2636	0.0220	1.965	0.164
2001	7	2	13	13	330,500	38,950,000	497,514	2.118	0.605	0.1797	0.0513	1.407	0.402

NOTES:
• 2001 data are preliminary.
• Since March 20, 1997, aircraft with 10 or more seats used in scheduled passenger service have been operated under 14 CFR 121.
• Flight hours, miles, and departures are compiled by the Federal Aviation Administration.
• Years marked with * are those in which an illegal act was responsible for an occurrence in this category. These acts, such as suicide and sabotage are included in the totals for accidents and fatalities but are excluded for the purpose of accident rate computation.

SOURCE: NTSB.

TABLE 5-5 Accidents, Fatalities, and Rates, 1982 through 2001, for
U.S. Air Carriers Operating Under 14 CFR 135, Nonscheduled
Service (On-Demand Air Taxis)

	ACCIDENTS		FATALITIES			ACCIDENTS PER 100,000 FLIGHT HOURS	
YEAR	ALL	FATAL	TOTAL	ABOARD	FLIGHT HOURS	ALL	FATAL
1982	132	31	72	72	3,008,000	4.39	1.03
1983	142	27	62	57	2,378,000	5.97	1.14
1984	146	23	52	52	2,843,000	5.14	0.81
1985	157	35	76	75	2,570,000	6.11	1.36
1986	118	31	65	61	2,690,000	4.39	1.15
1987	96	30	65	63	2,657,000	3.61	1.13
1988	102	28	59	55	2,632,000	3.88	1.06
1989	110	25	83	81	3,020,000	3.64	0.83
1990	107	29	51	49	2,249,000	4.76	1.29
1991	88	28	78	74	2,241,000	3.93	1.25
1992	76	24	68	65	*2,844,000	2.67	0.84
1993	69	19	42	42	*2,324,000	2.97	0.82
1994	85	26	63	62	*2,465,000	3.45	1.05
1995	75	24	52	52	*2,486,000	3.02	0.97
1996	90	29	63	63	*3,220,000	2.80	0.90
1997	82	15	39	39	*3,098,000	2.65	0.48
1998	77	17	45	41	*3,802,000	2.03	0.45
1999	73	12	38	38	*3,298,000	2.21	0.36
2000	81	22	71	68	*3,553,000	2.28	0.62
2001	72	18	60	59	3,400,000	2.12	0.53

NOTES:
• 2001 data are preliminary.
• Flight hours are estimated by the Federal Aviation Administration (FAA).
• Flight hours marked with *were substantially revised in February 2002 based on
  a change in the methodology used by FAA to estimate air taxi activity.
SOURCE: NTSB.

reportable accidents involving scheduled and nonscheduled air service each year.
However, serious accidents, those involving fatalities, are much rarer. The average
number of fatal accidents from 1982 to 2001 has been about 4 per year for scheduled and nonscheduled (Part 121) operations. Over this same period scheduled and
nonscheduled (Part 121) departures increased from 5.3 million to 10.5 million.
Over the past 10 years fatal accidents have averaged under 3 accidents per year
(this excludes the 4 fatal accidents due to terrorism in 2001). Clearly, the trend in
fatal accident rates in the United States is in the desired direction.

THE 1980s.   In the NTSB's first 15 years, airline safety improved steadily. In 1967,
the airline fatal accident rate was 0.006 for every 1 million aircraft miles flown. By
1980, this rate was down to 0.001, a reduction of 83 percent. And on January 1,

1982, U.S. airlines had completed 26 months without a catastrophic crash of a pure-jet transport. The period spanned 1980 and 1981; never before had there been even one calendar year without such an accident. The previous record had been 15 months, from September 1971 to December 1972. The airlines flew more than half a billion passengers on more than 10 million flights in the 26-month period—more than half a trillion passenger miles. The aerial transportation involved would have taken every man, woman, and child in the country on a flight of more than 2000 miles.

Just 13 days into 1982, the air carriers' remarkable record came to a shattering end when an Air Florida Boeing 737, taking off from Washington National Airport, crashed onto a bridge in a snowstorm. Seventy passengers, four crewmembers, and four persons on the bridge were killed. There were five fatal accidents in 1982 that produced a fatal accident rate of 0.057 per 100,000 aircraft hours. This rate was down 7 percent from the 0.061 recorded in 1981, when there were four bizarre single-fatality accidents. The fatality total, however, was 235, the third highest in a decade, compared with 4 in 1981 and none in 1980.

Commuter airlines achieved dramatically lower accident rates in 1982, which was the commuters' safest year in the 3 years for which their accident statistics were available. Total commuter accidents dropped from 31 to 26, and fatal accidents dropped from 9 to 5. Total and fatal accident rates were 1.283 and 0.247 in every 100,000 departures, reductions of 24 and 50 percent, respectively, from 1981 rates. This year was the fourth successive year of decrease in the commuters' total accident rate.

In 1983, the commuter airlines achieved sharply lower accident rates for the second successive year. The commuters had only 16 total accidents, 2 of them fatal, compared to 26 total accidents in 1982, of which 5 were fatal. Fatalities dropped from 14 to 11. The rate of 0.687 total accident per 100,000 departures, the rate most often used to measure commuter safety, was 46 percent lower than that in 1982.

The scheduled airlines had 4 fatal accidents in 1983 that produced 15 fatalities. There were 22 total accidents in 1983 compared with 16 in 1982 for a 1983 rate of 0.318 total accident per 100,000 aircraft hours.

The downward trend in accident statistics for the scheduled and nonscheduled air carriers continued in 1984. There was only 1 fatal airline accident, an en route crash of a Zantop International Airlines Lockheed Electra on a cargo flight. Three crewmembers and a lone nonrevenue passenger died in the accident at Chalkhill, Pa. The result was a near-record-low rate of 0.018 fatal accident in every 100,000 scheduled departures. This rate had been bettered only in 1980, when there were no fatal accidents in scheduled airline service.

Only 13 accidents, fatal and nonfatal, occurred in scheduled airline service in 1984. A record low, this produced a total accident rate of 0.229 per 100,000 departures. By comparison, the same rate was almost 3 times higher, 0.659, in 1975.

By the end of 1984, there had not been a catastrophic crash of a U.S. turbojet airliner since the July 9, 1982, takeoff crash of a Pan American B-727 at New Orleans.

That period of nearly 30 months marked the second time since 1979 that the airlines had operated for more than 2 years without a catastrophic crash of a jet airliner.

The earlier record, unprecedented at the time, had been established in the 26 months between a Western Airlines DC-10 landing crash at Mexico City on October 31, 1979, and an Air Florida takeoff crash at Washington, D.C., on January 13, 1982. Before that time, there had never been a single calendar year without such an accident. Unfortunately, the airlines' 1984 achievement was less than one day old when an Eastern Airlines B-727 crashed on New Year's Day near the top of an Andes mountain outside La Paz, Bolivia. This crash wrote a tragic finish to the 30-month record.

All other segments of U.S. civil aviation, except commuters, also recorded improved accident records in 1984. With 7 fatal accidents, the commuters' rate of 0.262 fatal accident in every 100,000 departures was a sharp increase over the 1983 rate, which had been a record low. Yet the 1984 rate still was well below the commuters' fatal accident rates in 1975 to 1981. The fatality toll in 1984 was 48; it was 11 in 1983.

On-demand air taxis in 1984 achieved substantial reductions in both total and fatal accident rates. Their 146 accidents produced a total accident rate of 5.14 per 100,000 hours, a 13 percent reduction from 1983. With 23 fatal accidents, the fatal accident rate was 0.81 per 100,000 hours, down 29 percent. Fatalities totaled 52, compared with 62 in 1983.

The year 1985 began in tragedy. On January 1, an Eastern Airlines B-727 smashed into the side of Mount Illimani, Bolivia, 22,000 feet above sea level during a descent for landing into La Paz. All 29 passengers and crew were killed. Although an NTSB-led expedition reached the accident site, deep snow and severe cold made recovering any of the bodies or the cockpit voice and flight data recorders impossible.

The Bolivian accident was the first of 13 fatal U.S. commuter and major airline accidents in 1985, in which 563 people died. In addition to the Bolivian accident, accidents involving substantial loss of life (8 or more fatalities) included these:

- A Galaxy Airlines charter flight on takeoff from the Reno, Nev., airport on January 21. *Fatalities: 70.*

- A North Pacific Airways commuter flight during landing approach into the Soldotna, Alaska, airport on February 4. *Fatalities: 9.*

- Delta 191 as it encountered a wind shear on final approach into the Dallas–Fort Worth airport on August 2. *Fatalities: 135.*

- A Bar Harbor Airlines commuter flight with child celebrity Samantha Smith on board during an instrument approach at Lewiston, Maine, on August 25. *Fatalities: 8.*

- A Midwest Express flight following an engine failure after takeoff from the Milwaukee airport on September 6. *Fatalities: 31.*

- A Henson Aviation commuter flight during an instrument approach into the Shenandoah Valley Airport near Grottoes, Va., on September 23. *Fatalities: 14.*

- A military-chartered Arrow Airways flight on takeoff from Gander, Canada, on December 12. *Fatalities: 256.*

Although the loss of life in commercial passenger flights was higher in 1985 than in any previous year in the 1980s, the overall record for U.S. civil aviation, as judged by accident rates and fatal accidents, was mixed.

For the scheduled (Part 121) airlines, the fatal accident rate of 0.066 per 100,000 departures in 1985 was more than 3.5 times that of 1984. For unscheduled (Part 121) airline service (the charters), the fatal accident rate of 1.261 per 100,000 departures was more than twice that of 1982, the last year before 1985 where there had been a fatal charter airline accident. In 1985, the total number of fatalities for both the scheduled and nonscheduled (chartered, Part 121) airlines was 526, making it the second-worst year in history for accidents involving U.S. air carriers. Only in 1977, when two charter B-747s collided on the runway at Tenerife Airport in the Canary Islands, were fatalities higher.

For the on-demand air taxis (Part 135), the fatal accident rate per 100,000 hours rose 68 percent in 1985 to 1.36 from 0.81 in 1984. And for the commuters, the fatal accident rate per 100,000 departures rose slightly from 0.262 (in 1984) to 0.273 (in 1985).

After 21 accidents and 5 fatalities in 1986, major mishaps of large, scheduled (Part 121) U.S. airlines hit a 13-year high in 1987 with 32 accidents, resulting in 231 deaths. This number was the highest since 42 accidents occurred in 1974 when departures were almost one-half of those in 1987. The 32 accidents for scheduled airlines resulted in an accident rate of 0.425 per 100,000 departures, up from the 0.289 (a 47 percent increase) recorded in 1986. However, the fatal accident rate of 0.041 represented an almost 200 percent increase from the year before. Among the fatal accidents in 1987 were these:

- Northwest Airlines flight 255, an MD-82, crashed after liftoff from Detroit Metropolitan Wayne County International Airport. *Fatalities: 156,* including 2 on the ground.

- Continental Airlines flight 1713, a DC-9-10, crashed shortly after takeoff from Denver's Stapleton International. *Fatalities: 28.*

- A Buffalo Airways B-707 crashed in Kansas City. *Fatalities: 4.*

l A Pacific Southwest Airlines BAe-146 plummeted to earth in California. *Fatalities: 43.*

In the commuter or regional airlines, there also was a sharp turnabout from 1986. There were 33 accidents in 1987, of which 10 were fatal—the highest since 38 accidents were recorded in 1980. Fatalities hit an 8-year high of 59. There were

1.174 accidents per 100,000 departures, compared with 0.50 in 1986. The fatal accident rate of 0.356 rose from 0.071 in 1986 to the highest level since 1979.

There were 65 fatalities in 1987 for on-demand air taxis, small planes operating nonscheduled service, the same number as in 1986. Their accident rate was 3.61 per 100,000 aircraft hours, down from 4.39 the year before, while the fatal accident rate, at 1.13, was down slightly from the previous year's 1.15.

Accident rates for U.S. scheduled air carriers declined in 1988 from the year before. According to the NTSB's statistics, there were 29 major air carrier accidents in 1988, which is 3 less than in 1987. Of the 29 accidents, 3 were fatal, resulting in 285 fatalities. The accident rate dropped from 0.425 per 100,000 departures in 1987 to 0.381 in 1988. The fatal accident rate fell from 0.041 per 100,000 departures to 0.027.

Commuter airline accidents decreased from 33 in 1987 to 18 in 1988. There were only 2 fatal accidents, compared to the 10 that occurred the previous year, with fatalities down from 59 to 21. The accident rate was 0.619 per 100,000 departures, compared with 1.174 in 1987, for the second-lowest rate in this decade. On-demand air taxis had 102 accidents, up slightly from 1987, for a total of 59 fatalities, down 6 from the year earlier.

The NTSB investigated two unusual major aviation accidents in 1988: an in-flight airline cargohold fire over Tennessee caused by a hazardous material spill and the partial disintegration of a jet over Hawaii. The explosion of a B-747 in the sky over Lockerbie, Scotland, claiming all 259 persons on board and an estimated 11 on the ground, was the last major accident of the year. When British investigators announced a week later that the aircraft was brought down by a bomb, the tragedy became the most deadly act of sabotage ever perpetrated against a United States airliner.

Although the total accident rates for U.S. air carriers and commuters declined in 1989, the fatal accident rates rose. The actual number of fatal accidents for the air carriers was the highest since 1968.

There were 28 accidents involving major U.S. scheduled and charter airlines in 1989, down from 30 in 1988. Of those 28, eleven involved fatalities, which was the highest since 15 fatal accidents in 1968. The fatal accident rate of 0.144 per 100,000 departures was the highest of the decade, up from 0.026 in 1988.

The major U.S. scheduled airlines suffered 24 accidents in 1989, down from 29 in the previous year. Of those 24, eight involved fatalities, the most since 1973. Of the 131 fatalities registered in 1989, 111 of them occurred in the crash of a United Airlines DC-10 in Sioux City, Iowa, on July 19 (a passenger who died 31 days after the accident is not registered in NTSB statistics).

The scheduled airline accident rate per 100,000 departures was 0.330, down from 0.381 in 1988. The fatal accident rate rose from 0.027 to 0.110.

Charter airlines had 4 accidents, 3 of them fatal, resulting in 147 fatalities. All but 3 of those deaths occurred in the February 8 crash of an Independent Air B-707 in the Azores. The 1.058 accident rate was up from 0.271 in 1988. The fatal accident rate in 1989 was 0.793 per 100,000 departures; there were no fatal charter accidents in 1988.

Commuter air carriers had 19 accidents in 1989, which is 1 more than in the previous year. The 5 fatal accidents, up from 2 in 1988, resulted in 31 fatalities, 10 more than in 1988. The 0.674 accident rate per 100,000 departures was slightly higher than the 0.619 rate in 1988, but the fatal accident rate rose from 0.069 to 0.177.

Air taxi accidents rose from 102 in 1988 to 110 in 1989. Although the number of fatal accidents declined from 28 to 25, fatalities increased from 59 to 83. The accident rate declined from 3.88 to 3.64 per 100,000 aircraft hours, and the fatal accident rate declined from 1.06 to 0.83. These numbers reflected the increase in hours flown.

THE 1990s.   The decade of the 1990s started off favorably with a decline in all accident rates for the scheduled, nonscheduled, and commuter air carriers in 1990. Even the on-demand air taxis experienced fewer accidents, but rates increased, reflecting the sharp decline in hours flown.

The major U.S. scheduled airlines experienced 25 accidents in 1991, up slightly from the previous year. Of those 25, four involved fatalities. Of the total 62 fatalities in 1991, 59 involved 2 major accidents.

On February 1, 1991, a USAir B-737 collided with a Skywest Metroliner on landing at Los Angeles International Airport. Both planes were destroyed in the accident, which killed 34 persons.

On March 3, 1991 all 25 persons aboard a United Airlines B-737 were killed when the plane crashed during final approach to Colorado Springs. The aircraft was approximately 1000 feet above the ground when the upset occurred.

Commuter air carriers were involved in 23 accidents in 1991, resulting in 99 fatalities, the largest number in two decades. Fatal accidents per 100,000 departures rose from 0.095 in 1990 to 0.284, an increase of 200 percent. On the other hand, there were 88 accidents and 78 fatalities involving U.S. air taxis in 1991, compared to 107 accidents and 51 fatalities in the year before. The 88 accidents represented the lowest number of air taxi accidents since the NTSB began compiling air taxi records in 1975.

In 1992, there were 7 fatal commuter accidents, compared to 8 in 1991. The number of fatalities aboard commuter air carriers dropped to 21 from 77 a year earlier.

The first of these occurred January 3 near Gabriels, N.Y., when a USAir Express Beech 1900 crashed and killed 2 persons. The airplane was on a scheduled flight from Plattsburgh to Albany when it crashed while on approach to Saranac Lake.

Two persons were killed January 23 when an Air Sunshine Cessna 402 crashed in Clewiston, Fla. The commuter flight, bound for Sarasota from Fort Lauderdale, crashed in poor weather.

On June 7, an American Eagle Casa 212 crashed while on approach to the Eugenio de Hosto Airport at Mayaguez, Puerto Rico. The scheduled commuter flight, which had originated in San Juan, was operating under instrument flight rules. Five persons died in the accident.

Three persons were killed the following day, June 8, when a GP Express Beech C99 commuter airplane crashed at Fort McClellan, Ala., while on approach to Anniston. Weather at the time of the accident was overcast with fog and haze.

On June 19, an Adventure Airlines Cessna 402 lost power in one engine and crashed after takeoff from a landing strip on the Grand Canyon's south rim near Meadview, Ariz. Three persons were killed in the accident, which was captured on videotape by two passengers on the plane.

On October 26, a Pacific Island Air Cessna 402 commuter plane crashed on takeoff at Saipan in the Pacific Ocean, killing 3 persons. The aircraft had taken off behind a DC-10 jetliner when it banked to the right at an altitude of about 250 feet before it went into an uncontrolled descent.

An accident occurred in light rain and fog on October 31 at Grand Junction, Colo., when an Alpine Air Piper PA42 crashed into a mountain while on an ILS approach to Walker Field. Three persons died.

Scheduled air carriers recorded 4 fatal accidents and 33 fatalities in 1992, compared to the same number of accidents and 62 deaths in the year before. The 33 fatalities represented the lowest number of deaths since 1986, when 5 persons in that category died.

There were 76 accidents and 68 fatalities involving air taxis in 1992, compared to 88 and 78 fatalities in the year before. The 76 accidents were the fewest involving air taxis since the NTSB began compiling air taxi records in 1975.

In 1993 the major scheduled airlines experienced only 1 fatal accident, that involving a ground crewmember being struck by a propeller. The fatal accident rate of 0.013 per 100,000 departures was the lowest since 1980, when there were no fatal accidents among the scheduled airlines.

Paradoxically, the scheduled air carriers experienced more accidents in 1993 (22) than in the previous year (16), resulting in a higher total accident rate, 0.285 versus 0.213.

The fatal accident rate for commuter airlines dropped from 0.225 to 0.111 per 100,000 departures, but fatalities rose from 21 in 1992 to 24 in 1993. The total accident rate dropped from 0.706 to 0.444.

On-demand air taxis recorded their lowest number of fatalities in the NTSB's history with 42. In 1992 there were 68. The fatal accident rate dropped from 0.84 per 100,000 hours to 0.82.

In 1994 the scheduled air carriers had 19 accidents, 4 of them fatal, for a total of 239 deaths, versus 22 accidents and 1 fatality in 1993.

On July 2, a USAir DC-9 approaching Charlotte/Douglas International Airport, with thunderstorm activity in the area, crashed when the crew attempted to abort the landing. There were 37 fatalities; 20 people on board survived the accident. This brought to an end a 27-month period in which the major U.S. scheduled airlines did not suffer a passenger fatality.

On September 8, a USAir Boeing 737-300 aircraft crashed while on approach to the Greater Pittsburgh International Airport. All 132 people on board were killed, and the plane was destroyed by impact, making it one of the worst aviation accidents in U.S. history.

Finally, on October 31, a Simmons Airlines ATR-72, operating as American Eagle flight 4184, crashed south of Roselawn, Ind. The flight, en route from Indianapolis to Chicago's O'Hare Airport, had been placed in a holding pattern for about 32 minutes because of traffic delays. The weather conditions during the period of holding were characterized by a temperature near freezing and visible moisture. All 64 passengers and 4 crewmembers were killed in the accident.

The 1994 fatal accident rate per 1 million miles flown for scheduled service rose to 0.0008, compared with 0.0002 in 1993. Per 100,000 aircraft departures, the fatal rate was up to 0.051 from 0.013.

Nonscheduled air carriers experienced their fifth consecutive year without a fatality. There were 4 accidents, compared with 1 in 1993. The 1994 total accident rate per 1 million miles flown was 0.0109 versus 0.0032 in 1993; 0.481 per 100,000 aircraft hours versus 0.138; and 0.967 per 100,000 departures versus 0.285.

Commuter air carrier fatalities rose from 24 persons to 25 in 10 accidents in 1994, of which 3 were fatal. There were 16 accidents in 1993, including 4 fatal. The 1994 fatal accident rate per 1 million aircraft miles flown declined to 0.005 from 0.007 in 1993, while the rate per 100,000 departures fell to 0.084 from 0.111 the year before. It was the fourth consecutive annual decline in accident rates.

Air taxi accidents totaled 85, of which 26 were fatal, compared with 69 and 19 the year before. The total accident rate per 100,000 hours flown rose to 3.45 from 2.97, and the fatal rate advanced to 1.05 from 0.82 in 1993.

The 1995 fatal accident rate per 1 million miles flown for the scheduled airlines declined to 0.0004 from 0.0008 the year before. Based on 100,000 departures, the fatal rate was 0.025, down from 0.051 in 1994.

Aside from the crash of an American Airlines B-757 in Colombia in December with 160 fatalities, the scheduled carriers experienced just one other fatal accident. This accident involved a U.S. cargo plane that went off a runway in Guatemala, killing 6 persons.

The Part 121 charter airlines, after 5 years without fatalities, had 2 deaths in 1995 for a fatal accident rate per 1 million miles flown of 0.0031, and 0.284 in terms of 100,000 departures.

Commuter or regional airline fatalities dropped to 9 persons from 25 in 1994. The fatal accident rate fell both in terms of million miles flown to 0.0036 from 0.005 in 1994, and from 0.084 to 0.062 in terms of 100,000 departures. It was the fourth consecutive annual decline in fatal accident rates.

On-demand air taxis, with a total of 75 accidents, had 52 fatalities, down from 63 in 1994. The accident rate per 100,000 hours dropped to 3.02 from 3.45 in 1994, and the fatal accident rate was 0.97, down slightly from 1.05 the year earlier. This was primarily the result of a decline in aircraft hours flown.

Total accidents for the scheduled airlines declined slightly in 1996 from 34 to 32, but the three fatal accidents resulted in the loss of 342 persons. On May 11, ValuJet flight 592, a DC-9, crashed into the Everglades shortly after takeoff from Miami International Airport, en route to Atlanta. All 105 passengers and 5 crewmembers aboard were killed. Only 2 months later, on July 17, TWA flight 800, a B-747 on a regularly scheduled flight to Paris, France, crashed into the Atlantic

Ocean off the coast of Long Island, N.Y., shortly after takeoff from John F. Kennedy International Airport. All 230 people on board the aircraft were killed. These two crashes resulted in the highest number of fatalities for a single year during the past two decades. The 1996 fatal accident rate per 100,000 departures for scheduled service increased from 0.025 to 0.038, and the rate per 100,000 flight hours rose from 0.016 to 0.023.

The nonscheduled carriers experienced 5 accidents, including 2 fatal ones, resulting in 38 fatalities. The total accident rate per 1 million aircraft miles almost doubled, from 0.0062 in 1995 to 0.0118 in 1996. The fatal accident rate per 100,000 departures increased from 0.284 to 0.530.

Commuter airlines had only 1 fatal accident in 1996. The fatal accident rate per miles, hours, and departures all declined for the fifth consecutive year. There were 90 accidents, including 29 fatal ones, for the air taxis during the same year, resulting in 63 fatalities. The accident rate per 100,000 hours declined to 2.80 from 3.02 in 1995, while the fatal accident rate declined to 0.90 from 0.97 in 1995.

The scheduled carriers experienced 44 accidents in 1997, an increase of 12 over the 32 in 1996. There were 3 fatal accidents, the same number as in 1996, which resulted in 3 fatalities. Given the increase in miles and hours flown and departures, the accident rates stayed about the same for the years 1996 and 1997.

The charter airlines had 1 fatal accident during 1997. On August 7, Fine Air flight 101, a DC-8 cargo plane, crashed shortly after takeoff from Miami International Airport. The airplane had a crew of 4. All on board and a driver on the ground were killed. This was a drop from the 38 fatalities in 1996 and brought the fatal accident rates down considerably.

Effective March 20, 1997, aircraft with 10 or more seats were required to conduct scheduled passenger operations under 14 CFR 121, which resulted in a significant reduction in miles and hours flown and departures for the commuter air carriers that operate under 14 CFR 135. Total accidents increased from 11 in 1996 to 16 in 1997, while fatal accidents rose during the same period from 1 to 5. As a result, the total accident rate per 1 million miles increased from 0.019 in 1996 to 0.065 in 1997. The rates per 100,000 aircraft hours went from 0.399 to 1.628 and, per 100,000 departures, rose from 0.313 to 1.148 during the same period.

On-demand air taxis experienced a decrease in the number of accidents, from 90 in 1996 to 82 in 1997. Fatal accidents also declined, resulting in a total accident rate reduction from 2.80 per 100,000 aircraft hours in 1996 to 2.65 in 1997. The fatal accident rate per 100,000 aircraft hours almost halved from 0.90 to 0.48 during the same period.

In 1998 the scheduled air carriers experienced only 1 fatal accident and 2 fewer total accidents than in 1997. As a result, all accident rates declined. Charter air carriers had a total of 7 accidents, an increase of 2 over 1997. However, there were no fatal accidents in 1998 or 1999. Commuter airlines also had an exceptionally good year in 1998 with total accidents being halved, from 16 to 8 and no fatal accidents. This was the first year in two decades that the commuter air carriers had no fatal accidents. On-demand air taxis, with a total of 77 accidents, down from the previous year, experienced 45 fatalities, an increase from 39 in 1997. The total rate per 100,000 aircraft hours decreased from 2.65 to 2.03, and the fatal rate per 100,000 aircraft decreased from 0.48 to 0.45.

Scheduled commuter carriers recorded 13 total accidents, including 5 fatal ones with a loss of 12 persons in 1999. On June 1, 1999, an American Airlines MD-82 overran the end of the runway, went down an embankment, and hit approach light structures after landing at Little Rock Airport. There were 11 fatalities, including the aircraft captain.

The nonscheduled carriers experienced 5 total accidents in 1999, 2 less than in 1998. The total accident rate per 1 million aircraft miles fell from 0.0178 to 0.0121, while the rate per 100,000 departures was reduced from 1.575 to 1.116.

Commuter airline accidents increased to 13 in 1999 versus 8 in 1998, and fatal accidents increased from none in 1998 to 5 in 1999. Combined with the decline in total operating statistics caused by more commuters becoming Part 121 operators, accident rates increased by 58 to 71 percent across the board between 1998 and 1999.

On-demand air taxis experienced a decrease in accidents and fatalities during 1999, resulting in an increase in the total accident rate from 2.03 in 1998 to 2.21 in 1999 and a decrease in the fatal accident rate from 0.45 to 0.36 during the same period.

2000 AND BEYOND.   The start of the new millennium saw an increase in total accidents to 51 from 47 in 1999 for scheduled Part 121 carriers. In 2000, there were 3 fatal accidents resulting in 92 deaths, as opposed to 2 fatal accidents resulting in 12 deaths in 1999. This equated to a 50 percent increase in the rate of fatalities per 100,000 departures between the two years. Chartered Part 121 operations continued to have no fatal accidents in 2000. Commuter Part 135 operations had only 1 fatal accident (as opposed to 5 in 1999) that resulted in 5 deaths (as opposed to 12 in 1999). This translated to a 78 percent reduction in the 2000 fatal accident rate of 0.164 per 100,000 departures over the 1999 rate of 0.744. On-demand air taxis had 22 accidents, 10 more than in 1999. The number of fatalities almost doubled to 71 in 2000. Accidents per 100,000 flight hours also increased by 72 percent to 0.62.

- On November 20, 2000, American Airlines (Airbus) flight 1291, from Miami, Fla., to Port Au Prince, Haiti, had a flight attendant receive fatal injuries during an emergency evacuation after the flight returned to Miami. There was one other serious injury.

- On February 16, 2000, a Douglas DC-8-71F, Emery Worldwide Airlines flight 17, scheduled cargo service from Sacramento, Calif., to Dayton, Ohio, crashed shortly after takeoff in Rancho Cordova, Calif. All 3 flight crewmembers were killed when the airplane disintegrated upon impact.

- On January 31, 2000, Alaska Airlines flight 261 (McDonnell Douglas MD-83), on route from Puerto Vallarta, Mexico, to San Francisco, Calif., crashed into the Pacific Ocean near Point Mugu, Calif. There were 88 fatalities.

- On Monday, September 18, 2000, a Piper PA-31T3 (Part 135 scheduled) crashed in Arkansas, killing 5 and seriously injuring 5 others. The probable cause, according to the NTSB, was the pilot's failure to extend the landing gear, his improper aborted landing procedure, and inadvertent stall. Also cited as causes were improper maintenance and the failure of the pilot to utilize the prelanding checklist.

The year 2001 changed the landscape of aviation for good. On September 11, four U.S. commercial aircraft (Part 121 scheduled) were crashed in separate acts of terrorism that collectively killed 265 individuals on board (and several thousands on the ground). This, coupled with two other fatal crashes, resulted in a total of 531 fatalities for Part 121 operations—the worst on record since the 1977 runway collision between two 747s at Tenerife Airport in the Spanish Canary Islands which resulted in the loss of 583 lives. If one were to discount the 4 terrorist disasters, Part 121 operations had only 2 fatal accidents (1 less than in 2000), but this resulted in 266 deaths—almost a threefold increase over that in 2000. While the fatal rate per 1 million miles flown was not significant between the two years, the fatal rate per 100,000 departures actually decreased by 26 percent in 2001. The Part 121 nonscheduled operations continued to have no fatal accidents. Part 135 commuter operations had 2 fatal accidents (compared to 1 in 2000), resulting in 13 deaths—8 more than in 2000. This resulted in a 2.5-fold increase in the accident rate per 100,000 departures to 0.402. On-demand taxis had a drop in fatal accidents by 4, to 18. There was also a drop in fatalities by 11, to 60. This resulted in a 15 percent drop of fatal accident rate per 100,000 flight hours from 2000 to 2001.

- On November 12, 2001, an Airbus 300-600 experienced a loss of control upon initial climbout and crashed into a residential area in Belle Harbor, N.Y., killing 265.

- On August 5, 2001, a US Airways Express/Piedmont ramp worker was killed when he was struck by the right propeller blades of US Airways flight 3340. The airplane was taxiing for departure from Washington Ronald Reagan National Airport for White Plains, N.Y.

- On October 10, 2001, PenAir flight 350 (Part 135 scheduled), a Cessna 208, crashed shortly after takeoff from Dillingham Airport in Dillingham, Alaska, killing 10.

- On October 3, 2001, West Isle Air flight 125, a Cessna 172N (part 135 scheduled), was substantially damaged after colliding with forested terrain approximately 1300 feet north of the Decatur Shores Airstrip, Decatur Island, Wash.

The pilot and the 2 passengers aboard the airplane were killed. The aircraft's maximum gross takeoff weight was exceeded, and the pilot failed to maintain airspeed during a low-altitude turn, which caused the aircraft to stall.

CONCLUDING REMARKS

The safety record of the airline industry in the United States has improved steadily over the past two decades. Concerns that airline deregulation might have degraded safety have not been supported by careful analyses of the industry's safety performance. Both aircraft accident and passenger fatality rates have continued a downward trend throughout the postderegulation period.

Some concluding remarks can be made in summarizing Tables 5-1 through 5-5. Airline passenger risk is not gauged solely by numbers of fatalities. Rather, passenger injury or fatality rates and the rate at which flights end in accidents or crashed are considered the best indicators of past risk. Statistical comparisons for commercial aviation are skewed by differences in aircraft size and in-flight distances. For example, since the mid-1970s, Part 121 airline operators have had the fewest fatal accidents. However, because each plane carries many passengers, these operators have had the most passenger fatalities in commercial operations. To complicate analysis further, more than 70 percent of jetliner accidents occur during takeoff, initial climb, final approach, or landing, but this figure represents only 6 percent of the flight time and even less of the mileage. Therefore, departure information for aircraft and passengers is necessary to estimate risk, and other exposure data do not permit appropriate comparisons among the aviation categories.

Although accident data are considered generally accurate and complete, exposure data quality varies with the aviation segment. Even though most scheduled Part 121 carriers must report extensive traffic data under U.S. Department of Transportation (DOT) requirements, smaller charter, commuter, and air taxi operators need report few or no data. Commuter and air taxi statistics are derived from estimated data provided to the NTSB by the Regional Airline Association (RAA) and the National Air Transportation Association. Because of inherent inaccuracies in these data, the estimates have limited utility for trend analysis, but they are valid approximations of exposure magnitude.

The three types of risk measurements presented in the tables should be considered together. The data show no significant increase in past passenger risk since the enactment of the Airline Deregulation Act. While the accident rates for the large scheduled airlines have not changed appreciably through 2001, all other categories have improved their safety records substantially.

The relative infrequency of Part 121 nonscheduled operations and accidents makes trend analysis for that part of commercial aviation very difficult. Accident rates for scheduled Part 121 were slightly higher in recent years, although fatality and accident statistical trends for a single year must be viewed with caution. Because commercial aviation accidents are relatively rare, a single crash of a large jet can skew the statistics.

Large-aircraft fatal accidents usually result in either few fatalities or few survivors. From 1982 to 2000, there were 82 fatal accidents involving scheduled and nonscheduled Part 121 carriers. Of that number, less than 50 percent accounted for more than 90 percent of the fatalities.

Industry segments have distinctly different accident rates. For example, scheduled Part 121 airlines have significantly better records than other types of air transportation. In contrast, nonscheduled 121 airlines provide less than 3 percent of the Part 121 departures but account for more than 20 percent of the fatalities.

Commuter airlines have accident and fatality rates 2 to 10 times above those of the large scheduled airlines. These disparate levels of safety often reflect differences in safety regulations, equipment, and operating environments. For example, commuter airlines might have less advanced technologies or lower training levels than major airlines because they have fewer aircraft in their fleets and fewer passengers per flight over which to distribute the costs involved. The largest commuter airlines have the best safety records; indeed the 20 largest Part 135 commuters (and Part 121 regionals) have safety records similar to those of jet carriers. Aircraft type and airport characteristics have little influence on the safety record.

OCCUPATIONAL ACCIDENT STATISTICS

The death toll trend due to occupational injuries has been on the decline over the years because of new technology, stricter safety regulations, and a shift in the economy toward safer service-industry jobs. According to BLS data, there were 5270 workplace deaths in private industry in 2001. This count excludes the deaths that resulted from the September 11 terrorist attacks. These deaths were down from 5616 in 1997, a 5-year reduction of 6.6 percent. According to the National Safety Council's *Injury Facts,* 2001 edition, deaths at the workplace in 1912 were estimated to be between 18,000 and 21,000. Thus, there has been a 73 percent decrease (using the midrange of 19,500 deaths) in deaths over a period when the workforce size quadrupled and there was a 9-fold increase in goods and services produced.

The Bureau of Labor Statistics is also the authoritative source for nonfatal occupational (work-related) illnesses and injuries in the United States. As mentioned in Chap. 4, the BLS in cooperation with OSHA and the states uses scientifically developed sample surveys to collect data on nonfatal injuries and illnesses. All deaths are reported directly to OSHA. The surveys exclude the self-employed, farms with fewer than 11 employees, private households, federal government agencies, and employees in state and local governments. The results are published in the *Standard Industrial Classification (SIC) Manual,* which was first developed and published in 1987. Information in the *SIC Manual* is published by industry type. There are *10 divisions* (A through J), and each division is divided into *major groups*. Major groups are further divided into *industry groups* with each industry group having SIC codes within it.

One of the groups of relevance to aviation is Major Group 45, *Transportation by Air,* which is under *Division E* (Transportation Sector). This Major Group contains

Industry Group 451 (*Air Transportation, Scheduled, and Air Courier*), Industry Group 452 (*Air Transportation, Nonscheduled*), and Industry Group 458 (*Airports, Flying Fields, and Airport Terminal*). The other group of relevance to aviation is Major Group 37, *Transportation Equipment,* and it is listed under *Division D* (Manufacturing Sector). This Major Group contains Industry Group 372 (*Aircraft and Parts*).

The *SIC Manual* contains injury and illness rates by industry and *case type*. The base used for the rate calculations is 100 annual full-time worker-hours, which works out to be 200,000 hours (that is, 40 hours/week per worker × 50/year × 100 workers). As mentioned earlier, rates are a better statistic than injury and illness counts. The *case rate types* include:

- Total case rate (total number of injuries and illnesses per 100 equivalent full-time workers)
- Lost-workday case rate (number of injury and illness cases that involved days away from work, days of restricted work activity, or both)
- Days away from work (due to illnesses and injuries) rate
- Total injury case rate
- Cases without lost-workdays rate

Total cases: There were 5.2 million injuries and illnesses reported in U.S. workplaces during 2001. This translated to a rate of 5.7 cases per 100 equivalent full-time workers. This rate was the lowest since that in the early 1970s. The Transport Equipment group had 221,700 illnesses and injuries, while the transport by air group had 135,500 illnesses and injuries. When hours were factored into the rate calculations, the Transport by Air group had a rate of 13.3 cases per 100 equivalent full-time workers, while the Transportation Equipment group had a rate of 12.6 cases per 100 equivalent full-time workers. Both groups had more than twice the U.S. industry average.

Injuries: Of the total cases mentioned above, there were 4.9 million injuries in 2001. This resulted in a rate of 5.4 cases per 100 equivalent full-time workers. The Transport by Air group had a rate of 12.6 per 100 equivalent full-time workers (twice the industry average), while the Transportation Equipment group had a rate of 9.4 per 100 equivalent full-time workers (1.8 times the industry average). While general industry injury and illness trends have been at their lowest levels in years, the aviation industry rates leave much to be desired.

Like many collected data, the BLS statistics do have some limitations. As in most surveys, the data are subject to sampling error. There are also other sources of error to which the data are subject. Some of these include the inability to obtain information about all cases in the sample, mistakes in coding or recording data, and definition difficulties. These statistics provide an estimate of workplace injuries and illnesses based on logs kept by employers during the year. The employer's understanding of which cases are work-related under OSHA's record-keeping guidelines may not always be clear. Finally, the number of injuries and illnesses that get reported can be influenced by level of economic activity, working conditions and work practices, worker experience and training, and the number of hours worked.

KEY TERMS

Safety factors

Saftey indicator

(Safety) exposure data

Accident rate

Hull loss

Non-hull loss

NTSB accident definitions: major, serious, injury, damage

Standard Industrial Classification (SIC) Manual

OSHA injuries and illnesses

OSHA case rates

REVIEW QUESTIONS

1. Why is the accident rate a better measure of safety than accident counts are?
2. Discuss some of the issues to be aware of in analyzing and comparing commercial aviation accident statistics.
3. What are hull-loss accidents and what are their primary cause? Describe the hull-loss accident trend since the early 1970s.
4. What is the most critical phase of flight? What are some lessons that can be learned from the Boeing summary?
5. Describe the general trend in aviation accidents during the 1980s and 1990s.
6. How do the accident experiences of Part 121 and Part 135 carriers compare during the 1980s and 1990s?
7. Define an OSHA illness and injury.
8. Explain how OSHA injury and illness rates are calculated.
9. List SIC codes relevant to aviation.
10. What are SIC codes and what are their limitations?

REFERENCES

Boeing Commercial Airplane Co. 2001. *Statistical Summary of Commercial Jet Aircraft Accidents: Worldwide Operations 1959–2001.* Seattle, Wash., June 2001.

IATA Safety Record (Jet) 1994. 1995. Montreal, Canada: International Air Transport Association.

National Safety Council. *Injury Facts,* 2001 edition, Itasca, Ill.

National Transportation Safety Board. 1982–2001. Annual reports. Washington, D.C.: NTSB.

U.S. Department of Labor. 2002. *News: United States Department of Labor,* Washington, D.C.: Bureau of Labor Statistics.

U.S. Department of Transportation. 1987. *Annual Report on the Effect of the Airline Deregulation Act on the Level of Air Safety.* Washington, D.C.: U.S. Government Printing Office, February.

U.S. Department of Transportation, Federal Aviation Administration. 1995. *Statistical Handbook of Aviation, Calendar Year 1994.* Washington, D.C.: U.S. Government Printing Office, November.

WEB REFERENCES

BLS charts of workplace injury and illness incidence rates,1974–2001. http://www.bls.gov/iif/oshwc/osh/os/osch0024.pdf.

Boeing Commercial Airplane Co, 2001, http://www.boeing.com/news/techissues/pdf/statsum.pdf.

Major group 45: transportation by air, http://www.osha.gov/cgi-bin/sic/sicser4?45.

Major group 37: transportation equipment, http://www.osha.gov/cgi-bin/sic/sicser4?37.

NTSB aviation accident statistics web site: http://www.ntsb.gov/aviation/Stats.htm

OSHA statistics: http://www.osha.gov/oshstats/.

Summary table, injury and illness incidence rates per 100 full-time workers, 1973–2001, http://www.bls.gov/iif/oshwc/osh/os/osnr0016.pdf.

ACCIDENT CAUSATION MODELS

LEARNING OBJECTIVES

After completing this chapter, you should be able to

- List the reasons for accident modeling.

- Explain the Office of Technology and Assessment (OTA) model of accident causation.

- Discuss the primary, secondary, and tertiary factors of accident causation, giving examples of each.

- Give a brief historical sketch of the 5-M factors.

- Explain the five causal factors in examining the nature of accidents: man, machine, medium, mission, and management.

- Explain Reason's model of accident causation, giving examples where appropriate.
- Discuss the concept of risk management in relation to conserving assets and minimizing exposure to losses.
- Identify the basic elements in a risk management program.

INTRODUCTION

The purpose of accident investigation is to uncover pervasive, unrecognized causal factors of accidents. This can help prevent similar accidents from occurring in the future. However, since commercial aviation accidents are relatively rare, proactive measures for identifying short-term changes in safety are required. The goal of *nonaccident* data analysis and *modeling* is to conduct a preemptive strike on the very first accident to prevent it from occurring by addressing the root cause of the accident. While incidents and accidents provide after-the-fact evidence that safety was inadequate, accident *modeling* assists with understanding how accidents happen so that measures (policy decisions or changes in the aviation operating environment) can be taken to prevent potential hazards from materializing.

The main goal of accident investigations is to establish the probable causes of accidents and to recommend control measures. Because most aviation accidents involve a complex maze of diverse events and causes, classifying or categorizing these accidents by type or cause gets quite complicated and involved. Also, accidents that are similar may often require different preventive strategies, although at times a single solution can eliminate or reduce the rate of occurrence of a wide range of accidents. For example, ground-proximity warning devices addressed the wide range of issues involved with controlled flight into terrain accidents for jetliners and helped reduced its occurrence rate.

The NTSB classifies accidents by several methods, such as causes and factors, sequence of events, and phase of operation. While determining up to five distinct occurrences in the chain of events leading to an accident, the NTSB categorizes the accident by the first occurrence. While aircraft component failures and encounters with weather are easy to establish as first occurrences, failures due to human errors are harder to trace. Accidents frequently have multiple causes, hence developing causal categories is difficult. In a majority of the cases, each cause is independent of the others, and if one did not exist, the accident might not have occurred.

Boeing, the Flight Safety Foundation, and other organizations have categorized accidents by primary cause. However, determining which of the multiple causes is the most important is a subjective process. One analysis (Sears, 1985) of major accidents involving large jet transports worldwide found that only 28 percent of these accidents had a single probable cause. Another study conducted by the Office of Technology Assessment (OTA) on NTSB Part 121 briefs showed approximately 60 percent of the fatal accidents by scheduled passenger carriers were initiated by human error, and human error was a causal factor in more than 70 percent of these accidents. Aircraft component failure, severe weather, and miscellaneous factors were listed as initial causes for the remaining accidents. However, when nonfatal accidents are included, the influence of mechanical failure doubles. Mechanical failure was the main cause of more than 30 percent of the accidents and was marginally involved in almost 50 percent of the accidents.

As mentioned earlier, accident modeling helps us understand the nature of accidents. Accident models

- Help explain the relationship between hazards and accidents
- Assist with understanding and explaining reality
- Aid in visualizing things that cannot be directly observed
- Must approximate conditions that exist in reality to be useful

There are several accident models discussed in the literature. Three of these models that have been most frequently associated with aviation are discussed next.

THE OFFICE OF TECHNOLOGY AND ASSESSMENT MODEL

In its attempt to uncover pervasive conditions that can lead to accidents, the OTA model outlines measurable potential safety factors/indicators that are causally related to accidents, fatalities, or injuries. A block diagram of this models is shown in Fig. 6-1. As can be seen in Fig. 6-1, items closely associated with accidents appear near the "accident" box. These items offer the greatest potential as safety indicators. Factors that are further away from "accidents" have correspondingly longer causal links to them. These factors (e.g., industry policy) are measured against more subjective standards and might be more difficult to quantify. The OTA model divides its safety indicators into three categories: primary, secondary, and tertiary. These safety factors and indicators describe the relative "closeness" between the measured safety factors and the fatality, injury, and accident rates. Theoretically, primary indicators provide the best measures of changes in safety, followed by secondary and tertiary indicators. Each of these categories is now discussed.

PRIMARY SAFETY FACTORS

Primary safety factors are those most closely correlated with accidents and include causal factors such as

- Personnel (pilots, controllers, etc.) capabilities
- Air traffic environment

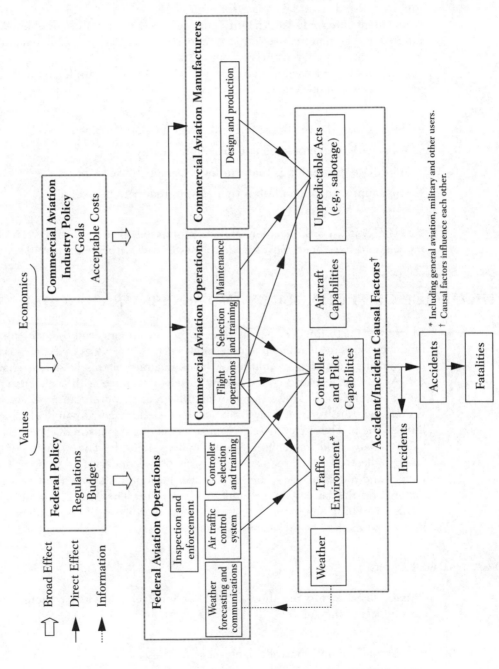

FIGURE 6-1 The Office of Technology and Assessment model.

Values   Economics

Broad Effect

Direct Effect

Information

Federal Policy

Regulations
Budget

Commercial Aviation Industry Policy

Goals
Acceptable Costs

Federal Aviation Operations

Inspection and enforcement

Weather forecasting and communications

Air traffic control system

Controller selection and training

Commercial Aviation Operations

Flight operations

Selection and training

Maintenance

Commercial Aviation Manufacturers

Design and production

Accident/Incident Causal Factors†

Weather

Traffic Environment*

Controller and Pilot Capabilities

Aircraft Capabilities

Unpredictable Acts (e.g., sabotage)

Incidents

Accidents

Fatalities

* Including general aviation, military and other users.
† Causal factors influence each other.

136

- Aircraft capabilities
- Weather
- Unpredictable acts

Incidents and measurable primary factors are primary indicators of safety. The five broad categories of primary factors are shown in Fig. 6-1. Lines and arrows are used to indicate interactions within primary factors and between *primary, secondary, and tertiary factors*. These interactions are important since few accidents or incidents result from a single, isolated cause. By examining these causal factors within an organization or system, safety levels can be monitored and proactive control strategies can be implemented to correct any deviations from established standards. A discussion of these five categories of primary factors now follows.

PERSONNEL CAPABILITIES.   Human errors account for more than two-thirds of commercial aviation accidents and include lapses in attention, judgment, or perception and deficiencies in knowledge or motor skills. These errors can be caused by vehicle, environmental, or health factors, including cockpit layout, workload, fatigue, or stress. Aviation personnel at greatest risk for committing these errors include flight crewmembers, dispatchers, mechanics, and air traffic controllers. Several reasons hamper the identification of human errors in the operating environment, some of which include privacy and sensitivity. However, human errors must be studied to be understood and prevented. Some indicators of personnel capabilities that are presently measured and used in either federal or industry standards include employee duty hours, age, training, and experience levels.

AIR TRAFFIC ENVIRONMENT.   Air traffic safety is heavily influenced by the infrastructure of the airways and airports and the volume and mix of air traffic. While most incidents are discovered before they become accidents, high traffic density puts continuous strains on many aspects of the air traffic control (ATC) system. For a given air traffic infrastructure, as traffic density increases, the risk of midair collisions can be expected to increase. While flight operation data can be readily determined with a high degree of accuracy, trending midair collision rates lack any kind of statistical validity as collisions occur too infrequently. Using midair collision as a safety indicator is also questionable as these statistics lack precision. Other potential measures of air traffic safety such as operational error, operational deviation, and pilot deviation statistics also have precision and consistency problems. Extended discussions of these measures can be found in Chap. 10 on airport safety. Definitions of some of these measures are repeated below to help explain some of the reasons for inconsistency in the data collected.

The FAA defines an *operational error* as "…an occurrence attributable to an element of the air traffic control system that results in less than applicable separation minima between two or more aircraft, or between an aircraft and terrain or obstacles and obstructions as required by FAA Handbook 7110.65 and supplemental instructions."

An *operational deviation* is "...an occurrence where applicable separation minima were maintained but loss in separation minima existed between an aircraft and protected airspace, an aircraft penetrated airspace that was delegated to another position of operation or another facility without prior approval, or an aircraft or controlled vehicle encroached upon a landing area that was delegated to another position of operation without prior approval."

A *pilot deviation* is "...the action of a pilot that results in the violation of a Federal Aviation Regulation or a North American Aerospace Defense Command (NORAD) Air Defense Identification Zone (ADIZ) tolerance."

Finally, the ratio of operations to controllers (controller workload) could be a factor of air traffic safety if one were to also consider the type of ATC equipment being used and the nature of the air traffic mix.

AIRCRAFT CAPABILITIES.   Aircraft component failures account for more than 40 percent of jetliner accidents. Components that fail include engines, structural members, landing gear, control systems, and instruments. Failures can result from improper maintenance, design flaws, or operator error. Replacement or repair trends and severity and the frequency of the component failure, especially for flight-critical components, are possible indicators of safety. The FAA, air carriers, and aircraft manufacturers maintain detailed databases of mechanical reliability data. Analysis and communication of observed trends prevent most problems from becoming critical. Other possible safety indicators include engine shutdown rates and unscheduled landings due to mechanical difficulties.

WEATHER.   An understanding and timely monitoring of weather conditions are required for safe operation of the aircraft. While modern aircraft can fly in most kinds of weather, unpredicted severe conditions, such as wind shear or heavy icing, can prove deadly. Poor weather can exacerbate existing unsafe conditions created by mechanical difficulties or errors in judgment. More on the effect of weather on air traffic can be found in later chapters on air traffic system technologies and aircraft system technologies.

UNPREDICTABLE EVENTS.   Unpredictable events are factors in the catchall category and include sabotage or terrorism. By definition, unpredictable or random events have no trends. Therefore, no unpredictable event indicators are possible except incidents and accidents that show levels of past risk. The thinking on terrorism and sabotage, however, has changed considerably after the terrorists events of September 11, 2001. The philosophy has shifted to possible prevention, and security has been stepped up considerably to minimize the occurrence of such acts in the future. More on this topic is discussed in Chap. 11 on aviation security.

SECONDARY SAFETY FACTORS

FAA operations, commercial aviation operations, and commercial aviation manufacturers are three groups of secondary factors that influence the primary factors. Examples within this secondary-factors group include

- Airline operating, maintenance, and personnel training practices
- Federal air traffic control management practices
- Aviation manufacturers' design and production practices

Quantifiable measures of these factors include aircraft or employee utilization rates. Flight operations and training practices of the airlines and the FAA (controller selection) are secondary factors that influence pilot and air traffic controller capabilities, which are primary factors. Similarly, flight operations and maintenance practices of the airlines, and design and production practices of aviation manufacturers (secondary factors), affect aircraft capabilities (a primary factor). Finally, inspection and enforcement policies of the FAA can affect the air traffic environment and influence pilot and controller capabilities.

Responsibility for commercial aviation safety lies with the FAA and the airlines. The Federal Aviation Regulations (FARs) establish the framework for commercial aviation operating practices. Specific practices tailored to individual carrier needs are permissible if approved by FAA principal inspectors and Flight Standards District Offices. Some FAA operating practices (through the FARs) require the reporting of some operational data like air carrier traffic, schedules, and financial information in an attempt to uncover problems that can compromise the safety practices of airlines. However, as currently reported and reviewed, no correlation with safety has been established. Some examples of potential secondary safety factors are now discussed.

(FAA) INSPECTION AND ENFORCEMENT.   Regulatory compliance audits that are conducted by the FAA can reveal airline management attitude, organizational skill, and operational safety. By using objective inspection guidelines and standards, conducting consistent and periodic audits with well-qualified auditors, and varying inspection team personnel, audits can deliver consistent, valid measures of safety trends in an organization. In addition, since FAA inspectors examine only a small percentage of an airline's records, aircraft, and operations, a measure of the quantity of inspection is needed to normalize the data used for analysis. A valid measure of regulatory compliance safety would be the number of violations per inspection hour. With appropriate guidelines, the quality of management practices can be measured by inspector audits.

FLIGHT OPERATIONS.   Increased air traffic volumes over the years have placed limits that have stressed the ATC system, and airline flightcrews and maintenance personnel have felt new demands. Factors worthy of trending under this category include aircraft daily utilization (number of hours per day an aircraft is used), departures per aircraft per day, percentage of fleet required for daily operations, and percentage of flights into high-density airspace.

MAINTENANCE.   The aircraft capability factors, including equipment design and manufacturing quality, are important measures of maintenance quality. Unit maintenance costs, if used as an estimate of safety, must be employed with caution as there are many reasons for variations among carriers, such as productivity and technological changes.

TRAINING.   Training factors include the number of hours of a type of instruction per applicable employee including the use of simulators, cockpit resource management, and wind shear training.

TERTIARY SAFETY FACTORS

Tertiary safety factors are furthest removed on the accident/incident causal chain and affect industry (airline and manufacturers) and FAA operating practices. These factors include federal policies and commercial aviation industry policies. Industry philosophy and policy differ among airlines and dictate operating decisions. Federal regulatory policy, in turn, influences industry policy and operating practices. Tertiary factors reflect the values (ethical business practices) of a society and its economic philosophy (e.g., deregulation).

THE 5-M MODEL

INTRODUCTION

The *man, machine, medium, mission*, and *management factors* represent a valuable model for examining the nature of accidents (Fig. 6-2). That is, when one seeks causal factors or preventive or remedial action, the diagram of the intertwined circles becomes a meaningful checklist for fact-finding and analysis to ensure that all factors are considered.

The five factors are closely interrelated, although it can be argued that management plays the predominant role. Mission is located as the target or objective to emphasize that effective mission accomplishment is implicit in professional system safety work.

This concept evolved when T. P. Wright of Cornell University first introduced the man-machine-environment (medium) triad into the aviation safety language during the late 1940s; he was influential in the development of the Cornell-Guggenheim Aviation Safety Division of the University College, University of Southern California (USC). Follow-on instructors used the 3-M (man-machine-medium) terminology.

The fourth M, management, was first introduced in 1965, at USC, when the school's initial advanced safety management and system safety courses were being developed. That emphasized the interrelationships between the man-machine-medium factors and their subset relationship to management.

The mission factor had been discussed at military-oriented USC courses but was not introduced into the diagram until 1976 at the suggestion of E. A. Jerome, consultant, writer, and former staff member with Flight Safety Foundation.

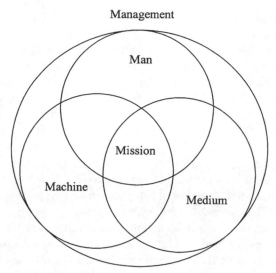

FIGURE 6-2   The 5-M model.

HISTORICAL SKETCH OF THE 5-M MODEL

Commercial aviation started after World War I. The great surplus of military aircraft and the large number of available trained pilots formed the basis for the new airlines that started to transport mail, passengers, and cargo on a commercial basis. In these early days of commercial aviation, the weather was a major factor determining whether a flight could be safely executed. The pilot required good visibility for safe takeoff; attitude control; navigation; terrain collision avoidance; avoidance of areas with adverse meteorological conditions, such as icing and heavy turbulence; avoidance of collision with other aircraft ("see and be seen"); as well as approach and landing.

The man-machine interface relating to the pilot and his or her aircraft, systems, and engine was of a simplistic nature, straightforward, and easy to understand and interpret. The pilot used most of what he or she had learned during day-to-day experiences during flights. The outside world as seen from the cockpit was very clear to interpret. Unfavorable weather and visibility conditions, however, could make safe flight impossible. Unpredictable weather thus jeopardized the economical necessity of predictable time schedules.

In the period from 1920 to 1950, the aircraft and system designers made great progress, which was reflected in increased economy as well as in improved safety of air travel. The introduction of gyroscopic blind-flying instruments enabled safe flight in low visibility. Radio navigation and approach systems allowed navigation and approach in bad visibility conditions. ATC, first based on *radio direction finding* (*RDF*) and later on radar in combination with radio communication, allowed for safe separation of all air traffic and a good sequencing for approach and landing.

Deicing and anti-icing systems reduced the dangers of ice accretion during flight. Weather radar enabled the pilot to detect hazardous meteorological conditions and circumnavigate these. Autopilots were introduced to automatically maintain attitude, altitude, and heading. Later on, these autopilots were provided with the capability to use both radio navigation and radio approach systems as sensors and thus could also reduce the pilot workload for navigation and approach.

The introduction of the turbine-powered, subsonic airliners such as the de Havilland Comet, Boeing 707, and Douglas DC-8 in the 1950–1960 period provided a quantum leap in the productivity of air transport. These aircraft were equipped with hydraulically powered flight controls and an increasing number of avionics—electrical and hydraulic systems—to enable the crew to fly these larger, heavier, and faster airliners. The Douglas DC-8 required a flightcrew of two pilots, one navigator, and one flight engineer. The operation of all systems became more complex, and the easy understanding of the early *visual meteorological condition* (VMC) flights with respect to the outside world in bad visibility was to a large extent lost. Due to the much higher cruising speeds of these turbine-powered subsonic airliners, even in VMC, safe separation between the aircraft had to be provided by ATC.

In the period from 1970 to 1990, three generations of wide-body airliners appeared. The first generation, introduced around 1970, was the wide-body, long-range airliner. Examples are the Boeing 747, the Lockheed 1011, and the Douglas DC-10. These wide-body airliners were equipped with three to four high-bypass-ratio engines, *inertial navigation systems* (INSs), and an *automatic landing system* (ALS). The cockpits of these airliners were equipped with many electromechanical instruments. These aircraft were flown by a flightcrew of two pilots and one flight engineer.

The second generation, introduced around 1980, was the long/medium-range, wide-body airliner with a new digital avionics system. Examples are the Airbus A-310 and the Boeing 757/767. These new wide-body airliners are equipped with an *electronic flight instrument system* (EFIS), a *flight management system* (FMS), and a *flight warning computer system* (FWCS). In greater detail, the EFIS, FMS, and FWCS provide the following functions:

- The EFIS *primary flight display* (PFD) provides a combined presentation of attitude, flight director, *instrument landing system* (ILS) deviation, flight mode annunciation, and speed and altitude information on a single *cathode-ray tube* (CRT) display, thus reducing the scanning cycle.

- The EFIS *navigation display* (ND) provides integrated map, horizontal flight path, weather radar, heading, and wind vector information, largely reducing the navigation task (in combination with the FMS) and improving the positional awareness of the pilot.

- The FMS provides integrated navigation and fuel management information, as well as a host of performance and navigation information, largely increasing the pilot's flight management and navigation capabilities.

• The FWCS provides alphanumeric and synoptic graphical system information, largely reducing the flightcrew workload with respect to systems operation and system malfunction handling.

The "glass" cockpit is equipped with six color CRT graphics displays for the EFIS and FWCS, as well as two alphanumeric monochrome *control display units* (*CDUs*) for the FMS; apart from the CRTs, a number of electromechanical instruments are still used to enable a safe continuation of flight in case all CRTs should fail. The introduction of the EFIS, FMS, and FWCS allowed the elimination of the flight engineer from the flight deck, providing a significant cost reduction for the operations with this type of airliner.

The third-generation airliner, introduced around 1990, is a long/medium-range aircraft with a revolutionary new digital flight control system, no longer using mechanical links between the pilot's control yoke and the hydraulic actuators of the flight control surfaces. This *fly-by-wire* (*FBW*) technology allows for new flight control concepts and envelope protection systems. Examples of these new FBW airliners are the Airbus A-320/330/340 and the Boeing 777. The engines of this new generation of airliners are controlled by *full-authority digital engine control* (*FADEC*) systems. In the cockpit, the CRTs have become larger, and the number of electromechanical instruments has strongly decreased.

For example, the Boeing 747-400 represents a significant change from earlier models by the reduction of flightcrew from three to two persons, with the role of the flight engineer being automatic and/or simplified and all controls fitted into the pilot's overhead panel, eliminating the flight engineer station. Boeing states that the design has reduced the number of cockpit lights, gauges, and switches from more than 970 in the basic B-747 to only 365 in the B-400, resulting in a major triumph of design and engineering skills. Good design of automatic systems can relieve pilots from the monotonous chore of monitoring systems and can free them for higher cognitive tasks that are beyond the capability of machines.

Modern scheduled aviation developed into a reliable and economical all-weather transport system. Through the use of ever-improving aerodynamics and engine technology, as well as the increasing use of lightweight composite materials since 1970, the fuel consumption per passenger-mile has been reduced by more than 30 percent. Radio navigation and approach systems, inertial navigation systems, weather radar, and ATC, in combination with ever-improving training and standardized procedures, allow safe flight, also in reduced-visibility conditions. However, in *instrument meteorological conditions* (*IMCs*), the pilot's situational awareness is relatively poor due to the nonexistence of outside visual attitude, navigation, weather, and terrain information.

Between 2000 and 2010, passenger enplanements on U.S. carriers are expected to grow by nearly 60 percent. It is expected that airline operations will increase by more than 30 percent over the same period. This growth translates to greater demand on the system—greater controller workload and increased pressure on airports, runways, terminal buildings, parking lots, and the airspace itself. However, the capacity of airports and airspace in western Europe and North America is limited, and already clear signs of saturation are visible.

The Boeing Company has projected a worldwide transport fleet of 23,000 airplanes by 2015, up from about 12,000 today. It has also reported that if the 1998 worldwide accident rate were held constant at the level of about 1 per 1 million departures, there could be a serious accident somewhere in the world every 1 or 2 weeks in 2015. This becomes even more significant with the recent announcement from Airbus Industrie of the proposed development of the double-deck A-3XX. The 555-seat standard A-3XX, with a range of 8150 miles, can be expanded to about 656 seats. That is well above the capacity of the 416-seat Boeing 747-400, the largest passenger jet now in service.

Given that today's accident rate is unacceptable, what have we to look forward to, given the constant increase in activity in the same finite blocks of airspace and real estate?

First, we can be encouraged by the progress to date. According to the Boeing Company, the accident rate for the newer generation of airplanes, such as the B-757, B-767, and the A-310, is considerably better than that for earlier designs. It is reasonable to expect that the current new models, such as the B-777, A-330, and A-340, will be safer yet, as a result of more sophisticated design and applied technology.

In summary, new technology will be available to the flightcrews and controllers as well:

- Better weather detection systems will provide information to airline dispatchers and pilots, allowing more efficient and safe flight around weather systems, both en route and near the airport.

- *Global positioning systems* (GPSs) are being used now, but will become the primary source for navigation and surveillance information, replacing ground-based, line-of-sight-limited VOR navigation facilities and radar facilities. GPS will also be the primary means of guidance for precision landings and departures at our nation's airports.

- Improved air traffic control tools are already being installed in FAA facilities to give the controller more reliable and efficient means to see and communicate with the airplanes under his or her control.

- Data link will allow clearances, weather, and traffic information to be provided in the cockpit in a fast, error-free, digital form. One of the big advantages of data link will be the elimination of "read-back" errors between the pilot and controller.

- Improved collision avoidance systems onboard the airplanes will reduce the number of collision scenarios.

- Flight decks will continue to improve, with added redundancy and integrated avionics giving the pilot more options and greater flexibility.

- Training of flightcrews will become more sophisticated. Flight data recorder information from "safe to destination" flights will be used by the airlines to improve training. The information, to the extent that it is generic, will be shared among the airlines, regulatory agencies, and manufacturers for improvements in many areas from operations to design.

- Human factors will be a major consideration from the onset of airplane design to ensure that the airplane can be operated and maintained easily within human limits.

Our national aviation system has evolved over the past seven decades to serve a vital role in the economy and our way of life. The system is complex, built on national standards with rigid quality control in all areas from the cockpit to the maintenance hangar to the air traffic control facility.

MAN

While many see the pilot as the only "man" in the system, others include all persons directly involved with the operation of aircraft—flightcrew, ground crew, ATC, meteorologists, etc. In its widest sense, the concept should include all human involvement in aviation, such as in design, construction, maintenance, operation, and management. This latter is the meaning intended in this discussion since accident prevention must aim at all hazards, regardless of their origin.

Unfortunately, the study of man (or human factors) usually does not receive sufficient emphasis. For example, during a pilot's training, he or she learns something of the mechanical aspects of the machine flown, the hazards of the weather, the operating environment, and so on. However, usually very little information is provided concerning his or her own behavior, limitations, vulnerabilities, and motivations.

As a result of refinements over the years, the number of accidents caused by the machine has declined, while those caused by man have risen proportionately. Because of this significant shift in the relationship between human and machine causes, a consensus has now emerged that accident prevention activities should be mainly directed toward the human.

People are naturally reluctant to admit to their limitations for a variety of reasons, such as loss of face among peers, self-incrimination, fear of job loss, or considerations of blame and liability. It is not surprising, therefore, that information on the human-factor aspects of accidents or incidents is not readily forthcoming. This is unfortunate since it is often these areas that hold the key to the "why" of a person's actions or inactions.

Many questions arise when one considers the why of human failures. Successful accident prevention, therefore, necessitates probing beyond the human failure to determine the underlying factors that led to this behavior. For example, was the individual physically and mentally capable of responding properly? If not, why not? Did the failure derive from a self-induced state, such as fatigue or alcohol intoxication? Had he or she been adequately trained to cope with the situation? If not, who was responsible for the training deficiency and why? Was he or she provided with adequate operational information on which to base decisions? If not, who failed to provide the information and why? Was the individual distracted so that he or she could not give proper care and attention to duties? If so, who or what created the distraction and why? These are but a few of the many why questions that should be asked during a human-factor investigation. The answers to these questions are vital for effective accident prevention.

In the past, the view that the man involved only the pilot led to the frequent use of the term *pilot error* as a cause of accidents, often to the exclusion of other human-related causes. As a consequence, any other hazards revealed by an investigation were often not addressed. Further, since the term tended to describe only *what* happened rather than *why*, it was of little value as a basis for preventive action. Fortunately, the term is now rarely used by investigation authorities.

The pilot is often seen as the last line of defense in preventing an accident. In fact, over the years, the skill and performance of pilots have prevented many accidents when the aircraft or its systems failed or when the environment posed a threat. Such occurrences usually do not receive the same attention and publicity as accidents, sometimes leading to an unbalanced perception of the skill and performance of pilots.

MACHINE

Although the *machine* (aviation technology) has made substantial advances, there are still occasions when hazards are found in the design, manufacture, or maintenance of aircraft. In fact, a number of accidents can be traced to errors in the conceptual, design, and development phases of an aircraft. Modern aircraft design, therefore, attempts to minimize the effect of any one hazard. For instance, good design should seek not only to make system failure unlikely, but also to ensure that should it nevertheless occur, a single failure will not result in an accident. This goal is usually accomplished by so-called fail-safe features and redundancy in critical components or systems. A designer must also attempt to minimize the possibility of a person using or working on the equipment committing errors or mistakes in accordance with the inevitability of Murphy's law: "If something can go wrong, it will." To meet these aims, some form of system safety program is often used during the development of a new aircraft type. Modern design must also take into account the limitations inherent in humans. Therefore, it includes systems that make the human's task easier and that aim to prevent mistakes and errors. The *ground-proximity warning system* (*GPWS*) is an example of such a system. It has significantly reduced the number of accidents in which airworthy aircraft collide with the ground or water while under the control of the pilot.

The level of safety of an aircraft and its equipment is initially set by the airworthiness standards to which it is designed and built. Maintenance is then performed

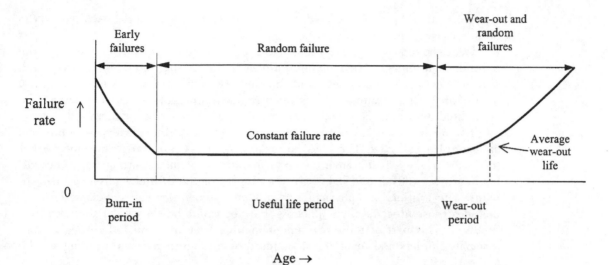

FIGURE 6-3 Aircraft system failure pattern.

to ensure that an acceptable level of safety is achieved throughout the life of the aircraft. Manufacturing, maintenance, and repair errors can negate design safety features and introduce hazards that may not be immediately apparent.

As the service experience with a particular aircraft type increases, the maintenance program needs to be monitored and its contents developed and updated where necessary to maintain the required levels of safety. Some form of reporting system is required to ensure that component or system malfunctions and defects are assessed and corrected in a timely manner.

The reliability of a component is an expression of the likelihood that it will perform to certain specifications for a defined length of time under prescribed conditions. Various methods can be used to express reliability. A common method for electronic components is the *mean time between failures* (MTBF), and the reliability of aircraft power plants is usually expressed as the number of shutdowns per 100,000 operating hours.

Failures normally arise in three distinct phases in the life of a component. Initial failures, caused by inadequate design or manufacture, usually occur early in its life. Modifications to the component or its use usually reduce these to a minimum during the main or useful life period. Random failures may occur during this period. Near the end of the life of a component, increased failures occur as the result of its wearing out. Graphic representation of this failure pattern gives rise to the typical "bathtub-shaped" curve (Fig. 6-3).

MEDIUM

The *medium* (environment) in which aircraft operations take place, equipment is used, and personnel work directly affects safety. From the accident prevention viewpoint, this discussion considers the environment to comprise two parts—the natural environment and the artificial environment.

Weather, topography, and other natural phenomena are thus elements of the natural environment. Their manifestations, in forms such as temperature, wind, rain, ice, lightning, mountains, and volcanic eruptions, are all beyond the control of humans. These manifestations may be hazardous, and since they cannot be eliminated, they must be avoided or allowances must be made for them.

The artificial portion of the environment can be further divided into physical and nonphysical parts. The physical portion includes those artificial objects that form part of the aviation environment. Air traffic control, airports, navigation aids, landing aids, and airfield lighting are examples of the artificial physical environment. The artificial nonphysical environment, sometimes called *system software,* includes those procedural components that determine how a system should or will function. This part of the environment includes national and federal legislation, associated orders and regulations, standard operating procedures, training syllabi, and so forth.

Many hazards continue to exist in the environment because the people responsible do not want to become involved in change, consider that nothing can be done, or are insufficiently motivated to take the necessary actions. Obstructions near runways, malfunctioning or nonexistent airport equipment, errors or omissions on aeronautical charts, faulty procedures, and so forth are examples of artificial environmental hazards that can have a direct effect on aviation safety.

MISSION

Notwithstanding the man-machine-medium concept, some safety experts consider the type of *mission,* or the purpose of the operation, to be equally important. Obviously the risks associated with different types of operation vary considerably. A commuter airline operating out of many small airports during the winter months in the New England area has a completely different mission than an all-cargo carrier flying extensive over-water flights to underdeveloped countries or a major carrier flying from New York to Los Angeles. Each category of operation (mission) has certain intrinsic hazards that have to be accepted. This fact is reflected in the accident rates of the different categories of operation and is the reason why such rates are usually calculated separately.

MANAGEMENT

The responsibility for safety and, thus, accident prevention in any organization ultimately rests with *management,* because only management controls the allocation of resources. For example, airline management selects the type of aircraft to be purchased, the personnel to fly and maintain them, the routes over which they operate, and the training and operating procedures used. Federal authorities promulgate airworthiness standards and personnel licensing criteria and provide air traffic and other services. Manufacturers are responsible for the design and manufacture of aircraft, components, and power plants as well as monitoring of their airworthiness.

The slogan "Safety is everybody's business" means that all persons should be aware of the consequences of their mistakes and strive to avoid them. Unfortunately, not everyone realizes this, even though most people want to do a good job and do it safely. Therefore, management is responsible for fostering this basic motivation so that each employee develops an awareness of safety. To do this, management must provide the proper working environment, adequate training and supervision, and the right facilities and equipment.

Management's involvement and the resources it allocates have a profound effect on the quality of the organization's prevention program. Sometimes, because of financial responsibilities, management is reluctant to spend money to improve safety. However, it can usually be shown that not only are accident prevention activities cost-effective, but also they tend to improve the performance of people, reduce waste, and increase the overall efficiency of the organization.

Management's responsibilities for safety go well beyond financial provisions. Encouragement and active support of accident prevention programs must be clearly visible to all staff, if such programs are to be effective. For example, in addition to determining who was responsible for an accident or incident, management's investigation also delve into the underlying factors that induced the human error. Such an investigation may well indicate faults in management's own policies and procedures.

Complacency or a false sense of security should not be allowed to develop as a result of long periods without an accident or serious incident. An organization with a good safety record is not necessarily a safe organization. Good fortune rather than good management practices may be responsible for what appears to be a safe operation.

On the whole, management attitudes and behavior have a profound effect on staff. For example, if management is willing to accept a lower standard of maintenance, then the lower standard can easily become the norm. Or, if the company is in serious financial difficulties, staff may be tempted or pressured into lowering their margins of safety by "cutting corners" as a gesture of loyalty to the company or even self-interest in retaining their jobs. Consequently, such practices can and often do lead to the introduction of hazards. Morale within an organization also affects safety. Low morale may develop for many reasons but nearly always leads to loss of pride in one's work, an erosion of self-discipline, and other hazard-creating conditions.

REASON'S MODEL

Dr. James Reason's model of accident causation focuses on understanding incidents and accidents and their contributing factors. Reason's model is widely used in the aviation industry and has been recommended by various organizations, such as the FAA, for use in investigating the role of management policies and procedures in aircraft accidents. Reason's model traces the root causes of accidents to errors that occur in the higher management levels of an organization. These errors are also referred to as *latent errors*. Reason contends that models are grossly inadequate if they attribute accidents solely to individual operator performance. Reason also proposes that human error is the end result rather than the cause of incidents or accidents. Today's technological systems involve complex multiple interacting factors that are distant in time and proximity from the immediate circumstance of an accident. Some of the salient features (Fig. 6-4) of Reason's model are the following:

- Systems are protected by multiple layers of defenses that are designed to prevent hazards or system failures from cascading into accidents.
- Each layer of protection, however, can develop "holes" or flaws through safety deficiencies, resembling Swiss cheese.
- As the number and size of these holes in the defenses increase, the chances of accidents also increase.
- When the holes in each of the layers of defenses line up, an accident occurs.

The model recommends focusing on events beyond the *active failures* of front-line employees to latent preexisting conditions that result from fallible decisions made by high-level decision makers. It is these latent failures that permit active failures to occur. Management should build defenses by creating a culture in which precursor events are detected and promptly corrected. This approach has the greatest potential for preventing accidents in an organization.

RISK MANAGEMENT

Risk management is a concept that has gained acceptance in many fields of business and industry. It stems largely from financial concerns and a realization that losses from different areas in a business must be either reduced or accepted. Safety losses stem from costs of accidents that result in fatalities, illnesses, injuries, product damage, and lost employee and production time. It is mentioned here because safety risk management is part of the overall business risk management process. Aircraft accident prevention attempts to minimize the risk posed by safety hazards. The application of risk management concepts, therefore, reinforces several of the accident prevention ideas in this chapter. Some basic definitions of (safety) risk and management are included below.

- *Risk management* is the overall process of identifying, evaluating, controlling or reducing, and accepting risks. It is the general term given to the process of making management decisions about risk that have been identified and analyzed.

•Warning systems disabled
•Absence of monitoring
•Safety regulations not enforced
•Over-reliance on automation

Defenses Inadequate

•Omission of checklist item
•Forgetting fuel, equipment or running gear checks
•Use of wrong procedure
•Violation of weather regulations

Unsafe Acts

•High workload
•Undue time pressures
•Acceptance of hazards
•Ignorance of the system

Preconditions

•Inadequate procedures
•Poor scheduling
•Neglect of hazards
•Insufficient training

Line Management Deficiencies

Decisions-Makers

•Lack of regulation
•Poorly planned deregulation
•Too rapid expansion of routes or services
•Lip service to safety

ACCIDENT

Defenses

Active Failures

Latent Failures

Latent Failures

Latent Failures

FIGURE 6-4 Reason's accident causation model.

- *Risk* is the combination of the probability of occurrence of the hazard (or loss potential) and the severity of its effects.
- Risk assessment is the process of evaluating risk.

Risk management involves conserving assets and minimizing exposure to losses. It means looking ahead to detect hazards before they lead to losses and taking appropriate action when these risks cannot be eliminated. Risks are usually categorized by the broad areas they threaten, such as assets, income, and legal liability. In the aviation industry, accidents usually involve all three areas. Since accidents can be considered as involuntary and unscheduled expenditures, managers are obliged to establish policies and procedures to attempt to eliminate or minimize them.

If risk management programs are to be successful, they need to be integrated into the various business functions within the organization. This requires strong visible support from the highest levels of management. A risk management program normally contains the following basic elements:

- A written safety policy (vision) statement signed off by the highest-ranking individual on site
- Goals and objectives that are integrated into employee evaluations
- Formal and informal accident/incident reporting systems
- An impartial review of incident and accident reports
- A systematic and continuous auditing (of site conditions and programs) process
- A process for prioritizing and addressing hazards
- A process for keeping senior management informed on a regular basis on the state of safety within the organization
- A feedback process to ensure that persons identifying hazards or submitting proposals for hazard reduction/elimination are informed of the outcome
- A process to continuously evaluate and update the risk management program

One of the major functions of airline management is to maintain economic viability while providing an acceptable service. This function requires that the cost benefits of expenditures be determined and includes the evaluation of risks or hazards as well as the consequences of accepting or not eliminating them. Often, the cost benefits of correcting hazards cannot be assessed in the short term because present expenditures are used to buy future safety. Risk taking is an accepted fact of air carrier life and can, therefore, influence management's attitudes toward safety.

CONCLUSIONS

In spite of the use of man, machine, medium, mission, and management as broad categories of hazards, a popular theory holds that most accidents or incidents can be traced to a human failure somewhere, not necessarily the person or thing immediately involved in the occurrence. For example, a machine is designed, built, and operated by humans. Thus a failure of the machine is really a failure of the human. Likewise, humans may not avoid or eliminate known environmental hazards, or they may create additional hazards. Thus, these could all be considered failures of humans rather than environmental failures. This interpretation, therefore, accounts for the wide discrepancy in the percentages of accidents attributed to human failure reported by different sources. Typically, these range from around 50 percent to close to 90 percent.

Fortunately, humans are adaptable and are able to compensate for many inadequacies in the design or construction of the machine. However, the closer the match between the human's capabilities and the machine's qualities, the greater will be the safety levels achieved. The larger the gap, the more likely it is that errors will occur or not be corrected. For example, operating the wrong lever or switch is more likely to occur if the handling of the aircraft is demanding or the flight deck is poorly designed.

The design of an aircraft should, therefore, aim at reducing the likelihood of human error. In other words, the machine should be forgiving and accommodating of human error. If errors are not self-evident, then their occurrence should be clearly signaled to the crew. As aircraft and procedures become more complex, the role of the human in the system deserves greater expert attention, particularly his or her workload in abnormal situations.

Risks associated with the mission can manifest themselves in any of the three basic categories. For example, one type of mission may place increased strain or pressure on the pilot, leading to his or her making errors or being placed in a situation for which the pilot was not adequately trained or prepared. Likewise, it may result in the aircraft or machine being used for a purpose for which it was not designed, possibly leading to premature failure of components that again could increase the pressures on the pilot and the likelihood of his or her making an error. Accordingly, an accident involving an aircraft being used on a mission for which it was not designed may appear to be caused by a crew error, while the underlying management error is not readily evident.

Safe aviation, therefore, involves the integration of the mission into the three basic elements of man, machine, and medium. Each element can influence the others to varying degrees, and they are often interdependent. A hazard in one can initiate a chain reaction leading to an accident in which all are involved. Likewise, when one eliminates a hazard in one element, the effect on the other elements needs to be considered.

Many aviation hazards are brought about by problems at the interface between these elements. As humans are involved in all three, it is vital that our inherent limitations be considered, necessitating increased emphasis on the study of the human involvement in aviation. In Chap. 7, we cover the subject of human factors in aviation safety.

KEY TERMS

Modeling

Operational error

Operational deviation

Pilot deviation

Primary, secondary, and tertiary accident causation factors

Man-machine-medium-mission-management factors

Inertial navigation systems (INSs)

Automatic landing system (ALS)

Electronic flight instrument system (EFIS)

Flight management system (FMS)

Flight warning computer system (FWCS)

Fly-by-wire (FBW)

Instrument meteorological conditions (IMC)

Full-authority digital engine control (FADEC)

Global positioning system (GPS)

Ground proximity warning system (GPWS)

Mean time between failures (MTBF)

Latent failures

Active failures

Risk management

Risk

Risk assessment

REVIEW QUESTIONS

1. Why do we model accidents?

2. List and discuss the three accident causation factors described in the OTA model. Give examples to explain concepts where appropriate.

3. Why can it be said that management plays a predominant role when examining the five-factors model?

4. Why do you think over the years the number of accidents caused by "machine" has declined, while those attributable to man have risen?

5. The medium or environment includes two parts—the natural environment and the artificial environment. Compare and contrast the two.

6. Describe the hazards inherent in the *mission* of several air carriers.

7. Give several examples of how management can influence the safety program. How can an effective safety program affect efficiency and cost-effectiveness?

8. Explain Reason's model of defensive layers to include the concept of active and latent failures.

9. What is the purpose of risk management?

10. Identify the basic elements in a risk management program.

REFERENCES

International Civil Aviation Organization. 1984. *Accident/Incident Reporting Manual* (Doc. 9422-AN/923). Montreal, Canada: ICAO.

———. 1987. *Accident/Incident Reporting Manual (ADREP Manual)*. Montreal, Canada: ICAO.

Phaneuf Associates, Inc. 1992. *Air Carrier Internal Evaluation Model Program Guide*. FAA Contract Report DTFA01-88-C-00064, February.

Reason, J. 1997. *Managing the Risks of Organizational Accidents*. Aldershot, U.K.: Ashgate.

Taylor, Laurie. 1997. *Air Travel: How Safe Is It?* 2d ed. London: Blackwell Science, Ltd.

Sears, R. L. 1985. "A New Look at Accident Contributions and the Implications of Operational & Training Procedures," *Proceedings of the Flight Safety Foundation's 38th Annual International Air Safety Seminar*. Arlington, Va.

U.S. Congress, Office of Technology Assessment. 1988. *Safe Skies for Tomorrow: Aviation Safety in a Competitive Environment*. OTA-SET-38. Washington, D.C.: U.S. Government Printing Office.

Warren, Dale. 1996. *A Perspective on Safety in Commercial Aviation*. Warrendale, Pa.: Society of Automotive Engineers, Inc.

Wiener, Earl L., and David C. Nagel. 1988. *Human Factors in Aviation*. San Diego, Calif.: Academic Press.

HUMAN FACTORS IN AVIATION SAFETY

LEARNING OBJECTIVES

After completing this chapter, you should be able to

- Discuss the significance of human error in major aircraft accidents.
- List and explain the factors that affect human performance, giving examples of each.
- Define human error and explain methods used to classify human error.
- Explain what is meant by error management.
- Discuss the philosophical difference between engineering and administrative human-error control strategies.

- Explain the role of cockpit standardization, cockpit automation, warning and alerting systems, display conspicuity and system recovery, the flight management computer, and air-to-ground communication in human-error management.
- List and explain the various administrative practices that airlines undertake to manage human error.
- Discuss how federal agencies' regulatory promulgations influence error management.
- Discuss the role of labor in human-error management.

INTRODUCTION

The people who operate and support the U.S. aviation system are crucial to its safety; the resourcefulness and skills of crewmembers, air traffic controllers, and mechanics help prevent countless mishaps each day. However, despite the excellent safety record, many studies attribute human error as a factor in at least two-thirds of commercial aviation accidents. Safety attention at present is, therefore, heavily focused on trying to understand the human decision-making process and how humans react to operational situations and interact with the new technology in aircraft and ground systems. The way in which human beings are managed affects their attitudes, which affects their performance of critical tasks. Their performance affects the efficiency and, therefore, the economic results of the operation. It is important to understand how people can be managed to yield the highest levels of error-free judgment and performance in critical situations, while at the same time providing them with a satisfactory work environment. Distractions, for whatever reason, must be minimized, especially during the performance of critical, rapidly time-varying tasks.

While the emphasis often focuses on the pilots, they are not the lone threat. They are, however, the last link in the chain and are usually in a position to identify and correct errors that result in accidents and incidents. The numbers vary somewhat, yet mismanagement by the flightcrew is a causal factor in anywhere from 57 to 90 percent of all major airline accidents. This is the single greatest threat to flight safety.

Basically, the problem is one of decision making. Essentially, three reasons explain why people make poor decisions: They have incomplete information, they use inaccurate information, or they process the information poorly. These are not mutually exclusive categories, nor are they limited to inexperienced pilots. Accidents occur all along the competency curve. In fact, a surprisingly high number of respected, proficient pilots, some of whom are check airmen and instructors, are involved in accidents. A highly respected pilot once said his biggest fear was that when he made a mistake, no one would tell him about it. On the other hand, if he had the reputation of being a marginal pilot, his crews would continually be on the alert for the inevitable error and would not hesitate to correct the situation. If crews consistently used a process that would assist in better decision making, while simultaneously providing a constant level of monitoring and backup, there would be fewer errors.

Postaccident investigations usually uncover the details of what happened. With mechanical failures, accident data analysis often leads logically to why the accident occurred. Determining the precise reason for human errors is much more difficult. Without an understanding of human behavior factors in the operation of a system, preventive or corrective actions are impossible.

Understanding human factors is especially important to systems where humans interact regularly with sophisticated machinery and in industries where human-error-induced accidents can have catastrophic consequences. However, human factors are not treated as a technology in commercial aviation. Technical decisions for aircraft design, regulation, production, and operation are based on "hard" sciences, such as aerodynamics, propulsion, and structures. Human capabilities do not lend themselves readily to consistent, precise measurements. And human-factors research requires much more time and cooperation than most other aeronautics research. Data on human performance and reliability are regarded by many technical experts as "soft" and receive little attention in some aviation system designs, testing, and certification. Data used in designs are often after the fact. This chapter explores areas of aviation in which human factors are especially important.

HUMAN FACTORS

Human factors is a science that attempts to optimize the interaction between people, machines, methods, and procedures that interface with one another within an environment in a defined system to achieve a set of systems goals. Human factors encompass multidisciplinary fields of study that include, but are not limited to, engineering, psychology, physiology, anthropometry, biomechanics, biology, and certain fields of medicine. Human factors science concentrates on studying the capabilities and limitations of the human in a system with the intent of using this knowledge to design systems that reduce the mismatch between what is required of the human and what the human is capable of doing. If this mismatch is minimized, errors (that could lead to accidents) will be minimized and human performance will be maximized. *Human performance* is a measure of human activity that expresses how well a human has carried out an assigned, well-defined task or a portion of a task (task element), and it is a function of *speed* and *accuracy*. This chapter looks at the accuracy component of human performance. If a task is not performed "accurately" in accordance with its requirements, an error has occurred. An expanded discussion on human error follows in the next section.

Accidents rarely involve a deliberate disregard of procedures. They are generally caused by situations in which a person's capabilities are inadequate or are overwhelmed in an adverse situation. Humans are subject to such a wide range of varying situations and circumstances that not all can be easily foreseen. Careful attention should therefore be given to all the factors that may have influenced the person involved. In other words, consideration must be given not only to the human error (failure to perform as required) but also to why the error occurred.

HUMAN PERFORMANCE

Variables that affect human performance can be grouped into physical factors, physiological factors, psychological factors, psychosocial factors, hardware factors, task factors, and environmental factors. These factors are now briefly reviewed.

Physical factors include body dimensions and size (anthropometric measurements), age, strength, aerobic capacity, motor skills, and body senses such as visual, auditory, olfactory, and vestibular.

Physiological factors include general health, mental, and medical conditions such as low blood sugar, irregular heart rates, incapacitation, illusions, and history of injury, disability, or disease. Also included in this category are human conditions brought on by lifestyle such as the use of drugs, alcohol, or medication; nutrition; exercise; sports; leisure activities; hobbies; physical stress; and fatigue.

Psychological factors include mental and emotional states, mental capacity to process information, and personality types (introverts and extroverts). Some human *personality traits* include the following:

- *Motivation* is a desire of an individual to complete the task at hand. Motivation affects one's ability to focus all the necessary faculties to carry out the task.

- *Memory* allows us to benefit from experience. It is the mental faculty that allows us to prepare and act upon plans. Memory can be improved through the processes of association, visualization, rehearsal, priming, mnemonics, heuristics, and chaining. Memory management organizes remembering skills in a structured procedure while considering time and criticality. It is a step-by-step process to increase the accuracy and completeness of remembering.

- *Complacency* can lead to a reduced awareness of danger. The high degree of automation and reliability present in today's aircraft and the routines involved in their operation are all factors that may cause complacency.

- *Attention* (or its deficit) determines what part of the world exists for you at the moment. Conscious control of attention is needed to balance the environment's pull on attention. An intrapersonal accident prevention approach would describe the hazardous states of attention as distraction, preoccupation, absorption, and attention mismanagement—the inability to cope with tasks requiring flexible attention and focused tracking and steering. The inability to concentrate can lead to lack of (situational) *awareness,* which has been identified as a contributing factor in many accidents and incidents. Awareness emerges from the processes of the mind. The component processes of the mind include memory, thought patterns, attention, reasoning, and physiological functions that affect mental state.

- *Attitude* strongly influences the functioning of attention and memory. Attitudes are built from thought patterns. An intrapersonal approach to the attitudes of crewmembers attempts to identify the desirable ranges between such hazardous thought patterns as macho–wimp, impulsive–indecisive, invulnerable–paranoid, resigned–compulsive, and antiauthority–brainwashed.

- *Perceptions* can be faulty. What we perceive is not always what we see or hear. Initial perceptions and perceptions based solely on intended actions are especially susceptible to error. An intrapersonal approach prescribes ways to make self-checking more efficient and reliable.

- *Judgment and decision making* are unique capabilities of humans. They enable us to evaluate data from a number of sources in the light of education or past experience and to come to a conclusion. Good judgment is vital for safe aircraft operations. Before a person can respond to a stimulus, he or she must make a judgment. Usually good judgment and sound decision making are the results of training, experience, and correct perceptions. Judgment, however, may be seriously affected by psychological pressures (or stress) or by other human traits, such as personality, emotion, ego, and temperament. An intrapersonal treatment of judgment and decision making presents an ideal decision-making model and then identifies departures from the ideal. These departures are due to cognitive shortcomings for which remedies can be described in the form of cognitive skills.

- *Self-discipline* is an important element of organized activities. Lack of self-discipline encourages negligence and poor performance.

- *Risk taking* is considered by some to be a fundamental trait of human behavior. It is present in all of us to varying extents since an element of risk is present in most normal daily activities. Risk will be present as long as aircraft fly and penalties for failure are high. Accordingly, the taking of risks needs to be carefully weighed against the perceived benefits.

Psychosocial factors include mental and emotional states due to death in the family or personal finances, mood swings, and stresses due to relations with family, friends, coworkers, and the work environment. Some of the factors that cause *stress* are inadequate rest, too much cognitive activity, noise, vibration and glare in the cockpit, anxiety over weather and traffic conditions, anger, frustration, and other emotions. Stress causes fatigue and degrades performance and decision making, and the overall effect of multiple stresses is cumulative. Interactions with coworkers are influenced by two important variables, namely, peer pressure and ego.

- *Peer pressure* can build to dangerous levels in competitive environments with high standards such as aviation in which a person's self-image is based on a high standard of performance relative to his or her peers. Such pressure can be beneficial in someone with the necessary competence and self-discipline, but it may be dangerous in a person with inferior skill, knowledge, or judgment. For example, a young, inexperienced pilot may feel the need to prove himself or herself and may, therefore, attempt tasks beyond his or her capability. Humans have many conflicting "needs," and the need to prove oneself is not limited to the young or inexperienced. Some persons, because of training or background, have a fear that others may consider them lacking in courage or ability. For such persons, the safe course of action may be perceived as involving an unacceptable "loss of face."

- *Ego* relates to a person's sense of individuality or self-esteem. In moderate doses, it has a positive effect on motivation and performance. A strong ego is usually associated with a domineering personality. For pilots in command, this trait may produce good leadership qualities in emergency situations, but it may also result in poor crew or resource management. The domineering personality may discourage advice from others or may disregard established procedures, previous training, or good airmanship. Piloting an aircraft is one situation in which an overriding ego or sense of pride is hazardous. Although usually not specifically identified as such in accident reports, these traits may often be hidden behind such statements as "pilot failed to overshoot," "descended below minima," "failed to divert to an alternate," "attempted operation beyond experience/ability level," "continued flight into known adverse weather," and so forth.

Hardware factors include the design of equipment, displays, controls, software, and the interface with humans in the system.

Task factors include the nature of the task being performed (vigilance and inspection tasks versus assembly operations), workload (work intensity, multitasking, and/or time constraints), and level of training.

Environmental factors include noise, temperature, humidity, partial pressure of oxygen, vibration, and motion/acceleration.

It is important to note that the factors discussed above can act alone or in combination with two or more other factors to further degrade human performance in the occupational setting. These factors can produce synergistic effects on human performance. Some examples include an air traffic controller monitoring air traffic during extremely low air traffic volume while on allergy medication or a quality control inspector monitoring low-defect-rate products while on cold medication.

HUMAN ERROR

It is human to err, as the saying goes, so it is important to accept the inevitability of human error. No individual can be expected to perform perfectly at all times. Also, acceptable performance is relative. What may be considered exceptional performance under one set of circumstances could be totally unacceptable under another set of conditions. Human error is an event that occurs whenever a task or a task element (portion of a task) is not performed in accordance with its specifications.

Occurrence of errors assumes a well-defined task with standards for comparison. An error occurs when a task element is

- Not performed when required—also referred to as the *error of omission* (e.g., wheels-up landing)
- Performed when not required—also referred to as the *error of commission* (e.g., landing gear deployed in flight)
- Performed incorrectly—also referred to as an *error of substitution* (e.g., shutting down the wrong engine during an in-flight engine failure)
- Performed out of sequence, e.g., not performing tasks in the order required by a checklist
- Performed late, e.g., braking that overshoots the runway

In addition to the source or cause of error outlined above, errors can be classified by the severity (consequences) they generate. For example, errors that cause death or total system loss are categorized as catastrophic errors.

MANAGEMENT OF HUMAN ERROR

Modern transport aircraft, for all their sophistication of design and manufacturing, are still highly vulnerable to errors caused by humans (e.g., crewmembers) who operate in the system. This section examines strategies for managing human error. Error management involves timely detection and elimination of errors and/or minimizing their effects once errors do occur. There are two basic strategies for controlling human error—engineering and administrative. *Engineering strategies* concentrate on the use of engineering (through automation, human-factors engineering, etc.) to eliminate the potential error source during the design phase of a system. Any residual error potential (during the operational phase) is managed by incorporating hardware (warning and alerting devices) and software (FMS) into a system that will eliminate, limit and compensate for, or warn of an error or error potential. *Administrative strategies* include employee selection and training, modifying or restricting human behavior through operating rules, work methods and procedures, workload management, regulatory non-design-related initiatives, and company management practices. Administrative strategies, while effective when used appropriately, are generally considered not to be as effective and permanent as engineering solutions. Administrative and engineering control strategies are discussed at length in the following sections.

Philosophically, any control strategy other than to use engineering to completely eliminate the error source merely serves to delay the occurrence and/or effects of an accident or incident. On another note, with the advent of modern automation, many industry analysts assumed that human error could be removed, replacing the fallible human with unerring devices. This view may be overly optimistic, and automation may merely change the nature of error and possibly increase the severity of its consequences.

ENGINEERING CONTROL STRATEGIES

While comprehensive coverage of engineering design strategies to control human errors is beyond the scope of this text, a few important features of this strategy will be broadly discussed.

On the forefront of engineering control strategies is *automation*. Automation, or assigning physical or mental tasks previously performed by the crew to machines or computers, is a frequently cited means of reducing human error. Although totally eliminating humans from the operational loop is not yet feasible or necessarily desirable, partial replacement is becoming increasingly common. Theoretically, automation minimizes or prevents operational human errors by reducing the physical or mental workload of the human operator or by eliminating the human from an operational control loop. Used appropriately, automation is a valuable tool; the autopilot, a flight path control device, is one such item.

In the past two decades, with the rapid growth in microprocessor technology, there has been a temptation on the part of some designers to build very complex systems based on the rationale that the systems could operate automatically. There are two fallacies in this argument. First, almost no major system on an aircraft truly operates fully automatically. Systems must be initialized or set up by the human, decisions about operating modes must be made, and then the systems must be monitored by humans for obvious reasons. Second, in the event of the failure of automation, it falls to the human to operate the system. This responsibility cannot be avoided or designed away. If the complexity of the system is unbridled, then the crew may not be able to perform their duties effectively or take over in the event of equipment failure.

In response, many design engineers with human-factors sophistication have recognized that simplification offers an alternative to automation. If the system can be simplified, there may be no need for complex automation, and the same goal can be achieved without placing the human in a potentially hazardous position. An example is the fuel system on a multiengine aircraft. Those favoring automation would find no problem with creating a complex tank-to-tank and tank-to-engine relationship, as long as its management could be automated. If, for example, a fuel imbalance were created, automatic devices would detect the imbalance, determine a remedy, open the required transfer valves, and turn on the appropriate pumps to restore the proper balance. No human intervention would be required.

This example represents a philosophical difference between two major aircraft manufacturers—Airbus and Boeing. The Airbus approach, as exemplified by the A-320, has been to remove the pilot from the loop and turn certain functions over to sophisticated automation. Compensation is automatic—the systems do not ask the crew's approval. Boeing's approach is to never bypass the crew: Sophisticated devices inform the crew of a need and, in some cases, a step-by-step procedure; but in the end, it is the crew who must authorize and conduct the procedure. Boeing is a strong advocate of simplification before automation. Their designers would look to a less complex relationship. An example would be fewer tanks to feed the engines, creating fewer tank-to-tank and tank-to-engine requirements, requiring less management by the crew and fewer opportunities for human error.

One of the potential difficulties with highly automated systems is that onboard computers may, unknown to the crew, automatically compensate for abnormal events. Efficient automatic compensation for abnormal events and conditions sounds attractive, but there is always a limit to the system's capacity to compensate. When automation is compensating for some worsening condition without the crew's knowledge, this can lead to a situation where it may be too late for the crew to override the system and prevent a catastrophe.

The other potential problem with automation is that automation itself is susceptible to errors. For example, digital navigation equipment is susceptible to keyboard entry or "finger" errors. Such errors can easily go unnoticed by the crew; it is believed that KAL 007 flew off course because of a keyboard error. A broader problem with automation is that automatic devices are often installed one item at a time, especially in older aircraft, without consideration of the overall pilot-cockpit system.

Ultimately, it may be a prudent approach to simplify any system to the extent possible; then and only then turn to automation, if it is still needed. There is no doubt that the use of automation will increase; and the role of the human in an increasingly automated environment needs to be studied and bases established for setting standards. Some error protection systems that come under the domain of engineering controls are discussed next.

COCKPIT STANDARDIZATION

Between-fleets standardization of hardware is considered desirable to reduce training and maintenance costs as well as to prevent human error that may occur as a result of the pilots moving from one aircraft to another. During periods of rapid expansion of aircraft inventories and pilot personnel, as the airline industry in the United States and elsewhere enjoyed in the late 1980s, there is frequent movement between aircraft as pilots bid for more lucrative assignments, more modern aircraft, or desirable bases. Some contracts limit the rapidity with which pilots may bid a new seat; others do not.

Most cockpit hardware is peculiar to the type of aircraft. However, certain cockpit hardware could be common to most or all models operated by a carrier; examples are radios, flight directors, certain displays, area navigation equipment, and weather radar. Other examples would be devices added after the original manufacture (e.g., TCAS, ACARS). When the carrier has the opportunity to purchase these add-on units, a common model will most likely be chosen for all the reasons stated above.

Where differences already exist between fleets, the airline may intervene by standardizing throughout the airline. For example, some airlines have invested in a common airlinewide model of the flight director.

Between-fleet standardization, if it involves retrofit rather than new equipment purchase, will be extremely costly, and its safety benefits may be modest compared to within-fleet standardization. Nonetheless, when pilots move rapidly through the seats of various aircraft or complete training for one aircraft and then return to another while awaiting assignment to the new aircraft, between-fleet standardization of cockpit hardware deserves inclusion in the list of intervention strategies.

Within-fleets standardization is far more critical. Long before the Airline Deregulation Act of 1978, carriers purchased aircraft from one another, thus generating mixed configurations within fleets. With the coming of deregulation, the pace of mergers and acquisitions, as well as used equipment purchases and leases, accelerated rapidly, producing fleets of traditional aircraft, such as B-727s, 737s, and DC-9s, that varied greatly with respect to cockpit configuration. These differences included different displays (e.g., various models of flight directors), warning and alerting systems (e.g., a host of altitude warning systems with various trigger points), every imaginable engine configuration, controls in different locations, various directions of movement of switches, and various operating limitations. One carrier, which had been through a number of mergers and acquisitions of other DC-9 operators, had eight different models or locations of altitude alerters. It later invested a very considerable sum to standardize the cockpits of its DC-9 fleet. Within-fleet standardization is considered a high-priority item by the line pilots and their safety committee.

In one rather strange example, a carrier with a large DC-9 fleet had seven DC-9-10 aircraft that it had purchased from another carrier. These aircraft had a 215-knot speed restriction for gear-down flight due to a modified gear door. For the rest of the fleet, it was 270 knots. These DC-9s were known as the "215 aircraft." Various informal "placards" appeared to remind the pilot that he or she was flying a 215 model.

COCKPIT AUTOMATION

Cockpit automation began in the mid-1930s with the introduction of crude autopilots. Autopilot development has enjoyed uninterrupted growth since the early models. By the 1950s, more sophisticated models could be found on aircraft of the Super Constellation and DC-6 era. Development continued into the jet age as autopilots and flight directors became components of flight guidance systems, which included area navigation (RNAV) and rudimentary autothrottles. Other devices such as autoslats, autospoilers, and autobrakes became part of the automation package.

It was not until the late 1970s that modern flight deck automation flourished, driven by the rapid development of the microprocessor. In 1982, Boeing introduced the B-767, the first of the "glass cockpit" (more correctly, *electronic flight instrument system,* or *EFIS*) passenger aircraft. Other Boeing models and those of other manufacturers followed. By the end of the decade, glass cockpits were offered to the airline industry by all major manufacturers of large airliners, as well as many manufacturers of smaller aircraft operated by the regional carriers. Glass cockpits can also be found in corporate and military aircraft.

The new cockpit designs combined many of the previously existing devices with new features based on a sophisticated *inertial reference system* (*IRS*) and color computer-graphic instrumentation. The computer graphic ("glass") displays not only replaced the traditional ("iron") electromechanical instruments, but also allowed a wide variety of information to be displayed that had not been available previously, e.g., a wind vector, a path predictor vector, navaids and airports, super-imposed color radar, and a moving map on the *horizontal situation indicator* (*HSI*). Color radar can be superimposed on the HSI map display, a capability also univer-sally praised by airline crews. These displays allowed pilot selection and deselection of information (e.g., emergency airfields on the map) and switch-selectable options for instrument configuration, a feature not possible prior to the EFIS era. The pilot has at her or his fingertips a vast storehouse of information that previously either was not available or had to be extracted from charts, tables, handheld (mostly ana-log) computers, and manuals.

WARNING AND ALERTING SYSTEMS

Warning and alerting systems provide another line of defense against human error. They may anticipate the possible error or condition (e.g., "insufficient fuel" mes-sage in a glass cockpit aircraft). They may warn the crew of an impending hazard (e.g., GPWS) or annunciate the error as it occurs (e.g., misconfiguration takeoff warnings). In many cases, the systems may be considered backups to human vigi-lance (e.g., out-of-balance fuel conditions) where the operator has the necessary information available before the system reaches the alarm condition. In other cases, warning and alerting systems are not, in the strictest sense, corrective actions against human error but are extensions of human sensory capability (e.g., engine fire warnings, baggage compartment doors not closed). These examples rep-resent not a lack of human vigilance but sensory limitations. Some systems are mixed—human capabilities may or may not be sufficient for detecting the alert condition (e.g., a potential conflict with another aircraft as annunciated by TCAS).

The *ground-proximity warning system* (*GPWS*) was mandated by the U.S. Congress as a solution to *controlled flight into terrain* (*CFIT*) accidents. The early models of the GPWS had their own operational problems, e.g., high false-alarm rates and alarm modes that were difficult to interpret. In spite of its shaky begin-ning, the merits of the GPWS have been well documented. In those countries in which the GPWS is required, CFIT accidents have been dramatically reduced, and they have virtually disappeared in the United States. Unfortunately, in other parts of the world, GPWS is not required, and each airline may decide whether to equip its fleet with the device. The crash of the Air Inter A-320 near Strasbourg, France, in January 1992 emphasized this regrettable fact.

No warning and alerting system is perfect. None can provide an absolute guarantee against the human error it was designed to prevent. The lamentable history of gear-up landings is a testimonial to this fact. A gear-up landing may seem a simple error to prevent, compared to a far more complex error such as a wrong-airport landing, for which no hardware/documentation intervention is obvious. Indeed, we have probably run out of corrective-action strategies to prevent gear-up landings. They still occur, even to highly experienced pilots.

The imperfection of warning devices is attributable to a variety of problems, from failure of human vigilance to internal failures of the device itself. To begin with, any alerting device is subject to both commissive and omissive errors. The designer attempts to balance these two types of inevitable errors. Deliberate disarming of the device or deliberate ignoring of the warning is very common.

Another weakness is crew-warning interaction. It is not unusual for crews to allow the alarm condition to alert them before they take action. The primary system (human vigilance) becomes the backup system, and the backup system (alerting device) becomes primary. The lines of defense are reversed, and human vigilance alone is an insufficient defense. An example of primary-backup inversions can be found in the common practice of an altitude callout 1000 feet prior to reaching target altitude. It is not at all unusual to see the responsible crewmember (usually the pilot not flying) allow the altitude alerter to sound and then make the callout. This practice relaxes a line of defense against altitude deviation. It is especially insidious since there are a great variety of possible trigger points for various models of the altitude alerter. Unfortunately, the practice described is very common.

There are other problems associated with warning systems or similar devices. Excessive false alarms unnecessarily distract operators and might lead to the device being ignored or disabled.

DISPLAY CONSPICUITY AND SYSTEM RECOVERY

Another line of defense against error is to make the error, once it enters the system, more conspicuous to the crew. This can be done through feedback and feedforward mechanisms. Feedback provides the operator with information on the impact of his or her control inputs. Feedforward mechanisms predict and display the future state of the system, which may provide guidance for control inputs. Feedforward is seldom an inherent part of the system; it must be inserted artificially. Feedback may be inherent to the system (e.g., prestall buffet) or may be artificially inserted or enhanced (e.g., electronic stall warning devices). Such mechanisms do not prevent the original error, nor do they ensure error tolerance. But it provides the crewmembers with a better opportunity to detect their own error and remove it before it affects the functioning of the system. The map mode of the HSI of the EFIS aircraft provides an excellent example. Lateral navigational errors show up very clearly in the map mode. Error-evident displays can be thought of as a form of feedback, at times employing feedforward.

The interrelationship between feedforward and feedback can be found in the *plan* mode of the HSI map display, which allows the crew to step through the lateral course after it has been entered. This is an error-display system at its best. In this mode, the crew steps through the lateral flight plan one waypoint at a time. The next waypoint in line appears to move toward the aircraft symbol. Thus, the crew would be alerted in the case of an illogical entry, a severe turn, or an inconsistent position on the course. With waypoint-to-waypoint navigation, an erroneously located waypoint would cause the course line to appear on the map with some highly suspect orientation, probably a sharp bend, which would alert the crew.

A good example of this capability occurred onboard a B-767 preparing to depart Atlanta for Miami. The clearance included as a waypoint the TEPEE (note spelling) intersection near Tampa. The captain entered TEEPE (note spelling) into the route page of the control display unit (CDU). Because there is a TEEPE intersection (near Waco, Texas), the CDU dutifully accepted the erroneous spelling and established it as a waypoint on the route from Atlanta to Miami. The sudden shift in course to the west-southwest toward TEEPE from the southward course toward TEPEE was immediately evident to the crew. A non-EFIS aircraft with the same CDU-FMS (such as some models of the B-737-300) would not have provided this form of error detection capability. The crew would have had to detect the error by some other check; but whatever the method used, it would lack the rich error display found on the HSI map.

Once the system properly displays an abnormal condition, there must be an effective means of removing it and allowing the system to recover. The system must not permit irreversible errors. With traditional aircraft, this was usually not a problem. Working with less sophisticated systems, the pilots were closely coupled to the machine; an error, once detected, could usually be reversed quite easily. The advent of highly sophisticated automation raises the question of escape from error and system recovery. Generally, the problem is not that the error is irreversible but that the recovery process can be difficult, time-consuming, and possibly error-inducing itself.

FLIGHT MANAGEMENT COMPUTER

The *flight management computer* (FMC) offers some novel features for protection against human error. The flight management computer can store and process a vast amount of information typically contained in manuals, checklists, performance charts, flight plans, weather reports, and documents and paperwork of all sorts. This information can be displayed to the crew in text, numeric, and graphic forms on selected pages of the control display unit, the glass instrument panels, and elsewhere. Some of the information is automatically displayed, requiring no request from the crew (e.g., the wind vector on the navigation display); other information is available in the FMC on demand through pilot selection of the correct CDU page. The display of certain valuable information, such as suitable emergency airfields, is switch-selectable. Finally, if the FMC detects an abnormal computer condition, a brief message can be displayed in the "scratch pad" line of the CDU, and the pilot is alerted on two other displays that an FMC message is waiting. An example would be a request for a waypoint "not in the database." The flight management computer can recognize and reject certain cases of input that were outside its domain. While the FMC of today can recognize and reject inputs because they are stylistically incorrect, it generally lacks the intelligence to detect inputs that are illogical or wildly incorrect but in the proper form.

Pilots are forced to enter information in a rigid format, which in one sense may be a defense against input errors, but in another sense creates a less user-friendly device. Why should a crew have to worry about whether a slash (/) is required between the latitude and longitude?

The B-767/757 and the glass cockpit aircraft that followed possessed rudimentary forms of computer-based error elimination and protection. The A-320, introduced in 1988, took error protection a step further. The fly-by-wire feature offered the opportunity to fly maneuvers, such as maximum safe angle of attack (AOA) for wind shear escape, with no danger of entering a stall. The computer would simply stop the aircraft's increase in pitch short of its computed safe AOA. If the pilot continued to pull back on the stick, no more nose-up pitch would be commanded. An intelligent computer interposes an electronic line of defense between the pilot's control and the aircraft's control surfaces. Incidentally, such a system that has the capability of controlling and correcting an error is referred to as an *error-resistant* or *error-tolerant system.*

Other EFIS aircraft, such as the 757/767, offer escape guidance on the *altitude directional indicator* (ADI) in the form of a target line for optimal nose-up pitch. In contrast with the approach taken in the A-320 design, the pilot remains in the loop. The pilot controls the pitch angle; the computer merely computes and displays the commanded nose-up pitch.

These two approaches emphasize not only disparate views of cockpit design but basic philosophical design differences: The A-320 essentially allows the pilot to pull the control stick all the way back and let the computer find the maximum angle of attack that will avoid a stall. Other EFIS aircraft depend on the pilot to follow the wind shear escape guidance cues. It is impossible to say which approach is more effective. Only time and experience will settle that question.

AIR-TO-GROUND COMMUNICATION

Verbal communication remains the weakest link in the modern aviation system; more than 70 percent of the reports to the Aviation Safety Reporting System involve some type of oral communication problem related to the operation of an aircraft. The ground collision between two B-747 aircraft in Tenerife in 1977, resulting in the greatest loss of life in an aviation accident, occurred because of a communication error. Technologies, such as airport traffic lights or data link, have been available for years to circumvent some of the problems inherent in ATC stemming from verbal information transfer.

One potential problem with ATC by data link is that the loss of the "party line" effect (hearing the instructions to other pilots) would remove an important source of information for pilots about the ATC environment. However, the party line is also a source of errors by pilots who act on instructions directed to other aircraft or who misunderstand instructions that differ from what they anticipated by listening to the party line. Switching ATC communication from hearing to visual also can increase pilot workload under some conditions. Further study is necessary to define the optimum uses of visual and voice communications. More details on communications are included in Chap. 8 on air traffic safety systems.

ADMINISTRATIVE CONTROL STRATEGIES

Administrative error management controls refer to the collection of practices and procedures that are developed, promulgated, and implemented by the airlines, regulatory agencies, and labor groups.

AIRLINE PRACTICES

Management is ultimately responsible for organizing and operating the complex aviation system, and its practices and decisions influence the judgment and skill of the pilots, mechanics, air traffic controllers, and other key people in the system. For example, airline management practices regarding pilot selection and training, as well as aircraft design, provide the underpinnings of pilot performance. Management practices that cause work overloads, duty shift conflicts, and labor/management rifts generate employee stress, which can lead to decreased performance and increased errors. Deregulation has resulted in a record increase in commercial traffic, increased use of hub-and-spoke systems, and crowded airspace and airport ground facilities. Resulting schedule pressures have stretched the limits of pilot, mechanic, and air traffic controller performance.

Schedule pressure is a function of the whole airspace system as well as of individual airline practices. Management attitudes, especially labor/management relations, determine how schedule pressure is interpreted in the cockpit and on the flight line. Additionally, airline mergers frequently have resulted in divisive seniority and pay-scale arguments between management and the merging workforces. Cockpit crews comprising pilots holding opposite views on unresolved merger issues bring additional stress to commercial flight operations.

Airline operating safety is based on well-rested and alert flightcrews. Good flightcrew communications, sound coordination patterns, and flightcrew familiarity are factors that can be controlled by management to help reduce crew fatigue and create positive effects on performance.

Airline management has the responsibility of addressing the human-factors problems that have arisen due to operating practices and management attitudes.

Some airlines have employee assistance and counseling programs and provide for good communication in both directions along the chain of command.

Employee Selection.   One strategy of minimizing error is employee selection— allowing into the system only those operators least likely to make mistakes. Airline pilots and air traffic controllers must meet prescribed health, age (mandatory retirement at age 60 for air carrier pilots), and training requirements and must pass written and operational tests of skills and knowledge. For the select group that survives the culling, continued quality is maintained through training and monitoring. Indeed, federal regulations require the periodic testing of flightcrew members to check results of training and operational experience, including flight proficiency and system knowledge. Pilots and controllers are also monitored through required periodic medical examinations, including for-cause drug and alcohol testing. The ability to monitor or test for certain physical and mental parameters of operator health that could affect operational safety is limited due to the sensitive issue of privacy. Selection may also occur through employing *experienced* pilots. Most airlines set minimum standards that are much higher than the FAR requirements. However, during times of labor shortages and high turnover rates, large and small carriers alike have been forced to lower their selection criteria for new hires. This can result in junior cockpit members advancing to captain without being "seasoned." Some commuter airlines, e.g., can find up to one-half of the captains in their first year of employment. This can compromise the strategy of error reduction through employee selection.

Employee selection can also occur through *type ratings,* a process that licenses airline pilots for a specific model. A pilot licensed to fly a B-737 is allowed to fly any version or derivative of the B-737, provided he or she is trained on their differences; but this pilot cannot fly a B-727 unless she or he first receives a full course of instruction, passes a written and flight examination, and is granted a type rating for the aircraft. *Type,* as used with respect to pilot ratings, means "a specific make and model of aircraft, including modifications thereto that do not change its handling or flight characteristics."

Training.   Pilot training is another strategy for reducing error. There is a practical and a regulatory requirement for training. Training can be *general-purpose* or *specific.* General-purpose training is more broad-based and covers statutory requirements of the Federal Aviation Regulations (FARs). These regulations must be interpreted by each company, consistent with its own philosophy and resources, as approved by the Federal Aviation Administration (FAA). *Crew resource management* (*CRM*) training is an example of a general-purpose training that offers a remedy for a broad, perhaps poorly defined class of problems, the origins of which are inadequate or inappropriate communication in the cockpit. The corrective action comes in the form of a training program for all pilots. At some carriers, the training is extended to other personnel, such as maintenance workers, cabin crews, and flight management. It is not remedial training for a handful of personnel who have been singled out as requiring corrective action, nor is it psychotherapy. CRM training is a broad-scale approach to social communication-based behaviors and attitudes. It attempts to change cockpit behavior, not personalities. CRM training at United Airlines, one of the pioneers in the field, was recognized by the captains in two fatal accidents—a B-747 door separation in flight near Hawaii and a DC-10 crash in Sioux City following total loss of hydraulics—as a major factor in their success in saving as many lives as they did. Such examples are difficult to come by, since it is usually problems and failures that get reported, not positive outcomes. CRM has generally been accepted by flightcrews as a worthwhile approach. *Line-oriented flight training* (LOFT)—full mission crew coordination training conducted in flight simulators—is another example of broad-based training.

Specific training designed to meet more specific problems usually fares better than that directed at less well-defined problems. When a specific problem has been identified, training can be directed toward a possible solution. An example of training to avoid foreseeable human error is wind shear escape maneuvers. There was a time when wind shear was a major safety concern, with little agreement on how pilots should maneuver their aircraft to avoid terrain while also avoiding low-altitude aerodynamic stalls. One procedure called for increasing the angle of attack until stick-shaker stall warnings were obtained and then "flying the shaker." Training programs for wind shear escape were formulated and introduced to the pilots at their next simulator check. The training requirement for glass cockpit aircraft is simplified by hardware. These aircraft have pitch-angle guidance for wind shear escape depicted directly on the attitude directional indicator. A yellow horizontal line commands the nose-up pitch angle to be followed, and the resulting angle of attack is kept just below the level for stick-shaker actuation.

Advanced simulators allow greater flexibility and safety and have become the preferred mode in training, and they also offer substantial cost savings. Training is not enough, however, if the environment is poorly designed. Another disadvantage is that training can get very expensive. During the "good times," the large commercial carriers increasingly recruit pilots from the smaller Part 121 regional and Part 135 commuter airlines, resulting in rapid turnovers in the regional pilot workforce, greater than 100 percent per year for some. The training burden on these smaller carriers can be enormous. In some cases, the flightcrew training costs at some regional airlines exceeded the pilots' salaries.

RESPONSIBILITY, ACCOUNTABILITY, AND ENFORCEMENT.   Once an individual has been properly trained and provided with a clear description of his or her task and the necessary tools to do the job, then that individual is responsible for her or his own actions. Such accountability applies in most professions. For example, management should be able to expect a pilot to comply with proper flight procedures or a technician to use a torque wrench when it is required. In short, the development and maintenance of a professional attitude and behavior are the responsibility of the individual. These attributes should be fostered by both management and the individual's professional association. Failure to perform to a designated standard should result in a person being held accountable. Responsibility and accountability should be accompanied by appropriate punishment or *enforcement* action for deliberate or repeated disregard of procedures, rules, or regulations. It should be remembered, however, that enforcement based on information obtained from the accident prevention process usually has a negative effect on subsequent accident prevention, because people will be reluctant to provide hazard information if it is going to be used against them. For unintentional errors or mistakes, punishment is inappropriate, and other corrective measures should be used, such as training, motivation, and so forth. Such measures create a climate of openness, foster safety awareness, and encourage hazard reporting.

PROCEDURES AND CHECKLISTS.   Procedures are step-by-step specifications drafted by management and provided to pilots. They are designed to dictate the manner in which tasks and subtasks are carried out and to provide a standardization of cockpit duties. Procedures and subtasks are most elemental steps by which pilots operate their craft and are fertile ground for human error. Procedures that are well documented and thoroughly thought through serve to increase crew performance while minimizing error. Most procedures involve running checklists. A *checklist* is a device (paper, mechanical, audio, or electronic format) that exists to ensure that procedures are carried out. Hence, running a checklist is a procedure in itself. The checklist is far more than a "laundry list" of items and tasks. The checklist serves to prevent error by stating what must be done, when, in what order, and by whom. It also provides the basis and sets the tone for cockpit discipline and standardization. This document, often a single piece of paper, is the very foundation of flight safety. Procedures, in turn, dictate to the crew how the tasks are done. Factors that influence the effectiveness of a checklist include content and its display, task sharing, continuation of steps after interruption, task allocation and sharing among the crew, and task transition and termination.

Electronic checklists may replace paper versions in future aircraft. Boeing has included such a device in its 777. Electronic checklists have many advantages over conventional versions, particularly when the checklist must be interrupted or items must be taken out of order. The electronic checklist will handle this very well; in the paper checklist, interrupting the process is an invitation to error.

PAPERWORK REDUCTION AND MANAGEMENT.   One area that is ubiquitous in methods improvement is paperwork reduction and management. We should distinguish between two types of paperwork: that which is necessary for any particular flight [flight plans, NOTAMs (Notices to Airmen), weather, weight-and-balance advisories, fuel slips, maintenance writeups, etc.] and that which is administrative (e.g., crew pay logs, engine performance logs, and discrepancy reports). Corrective-action strategies may consist of reducing cockpit workload by eliminating or simplifying the paperwork not needed for flight or by assigning it to other personnel in the cabin or on the ground. A related area in need of methods improvement is the design of the paperwork for compatibility in the cockpit. Many of the procedures and paperwork in use today by airlines were designed for traditional aircraft and have not been adapted to the advanced-technology cockpits. Carriers operating high-technology aircraft should examine every aspect of their operations and paperwork for incompatibility with the new aircraft.

WORKLOAD MANAGEMENT.   There are many opportunities for corrective action by managing (as contrasted with reduction) of workload. If workload cannot be eliminated or reduced, it can be managed. Management consists of reallocating workload to less flight-critical phases (e.g., programming that can be done at the gate rather than after takeoff) and reallocating duties (particularly in a three-pilot crew) to balance the demands on the individual crewmembers. For example, it has frequently been suggested that installing a transmitter-receiver, or an ACARS (Aircraft Communication Addressing and Reporting System), in the cabin for passenger-related communication by the flight attendants could reduce the radio communication load in the cockpit. This suggestion has been resisted by some pilots who hold a traditional view that all transmissions from a craft should emanate from the flight deck. (We note the prevalence of cellular telephones in the hands of passengers today.) Other pilots see the transfer of passenger-related communication duties to the cabin crew as good riddance. Another example is that a carrier may have a number of procedures that must be completed on climbing through 10,000 feet. In one aircraft, which has high climb performance, it is very possible to climb rapidly from 10,000 feet, where these duties are initiated, through FL180 (18,000 feet), where an altimeter adjustment is required (in U.S. airspace). This procedure could easily be ignored in the midst of a demanding workload. Good workload management suggests that some of these duties be reassigned to other points in the flight, possibly to FL180 or above, to avoid this potentially serious error.

COMMUNICATION.   The term *communication* usually includes all facets of information transfer. It is an essential part of teamwork, and language clarity is central to the communication process. Adequate communication requires that the recipient receive, understand, and can act on the information gained. For example, radio communication is one of the few areas of aviation in which complete redundancy is not incorporated. Consequently, particular care is required to ensure that the recipient receives and fully understands a radio communication.

The efficiency of communications within an organization is a management responsibility. Clearly written and easily understood directives, instructions, manuals, and so forth are required if staff members are to understand their responsibilities and duties and how they are expected to carry them out. The same applies to verbal communications, since an instruction that is not understood by the recipient may result in the wrong thing being done or in nothing being done at all.

There is more to communication than the use of clear, simple, and concise language. For instance, intelligent compliance with directions and instructions requires knowledge of why these are necessary in the first place. Therefore, management must first determine if an instruction is really necessary and, if so, ensure that the staff knows the reasons behind it, which enables the staff to respond more effectively.

Trust and confidence are essential ingredients of good communication. For instance, experience has shown that the discovery of hazards through incident or hazard reporting is only effective if the person communicating the information is confident that no retributory action will follow her or his reporting of a mistake.

Communications within the cockpit can be affected by what some psychologists call the *transcockpit authority gradient (TAG)*, which is an expression of the relative strength and forcefulness of the personalities involved. For safe operations, the gradient between the captain and copilot should be neither too steep nor too shallow, thus encouraging free communication between the pilots, leading to improved monitoring of the aircraft operation. When, e.g., the gradient is too steep, the copilot may be afraid to speak up, thereby failing in his or her role of monitoring the captain's actions. When it is too shallow, the captain may not adequately exercise his or her authority.

Miscommunication between aircrews and ATC controllers has been long recognized as a leading source of human error. It has also been an area rich in potential for interventions. Examples are the restricted or contrived lexicon (e.g., the phrase *say again* hails from military communications, where it was mandated to avoid confusing the words *repeat* and *retreat*); a phonetic alphabet ("alpha," "bravo," etc.); and stylized pronunciations (e.g., "niner" due to the confusion of the spoken words *nine* and *five*).

As a result of the tragic ground collision between two B-747s at Tenerife in 1977, blamed largely on miscommunications between the tower and the two aircraft, the FAA encouraged controllers to restrict the word *cleared* to two circumstances—*cleared to take off* and *cleared to land*—although other uses of the word are not prohibited. In the past, a pilot might have been cleared to start engines, cleared to push back, or cleared to cross a runway. Now the controller typically says, "Cross runway 27," and "Pushback approved," reserving the word *cleared* for its most flight-critical use.

Likewise, the term *cleared* was dropped from the "position and hold" instruction for the aircraft first in line for takeoff. Previously controllers said, "(Aircraft identifier) cleared to line up and hold." Because an aircraft in the number one position at the stop line was anticipating takeoff clearance, there were occasional incidents where a takeoff was initiated at this command. Now the controller simply instructs the number-one aircraft to *position and hold*.

The need for linguistic intervention never ends, as trouble can appear in unlikely places. For example, pilots reading back altimeter settings often abbreviate by omitting the first digit from the number of inches of barometric pressure. For example, 29.97 (inches of mercury) is read back "niner niner seven." Since barometric settings are given in millibars in many parts of the world, varying above and below the standard value of 1013, the readback "niner niner seven" might be interpreted reasonably but inaccurately as 997 millibars. The obvious corrective-action strategy would be to require full readback of all four digits when working in inches.

A long-range intervention and contribution to safety would be to accept the more common (in aviation) English system of measurement, eliminating meters, kilometers, and millibars once and for all. Whether English or metric forms should both be used in aviation, of course, is arguable and raises sensitive cultural issues. At this time, the English system clearly prevails, as does the English language. In some parts of the world, units are mixed: ATC instructions and instrumentation are in feet, yet weather is reported in meters. In 1983 an Air Canada B-767 ran out of fuel and made a successful dead-stick landing on a small, obscure airfield. The fuel instrumentation and calculations of most of the planes in the fleet were in pounds; this particular plane was instrumented in the metric system (kilograms of fuel). An error in conversion occurred that resulted in insufficient fuel on board by a factor of roughly 2.2, the conversion constant between kilograms and pounds.

TEAM CONCEPT (COGNITIVE REDUNDANCY).   This work method uses individuals in redundant combinations (backup) with one another so that the mistake of one will be caught by the other before it can lead to an accident. The *two-person concept* uses two or more persons capable of performing the required task. One completes the task, and the other independently checks his or her actions to ensure the task has been correctly performed. Crewmembers monitoring each other during aircraft operations and duplicate engineering inspections are examples of these procedures. Redundancy is designed into not only the aircraft and its systems but also the numerous maintenance and operating procedures involved. These procedures, as well as normal management functions, inevitably involve interactions among people. Viewing events from different perspectives is a cognitive technique for increasing awareness. Since different members of a crew naturally have different perspectives, the effectiveness and awareness of the crew can be dramatically increased through effective teamwork. Teamwork can, therefore, be thought of as a form of cognitive redundancy in which each member of the crew is backed up by the other members. An intrapersonal approach to teamwork presents each teamwork skill from the perspective of both the performer of the skill and the other team members.

PEER-PRESSURE CONTROL.   Accident prevention programs should clearly address the insidious nature of peer pressures. Far from resulting in a loss of face or appearing "scared," a person's decision to adopt the safe course of action clearly demonstrates strength of character or conviction. Peer pressure can also be helpful in eliminating aberrant behavior. For instance, review committees comprising pilots can be an effective means of modifying pilot behavior toward safe operating practices and could be used to complement normal management processes.

REGULATORY PRACTICES

Throughout the history of aviation, safety improvements have come primarily from technological developments, such as reliability and performance increases in aircraft, navigation devices, weather forecasting, and ATC. The federal government has the responsibility for setting appropriate standards for aircraft, airports, and navigation aids. Federal regulations address airline procedure, such as pilot flight time (e.g., the minimum rest period a crewmember must have during any consecutive 24-hour period is 8 hours), emergency operations, and the use of checklists. Air traffic rules, such as instrument approach and departure procedures, separation standards, and weather minimums, set operational limits for users of the national airspace system. Some examples of the roles that the FAA, NTSB and NASA play in addressing human error are now discussed.

FAA.   The FAA has the legal responsibility to promote air safety and the authority to do so through rulemaking and enforcement. As such, many of the FARs that exist today have originated as corrective-action strategies. The FAA influences human error, to a limited extent, through its certification process. Parts of FAR Part 25 attempt to deal with human-factors engineering by requiring the conduct of workload analyses to certify the aircraft for the size of the crew for which the design is submitted. There is no FAR requirement to analyze human-error potential, although it may take place informally during the certification process. In another example, the ground-proximity warning system, required under FARs in 1975, has essentially ended controlled flight into terrain accidents by U.S. carriers.

When errors are discovered by the FAA through accidents, incidents, check airmen, or the FAA's sponsorship of the Aviation Safety Reporting System (maintained by NASA), it may intervene through regulations or informally through emphasis on the matter in its various examinations and inspections of pilots and training centers. It can also intervene at airlines through its *principal operations inspectors* (*POIs*), who have considerable authority. It is the POI who must approve training programs, manuals, devices, procedures, checklists, etc.

Some corrective actions come as a result of a single accident. The speed limitation of 250 knots below 10,000 feet (FAR 91.70) followed the collision of a Constellation and a DC-8 over Staten Island in 1960. The DC-8 was flying at almost 500 knots on its way to Idlewild Airport (now Kennedy), navigating on a single VOR. The Constellation was flying to La Guardia.

The speed restriction was thought to make it less likely that an aircrew could overshoot its clearance limits. In addition, ATC modified its method of making handoffs from one controller to another. Previously, aircraft were cleared to a fix, at which the radar clearance actually terminated; then another radar controller would pick up the target and clear it to the next fix. Now the radar controller effects a position-handoff procedure, transferring authority for the target to the next controller. The handoff point does not terminate the clearance.

Another example of intervention through government regulation can be found in the *sterile cockpit rule.* In the 1970s, the airline industry was plagued by a rash of what came to be called CFIT accidents. In several cases, the cockpit voice recorder indicated a high degree of casual conversation and persiflage in the cockpit, implying a neglect of essential duties. As a result, the FAA promulgated FAR-121.542, which decreed that while moving on the ground under its own power or flying below 10,000 feet mean sea level (MSL), the cockpit was to be "sterile," meaning no nonpertinent conversation could take place. During sterile periods, there can be no entry into the cockpit by the cabin crew for nonessential reasons.

The sterile cockpit rule is largely unenforceable, but it does set the tone for a businesslike atmosphere in the critical phases of flight and sets a standard for cockpit behavior at critical times of operation. Some cockpit voice recorder read-outs in recent accidents have revealed less than assiduous devotion to the rule. It will always be controversial since it is invasive on the cockpit working atmosphere and self-expressions of the crew. Although no statistics support the efficacy of the sterile cockpit rule, it is generally seen as a plus for safety. Its benefits are not limited to CFIT accidents. With the growing concern over ground collisions at airports, the sterile cockpit rule probably plays a large part in preventing distractive behavior while taxiing.

The FAA enforces discipline over flightcrews through the levying of fines and/or suspension of licenses. For example, the FAA has cracked down on crews moving airplanes out of the gate or taxiing with a passenger standing in the aisle. Fines of $1000 can be levied against the captain for such an action, although it can be argued that passenger behavior is often beyond the pilot's control. The flightcrew depends on the cabin crew for information on passenger behavior, and often a passenger will stand up unexpectedly as the aircraft begins to move.

The FAA has the opportunity to intervene to prevent navigational errors in several ways. It can make changes in the system itself, procedures by which the system is operated by ATC personnel and cockpit procedures. For example, regarding cockpit procedures, it is not unheard of for an aircraft awaiting takeoff clearance to take the clearance of another aircraft and initiate a takeoff. This is particularly easy to do when parallel runways are being used for takeoff. As a corrective action to make this less likely, when more than one runway is in use, tower operators are now required to state the runway when issuing takeoff instructions (e.g., "American 123, runway zero-eight right, cleared for takeoff"). The aircraft crew usually acknowledges in kind, stating the runway along with its call sign and clearance, but it is not a requirement to do so.

Outside monitoring of airline flights is accomplished through the federal air traffic control system. Air traffic controllers detect gross navigation and guidance errors and provide useful information on weather and airport conditions to flight crews. En route controllers, in turn, are automatically monitored by ATC computers that record the separation between aircraft under positive control and sound an alert if the distance falls below minimum standards.

The air traffic control system is also likely to become increasingly automated. One aspect of the National Airspace System (NAS) Plan, the *Advanced En Route Automation (AERA)* system, could bring sweeping job changes for air traffic controllers through automation. AERA is software to be introduced in three stages as part of the *Advanced Automation System (AAS)*, the FAA's planned upgrade to the entire air traffic control system. The effectiveness of automation in accomplishing job tasks and the consequences of individual controller performance differences are being studied at the Civil Aeromedical Institute. The FAA plans to study controller selection and training requirements for AERA. More information on the future plans for the NAS can be found in Chap. 8 on air traffic safety systems.

There are no FARs relating cockpit automation to human performance. For example, the advanced cockpit electronic systems on the Boeing 757 and 767 airplanes required an *equivalent safety* deviation from current regulations to be certified. Automatic devices for the cockpit, which have subtle effects on human performance, are treated the same as other pieces of hardware in the regulations. Human-error hazard analyses are not required in the design, test, or certification stages. Some basic standards for cockpit design are included in FARs, but the standards do not address technological developments of the past decade, such as CRT displays and flight management systems. For example, although the use of color has increased in modern cockpit devices, the FAA has set standards only for warning, caution, and advisory lights. There are no rules governing other uses.

NTSB.   Human factors receive a great deal of emphasis in NTSB investigations of major accidents, the resulting determinations of probable cause, and recommendations for future accident prevention. The NTSB has a separate human performance division within its Bureau of Technology and usually includes a human-factors specialist on each major accident investigation team. Report forms, interviews, and analytical techniques are designed to elicit detailed information on the performance of the people involved in the mishap and the environmental and operating conditions that were present.

The NTSB accident database management and analyses are critically important because they provide the only valid statistical safety trends currently available to the federal government. Although lessons can be learned from individual accidents, the greatest understanding comes from analyses of clusters of accidents. For example, the frequent occurrence of flightcrew coordination problems in accidents has resulted in numerous NTSB recommendations urging the use of cockpit resource management training.

NASA.   NASA has traditionally provided a substantial amount of fundamental aviation research. For human factors in civil aviation, NASA contributes a major share of research, supplemented only by applied research programs in industry and basic research at a handful of universities. NASA is in a unique position that enhances its human-factors research efforts. While maintaining close working relationships with the FAA, NTSB, and military and commercial aviation industry, nonregulatory NASA is viewed as an impartial party, which gives NASA access to sensitive data unavailable to other federal groups.

Two research centers within NASA, Ames in California and Langley in Virginia, are responsible for most of the human-factors work. Generally, NASA-Langley investigates the physical aspects of human factors, while NASA-Ames studies the psychological elements. Physiological measures of pilot workload and advanced cockpit displays are among the topics addressed at Langley. The operational implications of human-factors research—cockpit resource management, information transfer, sleep cycle and fatigue, and the effects of advanced automation on flightcrew performance—are important fields of study at NASA-Ames. For example, line-oriented flight training was developed from the use of full-mission simulation as a research tool at Ames.

NASA-Ames also administers the Aviation Safety Reporting System, the only broad source of human-factors field data, other than NTSB investigations, available to the federal government. NASA-Ames increasingly has become the human-factors information clearinghouse. Although all databases have limitations, the Aviation Safety Reporting System analyses provide information unavailable to the FAA from other sources, such as the influence of new technologies or airline management practices on human performance.

LABOR PRACTICES

Organized labor has an important role in the resolution of management-related human-factors problems, and union contracts or initiatives often address issues not covered by federal policy. For example, some pilot contracts establish duty-time limits because FARs are not explicit in this area, and although FARs permit Part 121 pilots to fly 100 hours per month, few actually do. Additionally, labor organizations provide publications, training programs, counseling sessions, and communication channels to management for member employees. Unions also support independent studies and research efforts, such as the Air Lines Pilots Association's stress survey. ALPA has safety councils at each of its member domiciles.

CONCLUSION

People are pivotal to aviation safety. Although humans are largely responsible for commercial aviation's excellent safety record, human errors nonetheless cause or contribute to most accidents. Moreover, the rate of pilot-error accidents shows no sign of abating, while weather-related crashes are declining and aircraft component failures are rarely the sole factor in serious mishaps. Furthermore, accident and incident data analyses indicate that if only a portion of human-error problems can be solved, substantial reductions in accident risk can be attained.

It is clear that it is possible for those who design and operate aviation systems and other high-risk systems to erect lines of defense against error and to intervene in both general and specific ways to protect the systems. Furthermore, we have seen that this is possible in both traditional and advanced-technology aircraft.

It is equally important to recognize that the new technologies offer ways and means for management of error that are not available with traditional aircraft. Any limitations in the exploitation of this technology lie not in technology itself, but in the resourcefulness of persons who can effect corrective actions. Thus, we may conclude that the computers that drive the modern cockpit technology provide opportunities previously unknown for both the commission of and the control of human error.

Another concern among pilots and human-performance experts is that the increased level of cockpit automation may create a generation of pilots whose basic flying skills ("stick and rudder") deteriorate from lack of practice. A fact that reinforces this concern is that, from 1990 to 1999, the leading cause of air carrier accidents was loss of control of the aircraft and CFIT. These types of accidents account for 85 percent of large hull losses worldwide, according to the Boeing study, but 76 percent could have been averted had the pilots known how to respond to the situation. If manual skills ever become needed because of automation failure/degradation or unusual plane attitudes and conditions that automation cannot handle, the pilot may not be up to the challenge. Manual piloting skills may have degraded because of the (over)use of automatic flight systems in lieu of hand flying and/or because of the lack of training and practice on certain maneuvers and skills.

Human error on the flight deck can never be totally eliminated. However, through judicious design; constant monitoring of accidents, incidents, and internal reports; and the aggressive use of reporting systems such as NASA's ASRS, the warrants for and the means of corrective action can be found. Air transportation enjoys an excellent safety record today largely because no part of the system is ever allowed to rest.

Lack of awareness has been identified as a contributing factor in many accidents and incidents. Awareness emerges from the processes of the mind. The component processes of the mind are memory, thought patterns, attention, reasoning, and physiological functions that affect mental state. Intrapersonal skills can be developed from study and experience that can enhance these cognitive processes. An intrapersonal training approach emphasizes that improving the capabilities of the individual human mind is key to improving operational performance.

Viewing events from different perspectives is a cognitive technique for increasing awareness. Since different members of a crew naturally have different perspectives, the effectiveness and awareness of the crew can be dramatically increased through effective teamwork. Teamwork can, therefore, be thought of as a form of cognitive redundancy in which each member of the crew is backed up by the other members. The intrapersonal approach to teamwork is to present each teamwork skill from the perspective of both the performer and the other team members.

A great deal of study is presently underway in determining the root causes of human error and approaches needed to address the problem. However, this area is still in need of much work. Most of the Federal Aviation Regulations aimed at limiting human error are based primarily on past regulatory experience, not scientific evidence. Although previous experience is, of course, important, it is often insufficient or inappropriate in a changing environment. Recent technological developments, such as cockpit automation devices and displays, have outpaced the FAA regulatory process.

KEY TERMS

Human factors

Human performance

Motivation

Emotion

Awareness

Motivation

Memory

Complacency

Attention

Attitude

Perceptions

Judgment

Decision making

Self-discipline

Risk taking

Psychosocial factors

Transcockpit authority gradient (TAG)

Enforcement

Peer pressure

Ego and pride

Hardware factors

Task factors

Environmental factors

Engineering strategies

Administrative strategies

Fly-by-wire technology

Commuter Safety Initiative (CSI)

Cockpit resource management (CRM)

Line-oriented flight training (LOFT)

Type rating

Automation

Advanced En Route Automation (AERA) system

Advanced Automation System (AAS)

Electronic flight instrument system (EFIS)

Inertial reference system (IRS)

Horizontal situation indicator (HSI)

Flight management computer (FMC)

Control display unit (CDU)

Ground-proximity warning system (GPWS)

Controlled flight into terrain (CFIT)

Error-resistant

Error-tolerant

Attitude directional indicator (ADI)

Principal operations inspector (POI)

Sterile cockpit rule

REVIEW QUESTIONS

1. Discuss the significance of human error in major aircraft accidents.

2. What are the six major factors that affect human performance? Give examples of each.

3. Define human error, and explain methods used to classify human error.

4. Explain what is meant by error management.

5. What are the salient differences between engineering and administrative strategies used for controlling human error?

6. Discuss the role of the following in human error management:

   - Cockpit standardization

   - Cockpit automation

   - Warning and alerting systems

   - Display conspicuity and system recovery

   - Flight management computer

   - Air-to-ground communication

7. Explain how the airlines use each of the following approaches to manage human error.

   - Employee selection

   - Training

   - Responsibility

- Accountability and enforcement
- Procedures and checklists
- Paperwork reduction and management
- Workload management
- Communication
- Team concept (cognitive redundancy)
- Peer-pressure control

8. Discuss the role of federal agencies' regulatory practices in controlling human error.

9. What is the role of labor in human-error management?

REFERENCES

Alkov, Robert A. 1997. *Aviation Safety—The Human Factor*. Casper, Wyo.: Endeavor Books.

Barlay, Stephen. 1970. *The Search for Air Safety*. New York: William Morrow & Co.

Berlin, J. L., E. V. Gruber, P. K. Jensen, C. W. Holmes, J. R. Lair, and J. M. O'Kane. 1982. *Pilot Judgment Training and Evaluation*, vol. 1. Atlantic City, N.J.: FAA Technical Center, Report No. DOT/FAA CT-82/56.

Duke, Thomas A. 1991. "Just What Are Flight Crew Errors?" *Flight Safety Digest*. Arlington, Va.: Flight Safety Foundation, July 1–15.

Fiedler, F. E., and J. E. Garcia. 1987. *New Approaches to Effective Leadership: Cognitive Resources and Organizational Performance*. New York: Wiley.

Frazier, David. 1986. *The ABCs of Safe Flying*. Blue Ridge Summit, Pa.: TAB Books.

Garland, Daniel J., John A. Wise, and V. David Hopkin. 1999. *Handbook of Aviation Human Factors*. Mahwah, N.J.: Laurence Erlbaum Associates, Publishers.

Gibbons, A., S. R. Trollip, and M. Karim. 1990. *The Expert Flight Plan Critic: A Merger of Technologies*. San Diego, Calif.: Academic Press.

Green, G. G., H. Muir, and James M. Gradwell. 1991. *Human Factors for Pilots*. Aldershot, U.K.: Gower Publishing Co., Ltd.

Hawkins, F. H. 1987. *Human Factors in Flight*. Brookfield, Vt.: Gower Publishing Co.

Heller, William. 1982. *Airline Safety: A View from the Cockpit*. Half Moon Bay, Calif.: Rulorca Press.

Hollander, E. P. 1985. *Handbook of Social Psychology*, 3d ed. New York: Random House.

Hunt, Graham J. F., ed. 1997. *Designing Instruction for Human Factors Training in Aviation*. Brookfield, Vt.: Ashgate Publishing Co.

ICAO. 1984. *Accident Prevention Manual*, 1st ed. Doc. 9422-AN/923. Montreal, Canada: ICAO.

———. 1993. "Investigation of Human Factors in Accidents and Incidents." *Human Factors Digest*, no. 7.

———. 1989a. *Fundamental Human Factors Concepts*. ICAO Circular 216. Montreal, Canada.

———. 1989b. *Flight Crew Training: Cockpit Resource Management (CRM) and Line-Oriented Flight Training (LOFT)*. ICAO Circular 217. Montreal, Canada.

———. 1989c. *Training of Operations Personnel in Human Factors*. ICAO Circular 227. Montreal, Canada.

Jensen, A. D., and J. C. Chilberg. 1991. *Small Group Communication: Theory and Application*. Belmont, Calif.: Wadsworth Publishing Co.

Jensen, R. S. 1989. "Aeronautical Decision Making—Cockpit Resource Management." Washington, D.C.: DOT/FAA PM-86/46 Report.

Lauber, John K. 1989. "Human Performance and Aviation Safety—Some Issues and Some Solutions." *Air Line Pilot*. June 10–13.

Lintern, G., S. N. Roscoe, and J. Sivier. 1990. "Display Principles, Control Dynamics, and Environmental Factors in Pilot Performance and Transfer of Training." *Human Factors*. 32: 299–317.

Maurino, Daniel E. 1990. "Education Is Key to ICAO's Human Factor Program." *ICAO Journal*. Montreal. October, pp. 16–17.

Meister, David. 1971. *Human Factors: Theory and Practice*. New York: Wiley & Sons.

Melton, Carlton E. 1988. "Human Error in Aviation Can Be Deliberate, Inadvertent or Reflect Expertise." *ICAO Bulletin*. Montreal, Canada: ICAO Bulletin. October 23–25.

Nagel, David C. 1988. *Human Factors in Aviation*. San Diego: Calif.: Academic Press.

National Transportation Safety Board. 1967. "Aircraft Design Induced Pilot Error." Special Study PB 175629. Washington, D.C.

O'Hare, David, and Stanley Roscoe. 1990. *Flight Deck Performance: The Human Factor*. Ames, Iowa: State University Press.

Orlady, Harry W., and Linda M. Orlady 1999. *Human Factors in Multi-Crew Flight Operations*. Brookfield, Vt.: Ashgate Publishing Co.

Petersen, Dan. 1975. *Safety Management: A Human Approach*. Deer Park, N.Y.: Aloray Publishing Co.

Ramsden, J. M. 1976. *The Safe Airline*. London: Macdonald and Jane's Publishers, Ltd.

Sinaiko, H. Wallace, ed. 1961. *Selected Papers on Human Factors in the Design and Use of Control Systems*. New York: Dover Publications.

Villaire, Nathaniel E. 1994. *Aviation Safety: More Than Common Sense*. Casper, Wyo.: International Aviation Publishers.

Wiener, Earl L., and D. C. Nagel. 1987. *Human Factors in Aviation*. San Diego, Calif.: Academic Press.

AIR TRAFFIC SAFETY SYSTEMS

LEARNING OBJECTIVES

After completing this chapter, you should be able to

- Describe how air traffic is controlled.
- Describe the mission of the FAA and how deregulation has exerted pressures on the air traffic system.
- Distinguish between the airport, terminal, and en route service areas.
- Define the National Airspace System Plan (NASP).

- Highlight the key components of the communications modernization plan.
- Explain the purpose of the wide- and local-area augmentation systems.
- Discuss several advantages of satellite-based navigation.
- Compare and contrast automatic dependent surveillance radar with the current primary and secondary radar systems.
- Give several examples of new weather technologies in the modernization plan.
- Discuss some of the planned improvements in avionics.
- Explain the purpose of traffic flow management.
- Distinguish between the surface movement advisor and the airport movement area safety system.
- Explain how the STARS program will provide improvements for the TRACON.
- Identify the key components of the en route and oceanic modernization plan.
- Describe some of the emerging technologies in Free Flight Phase 1.
- Summarize the three phases of the NAS modernization plan.

INTRODUCTION

The FAA's mission for air traffic control is to promote the safe, orderly, and expeditious flow of aircraft through the nation's airspace. Today's national airspace is a complex system that has evolved over seven decades and is comprised of the following three major elements.

- *People*—individuals who operate and use the system, such as pilots, mechanics, regulators, instructors, technicians, air traffic controllers, and passengers
- *Equipment*—the aircraft that operate within the system
- *Infrastructure*—facilities and equipment used by individuals of the system, such as airports, air traffic control facilities, navigation aids, radar, communications, lighting aids, and the airspace itself.

The basic principles and operation of the ATC system have changed little since the 1970s. The airspace of the world is divided into *flight information regions* (*FIRs*). The country within the FIR is responsible for the ATC in that FIR. Around the major airports are located *control zones* (CZs) and *terminal maneuvering areas* (*TMAs*). The TMAs are connected by fixed airways, usually defined by radials of very high frequency (VHF) omnidirectional range (VOR) and distance-measuring equipment (DME) navigation aids.

Across the continents, radar is the main source of information for the surveillance of air traffic in the important parts of an FIR. Where possible, air traffic controllers communicate with the pilots by means of VHF voice radio links. Over the oceans and other areas where radar surveillance is not possible, procedural ATC is used. It requires large separations (10 minutes' flying time longitudinally, which is about 80 to 90 nautical miles, and 60 nautical miles laterally), and at regular intervals, a pilot position reports via the high-frequency radio link. Usually, aircraft are assigned fixed routes at a fixed altitude over these areas. To structure the traffic flow on these oceanic flights, the *organized track system* (*OTS*) was defined.

In the areas of the FIR where radar surveillance and VHF communication are possible, reduced separations can be used (3 to 10 nautical miles). Usually digital ATC computers are used to combine flight plan data with (secondary) radar transponder–derived identification and altitude information to provide an enhanced radar display showing aircraft positions labeled with flight identification, altitude, assigned flight level, ground speed, etc., to the air traffic controller. Furthermore, algorithms are used for some form of strategic planning and sometimes also for *short-term conflict alert* (*STCA*) in the event that two aircraft are likely to come into conflict. Around the airport, *departure procedures* (*DPs*) and *standard arrival routes* (*STARs*) are frequently used to structure traffic flow, to reduce radio communication, and to reduce the noise nuisance for the population. The flight can be completed by either a nonprecision approach or a precision approach. For a nonprecision approach, usually a *VOR* or *nondirectional beacon* (*NDB*) is used. From an altitude of 500 feet above terrain, the approach has necessarily to be continued in *visual meteorological conditions* (*VMCs*). The precision approach path is usually defined by an ILS, allowing a single straight three-dimensional flight path over the extended runway centerline, with a 3-degree glide slope. The *runway visual range* (*RVR*) and *decision height* (*DH*) that can be tolerated depend on the category of ILS ground equipment and the onboard *automatic flight control system* (*AFCS*) and auto land capability.

The air traffic controller uses traffic information from flight plans, radar information, and RDF information to monitor and maintain safe separation between aircraft and to facilitate the efficient movement of aircraft through the air traffic system. Through voice clearances, the air traffic controller can instruct the aircraft to maintain safe separations and can provide a safe sequencing of the aircraft for approach and landing. The planning process for the sequencing of aircraft for approach and landing normally starts when the aircraft enters the FIR.

Flying activities are conducted under two basic conditions—*instrument flight rules* (*IFR*) and *visual flight rules* (*VFR*). Air traffic control provides services for all instrument flying and some visual operations. While the tools available to the controller (radar displays, communications, and other aids) have advanced greatly, air traffic control, for the most part, is done manually, just as it has always been.

Controllers operate primarily from the following three types of facilities:

- Airport traffic control towers (ATCTs)
- Terminal radar approach control (TRACON) facilities
- Air route control centers (ARTCCs)

Controllers in the ATCTs control the movement of aircraft on the ground and in the vicinity (about 5 miles) of the airport. From TRACON facilities, controllers sequence and separate aircraft in terminal airspace, which extends from the point at which tower control ends to about 20 to 30 miles from the airport. From en route centers (ARTCCs), controllers assume control of the aircraft outside of terminal airspace and maintain control until the aircraft enters terminal airspace at its destination. *Central Flow Control* (CFC) uses weather technologies to predict airport capacity and hold aircraft on the ground when the predicted demand on a destination airport exceeds its capacity in bad weather. Figure 8-1 provides a general picture of how airspace is constructed and controlled.

In addition to these three types of air traffic control facilities and controllers, the FAA relies on other equipment and personnel to promote the safe and expeditious flow of aircraft. For example, flight service stations provide services such as disseminating weather information, primarily to general-aviation pilots. Throughout the country, long- and short-range radar tracks and identifies aircraft, and a variety of systems are used to detect and relay weather information to controllers and pilots. Communication equipment is used to exchange voice and other data between pilots and air traffic controllers.

Growth in commercial air traffic since deregulation has exerted pressures on several parts of the air traffic system for the following reasons:

- Air traffic levels have grown enormously since deregulation, but there have been no comparable increases in airport capacity.
- Many ATC centers are still using aging equipment, and many controllers still lack the preferred level of experience.
- The hub-and-spoke system of airline operations has caused an uneven distribution of traffic flow at hubs with major surges occurring during certain periods of the day. In addition to increased stress on the controllers, peaks in high volume traffic exacerbate schedule disruption when flights are canceled or delayed because of weather, equipment malfunction, or any other reason.

If air traffic continues to increase and no actions are taken to address the above concerns, there will be increased delays and the margins of safety now maintained by the ATC system will be compromised. This chapter reviews

- Technologies that improve the safety of the air traffic system
- Technologies or procedures that increase or better utilize the capacity of the system
- Technologies that improve communication between pilots and controllers in high-density airspace
- Technologies that detect and communicate weather conditions to pilots
- Training that helps pilots use presented information effectively
- Navigation and surveillance systems for controlling aircraft

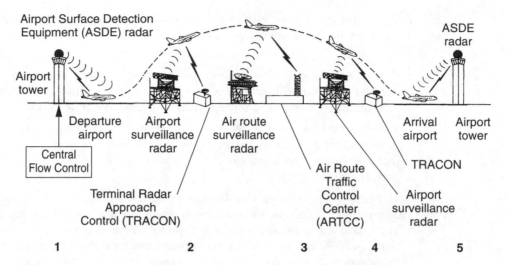

KEY:

1. The airport tower controls the aircraft on the ground before takeoff and then to about 5 miles from the tower, when the tower transfers aircraft control to a Terminal Radar Approach Control facility (TRACON). Controllers in the airport tower either watch the aircraft without technical aids or use radar–Airport Surface Detection Equipment for aircraft on the surface and airport surveillance radar for those in the air. Central Flow Control (in Washington, DC) can order the tower to hold flights on the ground if demand exceeds capacity at the arrival airport.

2. The TRACON, which may be located in the same building as the airport tower, controls aircraft from about 15 miles to about 30 miles from the airport, using aircraft position information from the aircraft surveillance radar. The TRACON then transfers control to an Air Route Traffic Control Center (ARTCC).

3. ARTCCs control aircraft that are en route between departure and arrival airports. Each ARTCC controls a specific region of airspace and control is handed off from one ARTCC to another when a boundary is crossed. Aircraft positions are detected by the air route to an Air Route Traffic Control Center (ARTCC).

4. The TRACON controls the arriving aircraft until it is within about 5 miles of the arrival airport tower, when control is transferred to the tower.

5. The airport tower controls the aircraft on the final portion of its approach to the airport and while it is on the ground.

FIGURE 8-1   General picture of how airspace is constructed and controlled.

NATIONAL AIRSPACE SYSTEM PLAN

To accommodate greater traffic with greater efficiency and automation, the FAA, in 1981, published a comprehensive plan to modernize and improve airways and aviation facilities. This was known as the *National Airspace System Plan* (*NASP*), and the crux of the NASP was the upgrading of the ATC system. There were major delays, however, in the execution of most of the major programs in the NASP. There were three reasons attributed to these delays.

- First, Congress was unwilling to appropriate the Airport and Airway Trust Fund because the unused fund monies could be applied against the federal deficit in an attempt to balance the budget within a reasonable time period.

- Second, the FAA was not able to spend money on NASP procurements because of engineering problems, particularly in software development, and changes in technology requirements caused by unanticipated developments in air transportation since 1981.

- Third, the FAA believed that its modernization problems were caused by federal acquisition regulations. To counter this, Congress enacted legislation, in November 1995, that exempted the agency from most federal procurement laws and regulations and directed the FAA to develop and implement a new acquisition management system that would suit its unique needs.

Although there was an element of truth to all three views, many individuals in the aviation community felt that there were delays in the NASP programs because the FAA underestimated the time needed to tailor existing technology to the ATC system requirements and did not provide adequate time for development and testing. However, delays and cost increases experienced for the NASP are not uncommon (especially for the federal government) for large and complex technological programs of this nature. The original version of the NASP called for substantial changes in ATC facilities, including automation to handle increased traffic by the early 1990s, but most of the major changes are now not expected to take place before 2015.

Automated tools for controllers can reduce workload, provide information to reduce the amount of potentially error-prone mental judgments controllers now must make, permit better teamwork, and enhance the working environment. Although automation can facilitate safe handling of higher traffic levels, automation changes the role of the air traffic controller and might in itself introduce new hazards.

NAS MODERNIZATION

By 2000, the *National Airspace System* (NAS) included more than 18,300 airports, 21 ARTCCs, 197 TRACON facilities, more than 460 ATCTs, 75 flight service stations, and approximately 4500 air navigation facilities. Several thousand pieces of maintainable equipment, including radar, communications switches, ground-based navigation aids, computer displays, and radios, are used in NAS operations. NAS components represent billions of dollars in investments by the government. Additionally, the aviation industry has invested significantly in ground facilities and avionics systems designed to use the NAS.

The U.S. aviation industry is soaring into the 21st century with projected increases in business, recreation, and personal travel. U.S. airlines alone expect that they will carry twice as many passengers by 2015 as they do today. To manage this increased load on the NAS, the air traffic control system and supporting services must be state of the art, led by a coordinated, long-term modernization effort.

To ensure that FAA services meet the increased demand, the FAA joined with the aviation community to develop an operational concept that identifies the capabilities that NAS users and service providers want for more efficient operations. Based on the operational concept, the FAA, with input from the aviation community, developed a modernization plan. The plan is a long-range blueprint for modernizing the NAS and improving NAS services and capabilities through 2015.

Over the past several years, hundreds of representatives—from airlines, general aviation, military, pilot associations, air traffic control associations, airports, product manufacturers, government contractors, and international organizations—participated in the future design of the NAS through government and industry working groups and many other feedback opportunities. Key influences on the modernization plan included the 1996 White House Commission on Aviation Safety and Security, which recommended that the FAA accelerate its modernization of the NAS, and the 1997 National Civil Aviation Review Committee, which recommended funding and performance management methods for improving NAS modernization.

COMPONENTS OF THE NAS PLAN

The modernization plan describes changes to the NAS in communications, navigation, surveillance, weather automation systems, and avionics, which will improve efficiency and enhance safety in our aviation system over the next two decades.

COMMUNICATIONS

Air traffic management depends on timely and accurate transmission of information during flight planning, in flight, and for airport operations. With the projected growth in air traffic, today's communications systems must be modernized to handle the additional demand and the need for faster and clear transmissions. This modernization will require replacing outdated hardware, better use of the available very high frequency (VHF) spectrum, and integrating systems into a seamless network using digital technology. During the transition, the FAA will continue to support analog communications. Key components of the communications modernization include these:

- *Controller-pilot datalink communication* (CPDLC). This introduces data exchange between controllers and pilots to reduce voice-channel congestion. CPDLC greatly expands current data capabilities and provides controllers and pilots with a means to communicate routine and repetitive messages. It will be available first from a commercial provider using VHF digital link (VDL) Mode-2 ground and airborne equipment for data only. It will eventually change to VDL Mode-3 for both voice and data. Digital communications over VDL Mode-3 are expected to provide the significant channel efficiency improvement needed to support projected aviation growth. VDL Mode-2 will continue to be available for use by airline operations centers and other users via a commercial service provider.

- *Integrated ground telecommunications infrastructure.* This digital infrastructure provides integrated voice, data, and video connectivity for air traffic control operations and administrative communications.

- *Digital voice and data communications via digital radios provided by next-generation air/ground communications* (NEXCOM). This program provides a system of ground radios and ground network interfaces that enable digital voice and data communications with aircraft and will replace more than 40,000 existing ground radios.

For the past 20 years, demands on the VHF spectrum for air traffic services have grown by an average of 4 percent each year, saturating the available spectrum in many locations. Transition to digital radios will effectively increase, by at least a factor of 2, the capacity of each VHF frequency through digital techniques.

Datalink is expected to reduce the chances of missed communications or misinterpretation of the message between controller and pilot. Transition to datalink communications will occur gradually as new applications are tested on a small scale prior to nationwide deployment and as users equip their aircraft with the necessary avionics. CPDLC initially will provide two-way exchange of air traffic control messages, such as transfer of communications and altimeter settings that are currently conveyed by voice. Voice communications will continue to be available. Oceanic and en route use of datalink will precede use in terminal airspace. Current airport datalink operations will expand.

NAVIGATION

The current aviation navigation system is comprised of more than 4300 ground-based systems whose signals are used by aircraft avionics for en route navigation and landing guidance. Despite the large number of ground systems, navigation signals do not cover all airports and airspace. Over the next 10 years, the navigation system is expected to use satellites, augmented by ground monitoring stations, to provide navigation signal coverage throughout the NAS. Reliance on ground-based navigation aids is expected to decline as satellite navigation provides equivalent or better levels of service.

A transition to satellite navigation significantly expands navigation and landing capabilities, improving safety and efficient use of airspace. In addition, it will reduce the FAA's need to replace many aging ground systems, decrease the amount of avionics required to be carried in aircraft, and simplify navigation and landing procedures. The transition to satellite-based navigation consists of the following:

- Use of the *Global Positioning System* (GPS) for en route/terminal navigation and nonprecision approaches (provided that another navigation system is onboard the aircraft). GPS is being described as the greatest aviation achievement since radios were introduced 50 years ago and is scheduled to revolutionize today's air traffic system. GPS is a radio navigation system composed of 24 orbiting satellites that provides extremely accurate three-dimensional position, velocity, track, and time, at low cost so that even general-aviation planes can use it. By picking up signals from four or more satellites, GPS receivers on the ground or in the aircraft can determine the location within approximately 330 feet of the exact position. By adding a datalink to the plane, its position, velocity, and track can be conveyed via satellite to communicate data to ground control units worldwide. Over-ocean surveillance can now be commonplace, providing additional flight paths and reduced separation for improved utility of the sky. Since all the traffic can now be determined accurately, each airplane's position can be uplinked or directly linked to planes, to display those within close proximity and with potential conflict. We will have not only communications, navigation, and surveillance technologies (CNS) but also the potential for collision avoidance at a price that all aviators can afford. GPS alone does not meet the accuracy, availability, and integrity requirements critical to safety of flight. Signal augmentation is required for most landing operations.

- Deployment of the *wide-area augmentation system* (*WAAS*) to provide en route/terminal navigation and category (CAT) I precision approaches. WAAS enhances GPS signals to provide more precise location information to an accuracy of approximately 25 feet. WAAS is designed to use reference stations covering wide areas throughout the United States to cross-check GPS signals and then relay integrity and correction information to aircraft via geostationary communication satellites. WAAS enhances availability by using these satellites to provide a GPS-like navigation signal (see Fig. 8-2).

- Deployment of the *local-area augmentation system* (*LAAS*) to augment GPS for CAT I, II, and III precision approaches. LAAS provides precise correction data to airborne and surface receivers that will result in navigation accuracy of less than 40 inches to distances of 20 miles or more from the airport (see Fig. 8-3).

Precision approaches are categorized in terms of decision heights (ceilings) and minimum visibility. Listed below are FAA (http://www2.faa.gov/programs/oep/AW -1a.htm) definitions and requirements for instrument approaches.

- *Nonprecision approach* (*NPA*). This is an instrument approach procedure based on a lateral path and no vertical guide path.

- *Precision approach* (*PA*). This is an instrument approach procedure based on lateral path and vertical guidance.

- *Category I.* Category I operation is a precision instrument approach and landing with a decision height that is not lower than 200 feet (60 meters) above the threshold and with either a visibility of not less than 1/2 statute mile (800 meters) or a runway visual range (RVR) of not less than 1800 feet (550 meters).

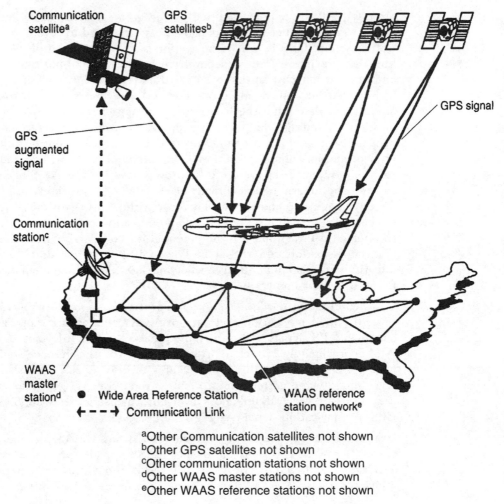

FIGURE 8-2   Wide-area augmentation system for all phases of flight.

- *Category II*. Category II operation is a precision instrument approach and landing with a DH lower than 200 feet (60 meters), but not lower than 100 feet (30 meters), and with a RVR of not less than 1200 feet (350 meters).
- *Category III*. Category III operation is a precision instrument approach and landing with a DH lower than 100 feet (30 meters) or no DH, and with a RVR less than 1200 feet (350 meters).

ADVANTAGES OF SATELLITE-BASED NAVIGATION

Satellite-based navigation enables significant operational and safety benefits. It meets the needs of growing operations because pilots will be able to navigate virtually

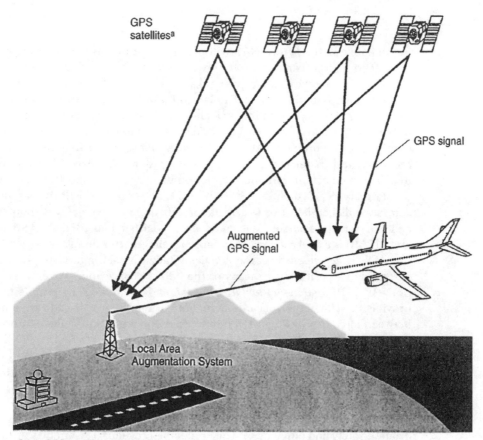

GPS
satellites[a]

GPS signal

Augmented
GPS signal

Local Area
Augmentation System

*Other GPS satellites not shown.

FIGURE 8-3 Local-area augmentation system for precision approach phase of flight. (*FAA.*)

anywhere in the NAS, including at airports that currently lack navigation and landing signal coverage. Satellite-based navigation will support direct routes.

With satellite navigation, the number of published precision approaches will increase, and more runways will be served by precision approaches, enhancing safety. Precision approaches provide the pilot with vertical descent and approach course guidance, while nonprecision approaches only offer course guidance. In addition, combining GPS with cockpit electronic terrain maps and ground-proximity warning systems can help pilots avoid controlled flight into terrain (CFIT).

Satellite-based navigation also decreases the number of ground-based navigation systems, thereby reducing infrastructure costs. For a precision approach today, each runway end needs a dedicated instrument landing system. WAAS can provide the precision approach guidance for most of the runways in the NAS. An LAAS system will provide CAT II/III precision approach guidance at all runways at an airport.

SURVEILLANCE

Surveillance in the future NAS will provide coverage in nonradar areas and include aircraft-to-aircraft capabilities for greater situational awareness and safety.

The NAS modernization plan calls for evolution from current primary and secondary radar systems to digital radar and *automatic dependent surveillance* (ADS). This change is designed to improve and extend surveillance coverage and provide the necessary flexibility for Free Flight (discussed later in this chapter). The FAA will continue to use primary and secondary surveillance radar to detect and track aircraft en route and in terminal airspace. New radar and surveillance systems will be installed to detect aircraft and vehicles on the airport surface at selected airports.

Primary radar (such as ASR-9, ASR-11, ASDE-3, and MPAR) is termed *independent* surveillance because it detects aircraft or motor vehicles without the need for enabling avionics and equipage. Secondary radar (such as Mode-S, ASR-11, ATCB1-6, and MPAR) is called *cooperative* because it relies on the aircraft to have a transponder.

ADS is *dependent* because it relies totally on each aircraft or surface vehicle to determine its position (by means of the onboard navigation system) and report position. An ADS system—*automatic dependent surveillance-addressable* (ADS-A)—will provide surveillance of oceanic aeronautical operations, based on communications between aircraft providing ADS information and a ground facility requiring receipt of ADS reports. Aircraft equipped with future air navigation system (FANS-1A) or aeronautical telecommunication network (ATN) avionics exchange identification, flight level, position, velocity, and/or intent data with ADS-A ground equipment through satellite communications. FANS-1A is an integrated suite of communication, navigation, and surveillance avionics. When installed in the aircraft and on the ground, possibly starting in 2003, these capabilities and accompanying procedures will increase aviation safety and efficiencies while reducing procedural separation distances.

A new avionics capability—*automatic dependent surveillance-broadcast* (ADS-B)—may be introduced by the users to provide pilots with an air-to-air surveillance capability in domestic and oceanic airspace. ADS-B uses satellite navigation and datalink to enable an aircraft to broadcast its identification, position, altitude, velocity, and intent to every other aircraft in the vicinity as well as to the ground tracking system. This broadcast information may be received and processed by other aircraft or ground systems for use in improved situational awareness, conflict avoidance, surveillance, and airspace management. This will ultimately reduce the dependence on human monitoring from the ground.

Throughout the evolution to ADS, surveillance and separation services in a mixed equipage environment will be provided. This means the FAA will continue to rely on aircraft transponders to verify aircraft position.

In the NAS modernization plan, primary and secondary radar will be retained in the terminal area, and only secondary radar will remain in the en route airspace.

WEATHER AUTOMATION SYSTEMS

Weather conditions interfere with flight operations and contribute to aviation accidents more than any other factor. Given these major impacts, the NAS modernization plan contains improved ways to collect, process, transmit, and display weather information to users and service providers, both during flight planning and in flight.

The key to reducing weather-related accidents is to improve pilot decision making through increased exchange of timely information. Service providers and users will receive depictions of hazardous weather, simultaneously enhancing common situational awareness.

The aviation weather plan will evolve from present-day separate, stand-alone systems to weather systems that are fully integrated into the NAS. The focus is on these two key capabilities:

• Improved processing and display, with the key systems being the *integrated terminal weather system* (*ITWS*) and *weather and radar processor* (*WARP*). ITWS is an automated weather system that provides near-term (0 to 30 minutes) prediction of significant terminal area weather for major terminal locations. ITWS integrates data from radar, sensors, National Weather Service models, and automated aircraft reports. It generates products, including wind shear and microburst predictions, storm cell hazards, lightning information, and terminal area winds. WARP is an integrated system that receives and processes real-time weather data from multiple sources and provides weather information for use by the ARTCCs and air traffic control system command center (ATCSCC) to support the en route environment. It also receives gridded forecast data from the National Weather Service and provides this information to other NAS automation systems. WARP has direct and indirect connections to the *next-generation weather radar* (*NEXRAD*) radars and prepares national and regional weather images for the controllers' displays.

• Improved sensors and data sources, featuring the NEXRAD, *terminal Doppler weather radar* (*TDWR*), and ground- and aircraft-based sensors. NEXRAD is a national network of Doppler weather radar to detect, process, distribute, and display hazardous weather, providing more accurate weather data for aviation safety and fuel efficiency. This radar has a 250-mile range, and the network covers the majority of the domestic en route airspace. This weather detection system provides information about wind speed and direction in the areas of precipitation, convective activity, tornadoes, hail, and turbulence. TDWR detects localized microbursts, gust fronts, wind shifts, and precipitation in the immediate terminal area at key locations. The radar provides alerts of hazardous weather conditions in the terminal area and advanced notice of changing wind conditions to permit timely change of active runways.

The FAA is conducting research to improve the capability to predict weather hazards and to communicate these to NAS users. New wind, temperature, icing, and weather hazard modeling will be used to help diminish weather-related delays and improve safety.

Aviation weather research is being focused on the following areas: in-flight icing, aviation gridded forecast system, ground deicing operations, convective weather, short-term ceiling and visibility predictions, turbulence, and wake vortices.

AVIONICS

Avionics will evolve to take advantage of the benefits found in the new *communications, navigation,* and *surveillance* (CNS)-related technologies in the NAS modernization. With the new avionics, users can access many enhanced technologies that will help them fly more safely and efficiently. Some of the planned improvements in avionics include

- Avionics for the GPS, WAAS, and LAAS to enable aircraft to navigate via direct routes and fly precision instrument approaches to virtually any runway
- New multimode digital radios for voice and data communications for pilots, controllers, and ground facilities
- ADS-B avionics that track the GPS-based position, velocity, and intent information to ground stations (air-to-ground) and other aircraft (air-to-air)
- Multifunctional cockpit displays to present information, such as weather, notices to aircrews (NOTAM), and moving maps to improve situational awareness

The FAA plans to evaluate a variety of ADS-B avionics and cockpit displays currently under development. The development of ADS-B avionics for air-to-air surveillance, an initiative of the user community, is being supported by FAA's standards and certification efforts. While ADS-B shows great promise for both air-to-air and air-to-ground surveillance, current aircraft transponders will continue to support surveillance operations in the NAS for the foreseeable future. If enough users equip their aircraft with ADS-B avionics, the FAA will install a compatible ADS ground system to provide more accurate surveillance information to air traffic controllers compared to radar-based surveillance.

ADS-B avionics will be linked to CDTI that shows the location of other ADS-B aircraft in the vicinity. It is envisioned that ADS-B with CDTI will be used by pilots for situational awareness relative to other ADS-B–equipped aircraft and, when controller concurrence is obtained, for tactical maneuvering, self-separations, and station keeping.

The *traffic and collision avoidance system* (TCAS), which provides pilots with advisory information to prevent midair collisions with other transponder-equipped aircraft, will be improved to accommodate changes such as reduced vertical separation above FL 290 and to track multiple targets at longer ranges. TCAS displays are used currently by pilots for collision avoidance and oceanic station keeping (maintaining miles-in-trail separation). TCAS will remain as an independent safety system to prevent air-to-air collisions.

OPERATIONAL PLANNING

To improve flight planning, the NAS modernization plan contains new and improved information services in the areas of *traffic flow management* (TFM) and flight services that enable collaboration-service providers and users sharing the same data and negotiating to find the best solutions to meet operational needs.

Traffic flow management capabilities are centralized at the ATCSCC. Some functionality is distributed to traffic management units at ARTCCs, TRACON facilities, and the highest-activity ATCTs.

At the center of this collaboration capability are integrated NAS information services, which include a systemwide computer network, use of standardized data formats, and interoperability across applications, to receive and share common data and jointly make planning decisions. The *NAS-wide information service* will evolve from today's current array of independent systems and varying standards to a shared environment connecting users and providers for traffic flow management, flight services, and aviation weather.

NAS-wide information service enables data exchange between users and the FAA to facilitate a collaborative response to changing NAS situations, rather than a local solution based on incomplete data. This gives users and service providers a common view of the NAS for improved decision making during all phases of flight, including flight planning.

In addition, the flight plan will be replaced by the flight object, which will be designed for dynamic updates and made available to authorized NAS service providers and users to manage flight operations collaboratively. The flight object will contain additional data such as the user's route and altitude preferences, the aircraft's weight, gate assignments, departure/arrival runway preferences, and location while in flight.

The goal of operational planning improvements is to integrate operational and business decisions to gain efficiency, predictability, and flexibility in flight operations. Each day, approximately 100,000 flights use the NAS, requiring many decisions to manage all the traffic. The NAS modernization plan provides tools to help users and service providers make collaborative decisions to prioritize and schedule flights and to better organize air traffic locally and nationally.

During flight planning, improved tools will be used to predict locations and the impact of traffic demand and weather along planned routes and at the destination. As the flight progresses, additional updates on weather, NAS status, and other user-specific data will be provided to the airline and other operational centers as appropriate. New tools will eventually help plan direct flight paths, sequence departures and arrivals, change routes, and balance capacity and demand throughout the NAS. The TFM plan focuses on modernizing the

- Infrastructure
- Process for data exchange and methods of collaboration
- Tools for NAS analysis and predictions

The *enhanced traffic management system* (ETMS) will be upgraded to replace hardware and software. ETMS is an existing traffic flow management computer system used by specialists to track, predict, and manage air traffic flows. The ETMS upgrade will replace controller workstations, computers, peripherals, and proprietary software to sustain current traffic flow management capability and meet the need for continued improvements in collaboration. For example, the control-by-time-of-arrival (CTA) toll will be implemented to manage arrival demand. With CTA, rather than the FAA, identifying departure times, users are given the authority to determine which flights and departure times are suitable for the capacity available at the destination airport. This allows the airlines additional flexibility to manage their arrivals.

FLIGHT SERVICE IMPROVEMENTS

FAA flight service stations (FSSs) provide flight planning assistance, aviation weather, and aeronautical information to commercial, general-aviation, and military pilots.

The FSS modernization plan replaces outdated automation systems that have limited capabilities with a new *operational and supportability implementation system* (*OASIS*). OASIS incorporates the functions provided by the direct user access terminal (DUAT) service and the graphics weather display system (GWDS). Like the DUAT service, OASIS allows pilots to self-brief and file flight plans.

In future enhancements, OASIS will be integrated into the NAS-wide information service to receive weather and NOTAM.

As this evolution occurs, OASIS will also be able to use flight object data and exchange additional information with the general-aviation pilots to improve flight planning. More accurate data also will allow for faster on-scene response times when it is necessary to initiate search and rescue services.

AIRPORT SURFACE OPERATIONS

At busy airports, numerous aircraft and surface vehicles—fuel trucks, service vehicles, luggage/cargo carriers—operate on the airport surface. To prevent accidents and maintain flight schedules, the personnel managing airport ground traffic and incoming/outgoing aircraft need accurate and complete information on aircraft and vehicle location and intentions, especially at night and in low-visibility conditions. This is possible through a combination of decision support tools, communications and surveillance technology, and new procedures and training.

The ATCT will evolve from having minimal automation support and relying on visual observations and voice communications between the tower and the users to the following:

- Information sharing between the FAA and the users via *Surface Movement Advisor* (*SMA*). SMA collects and shares ground movement data on the airport surface with the FAA, airline ramp control operations, and airport management.
- Expanded use of datalink to convey routine information.

- Surface surveillance tools that help expedite surface traffic and improve safety by reducing runway incursions by surface vehicles and aircraft.

- Improved radar displays.

- Improved traffic displays, weather information, and decision support tools to increase airport capacity utilization and mitigate the impact of adverse weather upon airport operations.

The initial SMA will be installed at selected airports to provide airline ramp control operators with arrival and departure information. Ramp control will be able to improve the sequence and metering of aircraft movement at gates and on ramp areas.

The *airport movement area safety system* (AMASS) will be deployed at the nation's busiest airports. AMASS complements airport surveillance by comparing the tracks of aircraft on final approach with the movement of vehicles/aircraft on the airport surface as detected by the ASDE-X (3) radar to predict conflicts and alert controllers. New runway incursion reduction capabilities will be implemented to help reduce the possibility of traffic conflicts. This includes additional surveillance, ATC tools, signage, lighting, new procedures, and increased training. In this connection, note that runway incursions are one of the FAA's top five safety concerns. An extended discussion of this topic is conducted in Chap. 10 on airport safety.

The *surface management system* (SMS) will evolve from the SMA prototype used in Atlanta. The SMS is designed to provide airport configuration, aircraft arrival/departure status, and airfield ground movement advisories to controllers, dispatchers, and traffic flow managers. SMS will interface with AMASS and the terminal automation system to help controllers coordinate arrival/departure flows with surface movements. SMS will increase surveillance information available to planning tools for the ATCT, TRACON, ramp control, and airport.

Enhanced SMS will enable users and providers to have access to flight planning, traffic management, arrival/departure, and weather information, giving a complete picture of airport operations. Using a perimeter "look-ahead" feature, the enhanced multifunctional displays will show conflict predictions between arriving aircraft and surface aircraft/vehicles. The goal is to have all airport operations, including air traffic control; aircraft, airline, and airport operations centers; ramp control; and airport emergency centers, receiving and exchanging common surface movement data.

DEPARTURES AND ARRIVALS

Resolving congestion at the busiest U.S. airports requires a combination of modern technology and additional runways. The FAA is working with airport operators to help plan and develop new runways to accommodate increased aircraft operations and use new technologies, while meeting environmental requirements.

Arriving and departing aircraft are sequenced in and out of the airport by traffic controllers at the TRACON facilities. Maintaining a steady flow of aircraft, particularly during peak periods, can be improved by providing controllers with tools for sequencing and spacing aircraft more precisely. The objective is to reduce variability in services and optimize use of airspace and available runways.

Focusing on maximizing airport capacity, the terminal modernization plan will include the installation of improved automation systems to provide the following enhancements:

- On-screen display of terminal weather to improve warnings of hazardous weather conditions
- Improved aircraft sequencing and spacing tools to improve efficiency and predictability of services
- Information sharing with users to improve safety and efficiency
- Information sharing between terminal and en route domains to improve flexibility
- Support for more flexible arrival and departure routes to maximize use of airport capacity

The terminal modernization plan specifically includes installation of the new *standard terminal automation replacement system* (STARS) and aircraft sequencing tools. STARS is an all-digital, integrated computer system with modern color displays and distributed processing networks. STARS can be easily upgraded and supports current and future surveillance technology, traffic and weather information, and sequencing and spacing tools. The new STARS workstation will display air traffic, weather overlays, and traffic flow management information for controllers. Future upgrades to STARS tower displays will add a capability to display airport surface traffic and runway incursion alerts and to provide the interface for terminal controller-pilot datalink communication (CPDLC).

STARS will interface with advanced communications, navigation, surveillance, and weather systems planned for the NAS modernization. It will replace the en route automated radar tracking system used in Alaska and at offshore locations. STARS will eventually interface with AMASS to improve the ability to manage airport surface traffic and prevent runway incursions.

The STARS program will provide other improvements for the TRACON. Controllers will have an integrated display of color weather from the ITWS and surveillance radar information. Also, the flight data management capability will allow terminal automation to perform flight data processing for aircraft within the terminal airspace. These data are currently processed by the en route automation system.

EN ROUTE AND OCEANIC

The evolution toward a Free Flight (discussed in the next section) environment requires significant improvements in en route and oceanic computer systems and controller decision support tools. The aging automation infrastructure must be replaced before new applications and improved services can be provided.

Currently, en route and oceanic facilities are colocated but do not share common systems, primarily because of the lack of surveillance and direct communications services over the ocean. The addition of oceanic surveillance and real-time direct communications will enable oceanic services to gradually become comparable with en route services, and oceanic and en route systems will evolve to a common hardware and software environment.

In the domestic airspace, aircraft are monitored by radar and typically follow the fixed route structure of airways, preventing pilots from flying the most direct route or taking advantage of favorable winds.

In oceanic airspace, aircraft follow "tracks" that are aligned each day with prevailing winds. Lack of radar surveillance and direct controller-pilot communications requires oceanic separation standards to be 20 times greater than those in domestic airspace. The large separations limit the number of available tracks. Therefore, some flights are assigned a less than optimum altitude, and there is insufficient opportunity to adjust altitudes to conserve fuel. Additional tracks and access to optimum altitudes would reduce fuel consumption and costs substantially.

The implementation of improved aircraft navigational performance, ADS in oceanic airspace, datalink communications, and better automation tools are expected to overcome these current limitations. The key components of the en route and oceanic modernization plan are as follows:

- Replacement of the en route controller displays

- Replacement of dissimilar en route and oceanic computer hardware systems with standard hardware

- New applications software that is compatible with the new standard hardware operating system to include
  - New and improved controller decision support tools
  - Utilization of advanced surveillance and communication information methods
  - Integration with NAS-wide information service to facilitate data sharing

FREE FLIGHT

Free Flight is defined as a safe and efficient operating capability under instrument flight rules in which the operators have the freedom to select their path and speed in real time. The primary driver behind Free Flight is economics. Airlines lose between $3.5 billion and $5 billion per year as a result of inefficiencies in today's ATC environment. Traditionally, air traffic restrictions are imposed to

- Ensure separation
- Prevent exceeding of airport capacity
- Deter unauthorized flight through special-use airspace
- Ensure safety of flight

The concept of Free Flight centers on allowing pilots, whenever safe and practical, to choose the optimum flight profile. Under the proposal for Free Flight, every aircraft would be surrounded by protective layers of airspace. Air traffic managers would intervene if critical boundaries were violated; otherwise, aircraft would be free to move in any direction and at any altitude and could change either or both at any time. Free Flight, which is expected to be implemented over three phases, will be more flexible than the current *National Route Program* (NRP), which enables users to choose predefined direct routes above 29,000 feet. The increased flexibility of Free Flight is expected to

- Decrease user costs
- Improve airspace flexibility
- Remove flight restrictions
- Save fuel
- Reduce delays
- Increase airport acceptance rates during rush periods
- Utilize runways and ramps more efficiently

During its transition through the three phases to final operation, Free Flight will adhere to the principles that it will

- Not compromise safety
- Be benefits-driven
- Emphasize the need for collaborative planning
- Address human-factors issues during all stages of development
- Accommodate users with various levels of equipage

New tools that give controllers, planners, and service operators more complete information about air traffic control and flight operations comprise a significant part of the NAS modernization near-term plan. Some of these tools are embodied in a program called *Free Flight Phase 1—Core Capabilities Limited Deployment*. Free Flight Phase 1 is the result of an agreement between the FAA and the aviation community to implement certain highly desired capabilities at selected locations by the end of 2002.

An important objective of Free Flight Phase 1 is to mitigate NAS modernization risks by deploying operational tools at a limited number of sites to evaluate performance, training procedures, human-factor requirements and solutions, and safety issues. Users and service providers will have the opportunity to assess system performance, operational benefits and acceptability, and safety before further deployment. With positive results, each Free Flight Phase 1 tool will be fully developed, integrated, and deployed at suitable locations.

Free Flight Phase 1 was officially launched in October 1998 and was chartered to deploy, on a limited basis, five tools or enabling technologies to areas of the country that have the highest traffic densities. These tools (see Fig. 8-4) are the User Request Evaluation Tool (URET), the Traffic Management Advisor (TMA), the Passive Final Approach Spacing Tool (pFAST), the Surface Movement Advisor (SMA), and the Collaborative Decision Making (CDM) support tool. The following is a brief description of the Free Flight Phase 1 tools.

- *User Request Evaluation Tool (URET)* is a conflict probe that enables en route controllers to manage user requests for route and altitude changes by alerting controllers of potential traffic conflicts up to 20 minutes ahead. It also checks for and alerts controllers to conflicts between routes and special-use airspace boundaries.

- *Traffic Management Advisor—Single Center (TMA-SC)* provides en route radar controllers with the capability to develop arrival sequence plans for selected airports, including assignment to the runway that best utilizes available airport capacity. TMA computes the aircraft's estimated arrival time at key arrival points to ensure aircraft meet flow constraints established by terminal traffic management coordinators.

- *Passive Final Approach Spacing Tool (pFAST)* is designed to help controllers balance runways and sequence aircraft according to user preferences and airport capacity. The pFAST computation uses aircraft descent performance characteristics, position/track data, user preferences, and controller inputs to generate a recommended arrival/landing sequence. pFAST works in conjunction with the Traffic Management Advisor tool in the en route center. TMA computes the times at which aircraft must arrive at terminal airspace, optimizing routines to achieve the runway arrival rate associated with the prevailing weather conditions. Note that TMA and pFAST are two Free Flight software tools of the *Center TRACON Automation System (CTAS)*.

- *Collaborative decision making (CDM)* support tools provide airline operations centers and the FAA with real-time access to current NAS status information, including infrastructure and operational factors, such as weather, schedules, equipment, and delay. This information enables users and the FAA to work collaboratively to better manage NAS traffic.

- *Surface Movement Advisor (SMA)* provides information sharing to airline and airport personnel who plan and manage the flow of traffic on the airline ramps (immediately adjacent to taxiways). SMA provides current aircraft arrival information (identification and distance from the airport) to ramp operators/managers. This improves efficiency by optimizing gate operations and ground support services while reducing taxi time and delays.

Free Flight Phase 1 requires no special avionics or special preclearance of its users. In operation, it will be very similar to the present NRP, where aircraft flying

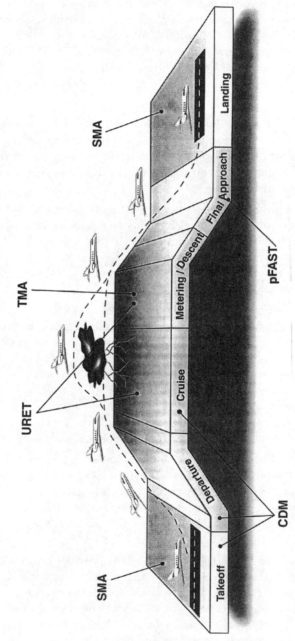

FIGURE 8-4   Free Flight Phase I tools (*FAA.*)

above FL 290 over routes of more than a few hundred miles can elect to fly off-airway least-time tracks between departure and arrival fixes. But under NRP rules, a constant altitude must be maintained unless a change is approved by ATC. In the eventual Free Flight regime, the filed plan will describe a true, four-dimensional *aircraft-preferred trajectory,* where altitude would progressively increase as fuel was burned.

FREE FLIGHT PHASES AND IMPLEMENTATION SCHEDULE

The NAS modernization plan is divided into three phases from 1998 to 2015. Each phase identifies the new capabilities/technologies, procedures, and training required.

Phase 1 (1998–2002)

This phase focuses on sustaining essential air traffic control services and delivering early user benefits. Controller workstations will begin major upgrades, and new controller automation tools will be used at selected sites. The WAAS will be deployed, and air-to air surveillance will be introduced.

Weather, flight planning, and tower datalink services will become increasingly available via commercial providers. Aircraft with Mode-S/TCAS transponders and cockpit displays will be able to monitor local traffic from data uplinked by FAA Mode-S radars. In oceanic airspace, two-way datalink air traffic service will become available via satcom and HF datalink transceivers.

The *host/oceanic computer system replacement* (HOCSR) has replaced 12-year-old host computers which, installed in every ARTCC, had been the backbone of the nation's overall control system. The new IBM systems, supplied by Lockheed Martin, operate at 32 million instructions per second.

The FAA's *display system replacement* (DSR) provides ARTCC en route controllers with large, full-color traffic presentations in place of 30-year-old circular monochrome displays. Built by Lockheed Martin under a $1 billion contract, DSR provides much greater capability via a Windows environment, and the controller interface was designed to ease transition from the older equipment. All ARTCCs became fully operational with their DSRs in the summer of 2000.

The first of 172 standard terminal automation replacement systems (STARS) entered FAA service in El Paso, Texas, in December 1999, which was followed by Syracuse, New York, in January 2000. STARS will provide new computers and controller displays to towers and other facilities handling terminal areas, roughly defined as the airspace within 50 miles of an airport. STARS is essentially the terminal equivalent of the en route DSR. Nationwide system installations are scheduled to be completed by 2007.

The URET (the conflict probe) has made the transition from test systems at Indianapolis and Memphis to partial operational status. Conflict probes initially will be installed at the Memphis; Indianapolis; Kansas City; Cleveland; Washington, D.C.; Chicago; and Atlanta centers to support Phase 1 of the FAA's Free Flight program.

Finally, the Air Traffic Systems Command Center (ATSCC) at Herndon, Va., was established in 1997 to oversee the total NAS traffic situation.

PHASE 2 (2003–2007)

This phase concentrates on deploying the next generation of CNS equipment and the automation upgrades necessary to accommodate new CNS capability. Some highlights of this phase include these:

- WAAS will be completed to provide more coverage and precision instrument approaches. LAAS CAT I (at airports outside WAAS coverage, such as those in Alaska) and LAAS CAT II/III (at most major airports) will be introduced.

- New digital radios that maximize the use of the VHF spectrum will be installed. ADS-B, initially used for air-to-air surveillance and self-separation via CDTI, will evolve over the period to become a key NAS technology as ADS-B ground stations are installed to provide nationwide, all-altitude surveillance. ADS-B stations and Mode-S (TCAS) surveillance radars will slowly transition to *selective interrogation* mode, allowing individual aircraft monitoring. Decommissioning of ground navigation aids will begin around 2007.

- Free Flight Phase 2 (FFP2) will include the geographic expansion and enhancement of FFP1 capabilities, including CDM, URET, and TMA-SC. Additionally, FFP2 includes controller-pilot datalink communication (CPDLC). As mentioned earlier, CPDLC provides for the exchange of digital messages between pilots and controllers as an alternative to voice. The benefits of CPDLC include the ability to transmit ATC information in a reliable, uncluttered medium and reduce voice communication flow on ATC radio frequencies.

- The desired benefit of CDM/Collaborative Routing Coordination Tools (CRCT) is to change the way air traffic is managed, prompting a shift from current practice, where the FAA is the service provider and is responsible for traffic decisions, to a collaborative paradigm in which users have input on decisions and all parties are simultaneously aware of NAS constraints. FFP2 will deploy CDM with CRCT functionality on the traffic flow management infrastructure to 21 ARTCCs and the Air Traffic Control System Command Center (ATCSCC) by 2005, as well as to participating Airline Operation Centers (AOCs).

- FFP2-planned URET enhancements include alternate flight plan processing, automation assisted dynamic rerouting, processing of International Civil Aviation Organization (ICAO) flight plans, nonradar airspace capability, and technology refresh of processors, routers, local-area networks, and operating systems. FFP2 will expand URET to all ARTCC facilities by 2004.

- TMA-SC enables en route controllers and traffic management specialists to develop complete arrival-scheduling plans for controlled aircraft. FFP2 will deploy TMA-SC to the Houston, Kansas City, Indianapolis, and Memphis centers.

FFP2 will also facilitate research and development (R&D) on certain key technologies. The following R&D projects will be monitored and evaluated for the purpose of making investment decisions in the 2003–2005 time frame:

- *Direct-To (D2)*. This tool is designed to assist en route controllers in identifying aircraft that can have their time en route reduced by flying directly to a downstream point closer to the destination airport. Also D2 provides conflict probe, trial planning, and flight plan amendment capabilities for en route R-side controllers.

- *Equitable allocation of limited resources (EALR)*. This capability extends the collaborative rerouting process by providing functionality to balance the assignment of entry into a congested airspace with assignment of reroutings. The capability ensures equity within an individual initiative as well as across several days.

- *Problem analysis resolution and ranking (PARR)*. This set of tools will assist the en route controller in managing flight data derived from URET. PARR will assist the controller in developing strategic resolution for aircraft-to-aircraft and aircraft-to-airspace conflicts. The integration of these tools will allow the entire sector team to access the full range of tactical and strategic tools and displays at each position.

- *TMA—Multicenter (TMA-MC)*. This decision support tool builds on TMA-SC to support efficient time-based metering of arrival traffic in arrival airspace spanning multiple centers. Research will provide TMA in the complex northeast corridor airspace for four centers that feed Philadelphia.

- *Surface management system (SMS)*. This is a system to reduce arrival and departure delays and inefficiencies that occur on the airport surface due to surface issues and downstream restrictions. SMS is a decision support tool that will help controllers and users of the NAS manage the movement of aircraft on the surface of busy airports, thereby improving capacity, efficiency, flexibility, and safety. SMS will support cooperative planning of other arrival and departure traffic management decision support tools to provide additional benefits.

The following R&D projects will be monitored for acceleration but are not expected to mature during the 2003–2005 time frame:

- *Expedite Departure Path (EDP)*. This decision support system is designed to assist controllers in TRACON facilities and ARTCCs with departure-related situations.

- *Advanced Vortex Spacing System (AVOSS)*. This is a completed research initiative by the National Aeronautics and Space Administration (NASA) that evaluated a ground-based system to make dynamic spacing recommendations for aircraft arriving in-trail to a runway, using current and projected atmospheric information to model wake vortex behavior. FFP2 is monitoring the work of the wake vortex research community for the possible acceleration of capacity-enhancing wake vortex-related technology.

- *En route/Descent Advisor* (*E/DA*). This decision support tool is used for the efficient handling of en route arrivals. It will have the capability to assist controllers in handling arrival aircraft in the descent phase of the sequencing process.

- *Active Final Approach Spacing Tool* (*aFAST*). This is a terminal component of the NASA-developed CTAS that provides not only runway assignment and sequence for arriving aircraft, but also a conflict-free approach path through a series of heading and speed adjustments.

PHASE 3 (2008–2015)

This phase completes the required infrastructure and integration of automation advancements with the new CNS technologies, enabling additional Free Flight capabilities throughout the NAS. Two important features will be NAS-wide information sharing among users and service providers and *four-dimensional* flight profiles that utilize longitudinal and lateral positions and trajectories as a function of time.

VHF datalink (VDL-3 or NEXCOM) radios, providing high-quality digital voice and data, will gradually become a requirement in domestic and international high-level airspace and high-density terminals. CPDLC will expand to provide flight and traffic information services (FIS/TIS). Today's GPS satellites will be replaced by higher-power, dual-frequency units offering improved performance but requiring new receivers. Ground navigation aid decommissioning will continue, but a backup network will remain. A sole-means GPS is not envisioned in the current FAA modernization plan.

Despite predictions that we will someday transition to a sole-means navigation source based on a global network of satellites, most observers agree that the nationwide infrastructure of ground-based navigation aids will be in place for several more years. Currently, the FAA operates about 1000 VOR/DMEs, 750 nondirectional beacon instrument landing systems (NDBs), and 1000 ILSs. Under the current plan for WAAS, the FAA would start phasing out ground-based navigation aids until a skeletal network of fewer than 200 VOR/DMEs, 200 ILSs, and 250 NDBs are left, with Loran-C possibly also providing backup to the satellites through 2008. Although recent studies have indicated that a sole-means satellite navigation system is feasible, few are willing to say that the world's air traffic control network can safely be based on a single space-based source. In a sole-means environment, any failure could be disastrous, whether in the satellite(s), the system control center, the aircraft's equipment, or interference from any source, including atmospheric or deliberate jamming. So, until these problems are adequately remedied, it seems clear that a backup system will be required.

Around 2011, after reaching the 50 percent reduction level, called the *minimum operating network* (*MON*), FAA will observe a temporary hold to assess the impact on airspace users, particularly general aviation. Should no difficulties be identified, decommissioning will continue down to 30 percent of the present numbers, which will be reached around 2013 and called the *basic backup network* (*BBN*). One factor affecting the decision to move down to the BBN level could be the further continuation of Loran-C, which is now expected to operate until at least 2008. In the MON and, later, the BBN, pilots will always have a backup safety net available to them. But this will also mean that very high frequency omnidirectional range (VOR) and instrument landing system (ILS) receivers will remain mandatory for IFR operations (Fig. 8-5).

NAS AND ATC FUNDING

Modernizing the NAS and the ATC system is a necessary but very expensive business. As examples, WAAS is estimated to cost around $3 billion, while equipping all 21 ARTCCs with display system replacement en route workstations will cost more than $1 billion, as will the future STARS installations nationwide. Even maintaining the current VOR/DME, NDB, and ILS network across the country runs around $200 million per year. Some claim that full ATC system modernization, including ongoing operations and maintenance and the staff to run it, could cost close to $100 billion over the next 15 years.

The FAA is dependent upon Congress to appropriate funds on a year-to-year basis. The agency's budget request, part of the Department of Transportation package, is included in the President's January budget submittal for the following fiscal year. Before the agency's request is included in the Presidential submittal, it has already gone through review by the Department of Transportation and the Office of Management and Budget.

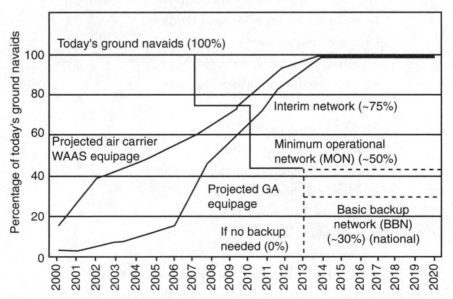

FIGURE 8-5   NAS Plan for the progressive reduction in ground navigation aids. (*FAA.*)

After Presidental submittal, the agency's request is submitted to congressional committee scrutiny through study, hearings, questions and responses, and finally to the full Senate and House for approval on a department-by-department basis. Unfortunately, the FAA has had some spectacular program failures in the past, and Congress has a long memory for such embarrassments. Another factor is that the FAA budget is considered in the context of the entire budget and national priorities, placing the agency in the position of competing with social, space, and defense programs. Only after the budget process is complete, taking nearly 2 years from initial work to approval, can the agency start procurement or construction of facilities.

The question at this point is, Will the FAA be successful under the present process in competing for airport improvement, system modernization, and operating funds while facing the challenge of servicing a near doubling of airline passengers in the next 15 years?

KEY TERMS

Flight information regions (FIRs)

Terminal maneuvering areas (TMAs)

Control zones(CZs)

Very high frequency (VHF)

Very high frequency omnidirectional range (VOR)

Distance-measuring equipment (DME)

Organized track system (OTS)

Short-term conflict alert (STCA)

Standard instrument departures (SIDs)

Departure procedures (DPs)

Standard arrival routes (STARs)

Nondirectional beacon (NDB)

Visual meteorological condutions (VMCs)

Automatic flight control system (AFCS)

Instrument flight rules (IFR)

Visual flight rules (VFR)

Airport traffic control towers (ATCTs)

Terminal radar approach control (TRACON)

Air route traffic control centers (ARTCCs)

Central Flow Control (CFC)

National Airspace System Plan (NASP)

National Airspace System (NAS)

Controller-pilot datalink communication (CPDLC)

Integrated ground telecommunications infrastructure

Next-generation air/ground communications (NEXCOM)

Datalink (DL)

Global Positioning System (GPS)

Wide-area augmentation system (WAAS)

Local-area augmentation system (LAAS)

Nonprecision approach (NPA)

Precision approach (PA)

Decision height (DH)

Runway visual range (RVR)

Category (CAT) I, II, and III approaches

Automatic dependent surveillance (ADS)

Automatic dependent surveillance—addressable (ADS-A)

Automatic dependent surveillance—broadcast (ADS-B)

Cockpit display of traffic information (CDTI)

Integrated terminal weather system (ITWS)

Weather and radar processor (WARP)

Next-generation weather radar (NEXRAD)

Terminal Doppler weather radar (TDWR)

Traffic and collision avoidance system (TCAS)

Traffic flow management (TFM)

NAS-wide information service

Enhanced traffic management system (ETMS)

Operational and supportability implementation system (OASIS)

Surface Movement Advisor (SMA)

Airport movement area safety system (AMASS)

Surface management system (SMS)

Standard terminal automation replacement system (STARS)

National Route Program (NRP)

Free Flight—Phases 1, 2, and 3

User request evaluation tool—core capability limited deployment (URET CCLD)

Traffic Management Advisor—Single Center (TMA-SC)

Center TRACON automation system traffic management advisor—single center (CTAS TMA SC)

Passive Final Approach Spacing Tool (pFAST)

Collaborative decision making (CDM)

Host/oceanic computer system replacement (HOCSR)

Display system replacement (DSR)

REVIEW QUESTIONS

1. Describe how air traffic is controlled.
2. What is the mission of the FAA's air traffic control system?
3. How has deregulation affected the ATC system?
4. What is the National Airspace System Plan (NASP)? And what were the reasons for its delayed schedule?
5. What are some of the key components of the communications modernization?
6. How is datalink expected to improve communications?
7. How does GPS work?
8. Compare and contrast WAAS and LAAS.
9. Distinguish between category I, II, and III approaches.
10. What are some of the advantages of satellite-based navigation?
11. Distinguish between primary and secondary radar.
12. What is automatic dependent surveillance (ADS)? Distinguish between ADS-A and -B.
13. How does the NAS modernization plan improve ways to collect, process, transmit, and display weather information to users and service providers?
14. How does NEXRAD differ from TDWR?
15. What is TCAS? What is the NAS-wide information service? What is the enhanced traffic management system (ETMS)? How will flight services be improved?
16. What is the function of the surface movement advisor (SMA)? How does it differ from the surface movement system (SMS)?
17. How will the STARS program improve TRACON?
18. What are the key components of the en route/oceanic modernization plan?
19. What is Free Flight and what are its goals?
20. Describe the five tools or enabling technologies of Free Flight Phase 1.

21. How will Free Flight Phase 1 differ from the present National Route Program (NRP)?

22. Distinguish between the three phases of the NAS modernization plan.

REFERENCES

Clausing, D. J. 1987. *The Aviator's Guide to Modern Navigation*. Blue Ridge Summit, Pa.: TAB Books.

———. 1996. *Aviation System Capital Investment Plan*. Washington, D.C.: U.S. Government Printing Office, January.

———. 1998. *Air Traffic Control: Evolution and Status of the FAA's Automation Program*. General Accounting Office. Washington, D.C.: U.S. Government Printing Office, March.

———. 1998. *Air Traffic Control: Observations on FAA's Modernization Program*. General Accounting Office. Washington, D.C.: U.S. Government Printing Office, February.

———. 1999. *Air Traffic Control: FAA's Modernization Investment Management Approach Could Be Strengthened*. General Accounting Office. Washington, D.C.: U.S. Government Printing Office, April.

FAA/DOT. 1985. *National Plan of Integrated Airport Systems 1984–1993*. Washington, D.C.: U.S. Government Printing Office, August.

———. 1989. *National Airspace System Plan*. Washington, D.C.: U.S. Government Printing Office, September.

———.1999. *National Airspace System (NAS) Architecture Version 4.0*. Washington, D.C.: U.S. Government Printing Office, August.

Loose, J. 1986. "ATC History: The Modern Era 1972–1985." *Journal of Air Traffic Control*. 8:2 (April–June, 1986), 9–12.

Nolan, Michael S. 1994. *Fundamentals of Air Traffic Control*, 2d ed. Belmont, Calif.: Wadsworth Publishing Co.

U.S. Congress. 1986. *Air Traffic Control: FAA's Advanced Automation System Acquisition Strategy*. General Accounting Office. Washington, D.C.: U.S. Government Printing Office, July.

———. 1994. *Air Traffic Control: Status of FAA's Modernization Program*. General Accounting Office. Washington, D.C.: U.S. Government Printing Office, April.

———. 1995a. *Aviation Research: Perspectives on FAA's Efforts to Develop New Technology*. General Accounting Office. Washington, D.C.: U.S. Government Printing Office, May.

———. 1995b. *National Airspace System: Comprehensive FAA Plan for Global Positioning System Is Needed*. U.S. Congress, General Accounting Office. Washington, D.C.: U.S. Government Printing Office, May.

———. 1995c. *Federal Aviation Administration: Issues Related to FAA Reform*. U.S. Congress, General Accounting Office. Washington, D.C.: U.S. Government Printing Office, August.

U.S. Congress, Office of Technology Assessment. 1988. *Safe Skies for Tomorrow: Aviation Safety in a Competitive Environment*, OTA-SET-38. Washington, D.C.: U.S. Government Printing Office.

AIRCRAFT SAFETY SYSTEMS

LEARNING OBJECTIVES

After completing this chapter, you should be able to

- Recognize the importance of jet engine development and advances in solving problems of fuel consumption, noise, reliability, durability, stability, and thrust.
- Describe several breakthroughs in technology evolving from development of the B-47.
- Explain how improvements in high-lift systems have resulted in safer aircraft.
- List and briefly describe advances that have taken place in five stopping systems.
- Discuss the importance of stability and control characteristics in relation to safety; give several examples of developments in powered controls and low-speed stall characteristics.
- Understand how criteria and procedures used in aircraft design over the years have produced long-life, damage-tolerant structures with excellent safety records.
- Explain the concept of fail-safe design.
- List and discuss the parameters that define aircraft aging.
- Discuss factors that affect cabin safety and crash survival.
- Describe several approaches and new technologies designed to address the problems of wind shear, volcanic ash, ice, and precipitation.
- Summarize some of the flight-deck technology changes that have made significant contributions to improving safety.
- Explain how computational fluid dynamics (CFD) has enhanced the study of wing design and engine/airframe integration.
- Discuss the importance of the following technologies to aircraft development and safety: wind tunnel, piloted simulator, structural tests, flight tests, flight data recorder, and crew voice recorder.

INTRODUCTION

Rapid advances in technology have led to the development of extremely complex and highly sophisticated commercial aircraft. Major portions of tasks that were demanding and performed manually are now automated. New fly-by-wire concepts, for example, sever the mechanical connection between the pilot and the aircraft wing and tail. *Fly-by-wire* refers to the electronic linkage from the sidestick (or control wheel) to flight control computers to activate flight controls. This type of system can come in a number of forms. In many aircraft, such as the Airbus A-320, it incorporates *protections,* which prevent the pilot from exceeding certain limits.

Pilots often perceive that they can fly closer to the margins with a conventional airplane and get the last ounce of performance. In reality, the bigger danger may lie in overreacting and getting into an unrecoverable attitude or being too timid and not getting the required performance. For example, close traffic may require rapid pitch and roll inputs. This is no time to consider limits. With a conventional aircraft, there is a real risk of excessive bank or overstressing the aircraft, while the fly-by-wire system will automatically limit bank to 67 degrees and prevent more than 2.5 G's as in the case of a wind shear escape maneuver during approach. Older, conventional aircraft may have no recovery tools, other than a target pitch attitude, with additional pitch increases up to stick-shaker actuation. Others have more accurate flight director guidance for pitch target but still have no stall protection. The fly-by-wire aircraft combination of an excellent speed reference mode for the flight director and stall protection, even with full aft stick, is a distinct advantage. This allows the pilot to pitch up to the command bars, yet not be concerned with overdoing it. This is a valuable feature. In other words, it really does not matter that a conventional airplane may have a theoretical extra margin of performance by pressing the limits. What matters is the level of performance that is readily available to the pilot.

Pilots soon gain confidence that fly-by-wire will not allow limits to be exceeded, which promotes maneuvering to the edge of the safe flight envelope, probably closer than they would dare with a non-fly-by-wire airplane.

Much of the technology of the airplane can be attributed to the many new and improved design tools that the engineer has available. Wind tunnels, simulators, and analog and digital computers have each contributed to the remarkable safety-record improvements achieved by today's jet transports. The Wright brothers' achievement was aided (maybe even made possible) by the wind tunnel they built to verify the density of air.

The purpose of this chapter is to relate how this wealth of technology and the technology of design tools have contributed to the excellent safety record of today's jet transport fleet. In addition, a discussion of new technologies that should further improve the safety and efficiency of tomorrow's airplane is included.

World War II hastened the development of several different technologies used directly in the development and use of the jet transport. The jet engine, radar, and wing sweep are three of those major technologies. The only airplanes available to the airlines in the late 1940s were propeller-driven, powered by large reciprocating engines. Efforts to increase their size and performance came at the expense of engine reliability. In many accidents, the loss of an engine was a contributing factor.

After initial development, it was realized that the jet engine offered improved reliability, safety, and performance. The jet engine's greatest challenge was its propensity for high-speed, high-altitude operation. This led to a swept-wing design, which favored a much higher wing loading. Early operation at high speed at high altitude near the airplane's critical Mach number was an adventure in pilot technique, clear air turbulence, and jetstreams. The impact of these factors and the technical solutions found to make the jet the safe and efficient mode of travel it is today make a very interesting story.

The jet transport era was ushered in with the advent of the Boeing B-47. It was followed in 7 years by the Boeing Dash 80 prototype. There were many safety-of-flight issues associated with these large swept-wing jet aircraft. This chapter focuses on those technologies that were pivotal to safety as the jet frontier was opened.

JET ENGINE DEVELOPMENT

Credit for the jet engine goes to two individuals who worked independently of each other just prior to and during World War II: Frank Whittle of England and Hans von Ohain of Germany. Through the efforts of General Hap Arnold, the British made Frank Whittle's work available to U.S. engine companies. General Electric (GE) started jet engine development for the Air Force, and Westinghouse started it for the Navy. The effort at GE first produced the axial flow J-47 engine, which was used in quantity in the F-86 and B-47 production airplanes. The work of Pratt & Whitney (P&W) led to the J-57 engine used on the B-52 and 707 series aircraft.

Very rapid progress has been made through the years by the engine manufacturers in solving problems of fuel consumption, noise, reliability, durability, stability, and thrust. The resulting improvements in jet engine performance and reliability rapidly overtook those of the piston engines which were experiencing significant problems with increased size. The reduced complexity and frequency of maintenance of the jet engine lessened the chances of human error. As is discussed later, simplification in controls improved layout and display conditions in the flight deck that subsequently served to reduce the chances for human error.

Jet airplane performance resulted directly from jet engine propulsion efficiency, increasing as airspeed increased. Propeller efficiency falls off as airspeed increases, making it impractical for high-speed flight. Efforts to improve jet engine capability have been nothing short of spectacular. Improvement in cruise specific fuel consumption as a function of the year the engine was introduced into commercial service is shown in Fig. 9-1. This progress has been made while meeting diverse engine-thrust levels asked for by the aircraft manufacturers to satisfy ever-changing payload-range requirements.

These engines also had to satisfy more stringent noise requirements. Much of the improvement in the last 30 years has been due to the engine and its installation when measured in terms of increased passenger-miles per pound of fuel burned. This jet engine technology was even more impressive when one notes that this performance improvement was achieved with a corresponding improvement in the engine's reliability and safety. But the *in-flight shutdown* (IFSD) rate and the resultant loss of thrust and systems were only a part of the safety concern, as the airplane was designed for this event. The other concern was the failure mode (passive failure or a disk rupture), which had to be treated differently. This was the primary reason for going to the pod-mounted engine installation.

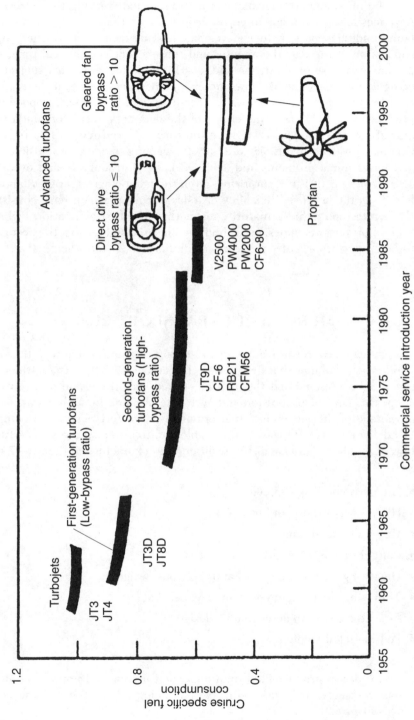

FIGURE 9-1 Turbojet/turbofan engine thrust specific fuel consumption (TSFC) improvement over the first 45 years.

223

The air transport industry has never stopped working to improve its record. Advances have been made in engine inspection and monitoring technology to detect fault conditions early in their development. This technology further improved with application of the digital computer and sensors that can, in real time, monitor the dynamic response of an engine and look for abnormalities. This subject is currently being addressed by the airplane and engine manufacturers and the regulatory agencies. In the past 15 years the rate of occurrence of IFSDs has dropped significantly. This record is a tribute to the efforts of the operators, regulatory authorities, engine manufacturers, and aircraft manufacturers to reduce IFSDs to a minimum. Compared with early turbojet and turbofan engines, modern propulsion systems have had fewer initial problems and achieved better introductory and overall reliability. This improvement is not surprising, given the technological progress and experience. It reflects the improved reliability of today's engines, which have consistently lower IFSD rates and achieve maturity faster than the first-generation turbine engines. With improvements in design, technology, analysis, and testing, the next generation of turbofan engines at entry into service will have IFSD rates better than mature rates for older engines.

LONG-RANGE COMMERCIAL JET TRANSPORT ERA

A great leap forward for commercial transport airplanes came with the Boeing B-47, a military jet bomber that first flew on December 17, 1947, the 44th anniversary of the Wright brothers' first flight. It all started with the vision of a few designers to adapt this new jet propulsion system to the airplane in such a way as to take full advantage of the jet engine's performance characteristics. The Boeing Transonic Wind Tunnel (BTWT) was an invaluable tool to establish the successful configuration. A number of radical airframe differences characterized the XB-47 design:

- Highly swept wing (35 degrees)
- High-aspect-ratio plan form (9.43)
- Very wide speed range
- Long-duration, high-altitude operation
- High wing loading (double that of previous designs)
- Thin wing (12 percent constant-thickness ratio)
- An extremely clean aerodynamic design
- Pod-mounted engines

This design produced a revolutionary performance advantage but also presented some real safety challenges requiring technological solutions. Some of these challenges involved

- How to take off and land
- Stopping-distance considerations
- Control system capability over a large speed range, and flutter
- Structural integrity for this wing plan form and speed range

These challenges were but a few encountered by the test program following the first flight on December 17, 1947. An experimental flight test program was undertaken, during which many design decisions were validated and valuable lessons learned (Fig. 9-2). Each of these contributed to today's jet transport safety. More than 2000 B-47s were manufactured, and the airplane remained operational until the late 1960s, again contributing much to today's airplane safety.

Another great leap forward came with the Boeing B-367-80 prototype. The Dash 80 took full advantage of the B-47 experience. One significant difference from the B-47 was the return to a tricycle gear, which made it possible to tailor its high-lift system for takeoff and landing and to increase the weight on its wheels for improved stopping capability. These changes are discussed along with improvements in the control system, structure, and other technologies that evolved from the Dash 80 prototype.

HIGH-LIFT SYSTEMS

Development of *high-lift systems* had to keep pace with the transition from piston engine to jetliner operations. Early jet bombers used simple trailing-edge flaps and explored but did not employ leading-edge flaps. These trailing-edge flaps provided a low-drag solution to the takeoff climbout problem but created another problem for the approach to landing. Since flap drag was low in the approach phase, the power setting of the engines was low, and it became difficult to quickly accelerate the engines to go-around power for a missed approach. Glide path control was also difficult with these low-drag flaps. With the B-47 in the landing configuration, the engines took 13 seconds to accelerate to 60 percent rpm. Part of the solution was to incorporate multielement slotted trailing-edge flaps, which provided the required higher drag and power settings. The remainder of the solution was to reduce the acceleration time for the engines.

Incorporation of leading-edge flaps resulted in greater safety due to better takeoff and landing field length performance and greater stall margin. Leading-edge flaps were incorporated on some of the early jet transports primarily because of the different takeoff characteristics of jetliners versus piston-powered airplanes. Prop wash over the wing of propeller aircraft provides a built-in factor of safety at low speeds, enabling the airplane to continue to climb at airspeeds less than the power-off stall speeds. Leading-edge flaps on jets extend the margin of safety to even lower speeds and permit takeoff and climbout margins similar to those in prop airplanes. Some overrun accidents with early jets not equipped with leading-edge devices were attributed to early rotation of the airplane.

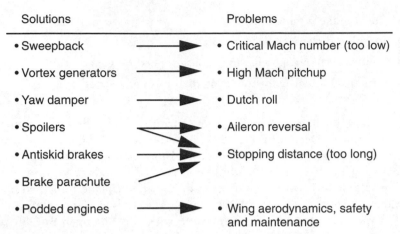

Solutions	Problems
• Sweepback	• Critical Mach number (too low)
• Vortex generators	• High Mach pitchup
• Yaw damper	• Dutch roll
• Spoilers	• Aileron reversal
• Antiskid brakes	• Stopping distance (too long)
• Brake parachute	
• Podded engines	• Wing aerodynamics, safety and maintenance

FIGURE 9-2 Lessons learned from development of the B-47.

The early jetliners operated into and out of a limited number of airports, and their high-lift systems were adequate for the range of altitudes and temperatures encountered at these airports. As jet travel became more economical, an increased diversity of airports was served, and the technology of flap systems changed to accommodate them. In the interest of shorter field-length performance, the 727 program developed a high-camber airfoil by having a highly deflected slat and a three-element trailing-edge flap system. This system provided the capability to operate at lower approach and landing speeds, giving the 727 the ability to land on shorter runways. Today's jetliners incorporate two-position leading-edge slats to provide low-deflection setting for low-drag takeoffs, a higher-deflection position for landing at lower speeds, and better visibility from a lower deck angle.

A further merit of a good high-lift system is to have higher leading-edge-device camber on the outboard wing to provide a more stable stall plus better stall characteristics, which provide enhanced safety by making for a more forgiving airplane. To enhance safety and improve stall characteristics while maintaining the low-camber leading edge for takeoff, the 757 added the auto slat to its configuration. This feature provides an extra margin from stall when the auto slat actuation angle of attack is reached. This feature also provides additional maneuvering margin in the case of a wind shear encounter.

STOPPING SYSTEMS

A number of improvements in aircraft stopping systems have taken place over the years, greatly enhancing safety. They include antiskid, fuse plugs, autobrakes, speed brakes, and thrust reversers.

ANTISKID. The higher takeoff and landing speeds for jetliner operations compared with prop airliners created the need for more efficient stopping devices. An *antiskid* *system* was recognized as a necessity early in the jet age. Most of the time pilots do not need antiskid at all. The runway is dry and long, and the brakes are not applied hard enough to skid the tires. However, when the runway is slippery and short, the ability of the antiskid system to maximize braking effectiveness becomes very important. Experience with large propeller aircraft indicated that it was very difficult for the pilot to sense skidding of the tires on the runway during a stop. Some cockpit crews on propeller-driven bomber crews even relied on the crew in the back of the airplane to call out over the intercom if their tires were skidding!

Early antiskid systems were developed and tested successfully on the dual-wheel main landing gear assemblies of large propeller-driven aircraft. However, the design of antiskid systems soon became a new technological challenge because of the incorporation of multiple-wheel main landing gears on the first jetliners. The early antiskid systems for the four-wheel truck main gears controlled tandem pairs of wheels. Later the technology to control each wheel independently was incorporated to maximize the effectiveness of the antiskid system. Digital antiskid systems incorporating microprocessors enhance the reliability and effectiveness of today's antiskid systems.

FUSE PLUGS. Air pressure buildup in tires after braking during early high-speed refused-takeoff tests caused the tires to explode, sending chunks of rubber flying into the airframe. Incorporation of low-melting-point fuse plugs in the wheels allowed the tires to deflate before severely heating and exploding.

AUTOBRAKES. The incorporation of an *automatic braking system* is a recent enhancement to safety. This system enables automatic brake application on landing or during a *refused takeoff* (RTO). The landing autobrake system controls brake pressure to maintain aircraft deceleration at one of five pilot-selected values, provided that sufficient runway friction is available to maintain this level. The RTO autobrake system applies full braking upon closing throttles above a fixed speed (for example, 85 knots). Using autobrakes frees the pilot to concentrate on other activities, such as applying reverse thrust and guiding the airplane to a smooth, safe stop.

SPEED BRAKES. While main wheel brakes remain the primary method of stopping aircraft on the runway, technological development of ancillary stopping devices has kept pace with development of wheel brake systems. The need to apply a download on the wheels was recognized as a requirement to enhance the effectiveness of wheel brakes. The use of *wing spoilers* (*speed brakes*) to increase download on the main wheels was incorporated on the early jets. The percentage of wingspan covered by these devices has increased on later models, increasing their effectiveness. Implementation of automatically deployed spoilers on later models has enhanced stopping capability significantly. Sequencing of the spanwise spoilers after touchdown eliminates pitchup, which can occur when all the spoilers are deployed simultaneously.

THRUST REVERSERS. Commercial jet transport operation into regional airports would not have been possible without *thrust reversers*. Although FAA regulations did not require thrust reversers, they were a must for most airline customers. It was very desirable that jetliners be able to land on slippery runways within the FAR-required field length. Early experience revealed that some types of thrust reversers lost their effectiveness at high landing speeds, which required the airframe manufacturers to conduct tests to verify early reverser concepts.

Designs of the 737-100 and -200 presented The Boeing Company with a difficult challenge in integrating the thrust reversers, because the engines were mounted very close to the lower surface of the wing and very close to the ground. The initial design decision was an economic one: Use the entire power package and nacelle from the Boeing 727. The resulting reverser configuration partially trapped the reverser efflux between the trailing-edge flaps and the leading-edge devices, creating a "bubble" of air on which the airplane floated in ground effect, greatly reducing stopping effectiveness. A complete redesign of the reverser, including a lengthening of the nacelle to accommodate a target reverser aft of the flaps, resulted in a tremendous improvement in stopping capability. In fact, the 737-200 has such an effective reverser that the airplane can stop within its FAR scheduled wet distance using only thrust reversers.

FLYING QUALITIES

Before it flew, the flying qualities of the B-47 jet airplane were known to be significantly affected by its swept wing, high speed, and high wing loading. A number of technological advances enhanced the handling qualities of this configuration so that today the flying qualities of the Boeing 757, 767, and 777 and Airbus 320, 330, and 340 are the standards of the industry.

The stability and the control characteristics designed into an airplane have one of the most significant impacts on the airplane's flight safety. The airplane's response to an engine-out, system failure, or atmospheric disturbance has to be controllable within the trained commercial pilot's ability. To ensure this controllability, the technology for modeling airplane dynamic response has been continually advanced along with improved modeling of atmospheric disturbances. These models are then incorporated into flight simulators, and piloted evaluations of the designs are undertaken to give direction as to the best flying qualities and pilot techniques. Out of these simulations, technology was developed for determining airplane configuration, including the best control laws for automatic flight control and stability augmentation systems. The computer's and simulator's parts in enabling this progress, and thus improving aircraft safety, are discussed later.

POWERED CONTROLS.   The B-47 was ahead of its time with closed-loop hydraulic position servos on all control surfaces. Each axis had manual backup through servo tabs and aerodynamic balance cavities. Hydraulic-powered elevator force feel was provided with a Q-spring. It sensed an increase in dynamic pressure and adjusted the pilot's force feel system to protect against an inadvertent maneuver that could exceed the airplane's structural limit. The technology that followed on jetliners greatly improved the state of the art of hydraulic systems and actuators by providing more reliable, redundant hydraulic supply and smooth transition from boost to manual control. It also prevented a jammed control failure mode.

The B-52 design proceeded before the B-47 hydraulic power control system could be perfected. The B-52 was designed with manual controls on all axes. The elevator and rudder controls had 10 percent chord surfaces, and a variable-incidence horizontal stabilizer was used for longitudinal trim and control. Aileron control was supplemented by wing spoiler controls that were flight-tested on the B-47. On commercial designs, the rudder and elevator sizes were increased to provide more engine-out control and maneuvering capability, stabilizer mistrim, and dive recovery capability on the longitudinal axes.

Spoiler lateral controls were explored on the B-47 after it was discovered that at high airspeeds aileron control reversed due to the resultant elastic torsion of the wing from an aileron-developed lift component opposite to and of greater magnitude than that of the deflected aileron.

The Boeing Dash 80 prototype perfected the combination aileron and spoiler control further so that essentially no yaw or pitch coupling occurred following a control input. The adverse yaw associated with a lateral control input is balanced by yaw from the spoilers.

The B-47 also encountered adverse shock interactions at high speeds. The swept wing greatly delays this onset, but as Mach 1 is approached, they occur. This problem was investigated thoroughly in flight. The solution turned out to be rather simple. Small vanes called vortex generators were installed on the wing, altering the boundary-layer shock interaction to reduce pitchup and increasing the buffet margin to allow greater turn performance at altitude. These devices have proved valuable on subsequent transport designs.

The other flying quality discovery of the B-47 program was that the swept wing has significantly different Dutch-roll characteristics. In a straight wing, it is primarily a nose oscillation, but with wing sweep, it has significant roll that is slightly out of phase with the yaw. It was computed prior to flight that with no dihedral the Dutch-roll mode would be sufficiently damped, but flight test showed that not to be the case. Again, technology advanced quickly with the invention of the electromechanical yaw damper to solve this problem. It senses the yaw motion that occurs in this mode and applies opposite rudder, thus relieving the pilot of extra workload or the possibility of the aircraft becoming upset in turbulence. The electromechanical yaw damper proved to be a solution with a long life. Electronic yaw dampers are in all present-day commercial jet transports.

LOW-SPEED STALL CHARACTERISTICS.   Design of the high-lift systems was a challenge in that these systems had to provide commercial jets with safe takeoff and landing margins, in all places and conditions. The system also has to exhibit satisfactory flying qualities in the extremely unlikely event of a stall.

Early swept-wing designs exhibited poor stall characteristics because the wing tip would stall first. The remaining inboard wing lift, being ahead of the center of gravity, would give the aircraft an undesirable pitchup tendency. Wing technology involving the selection of airfoils, wing twist, and tailoring of the leading-edge and trailing-edge flaps was initially advanced through the use of wind tunnels, then flight testing. Later, the understanding of this problem was greatly advanced through the use of *computational fluid dynamics* (*CFD*).

Two technologies, the stick-shaker and auto slat gapper, have been effective in providing good stall characteristics for both takeoff and landing. The best design philosophy is to first work on avoiding the stall. Some airfoil-wing combinations provide adequate stall warning to the pilot by virtue of buffet that occurs as the stall angle of attack is reached. As high-performance wings were developed, this buffet margin was found inadequate and a "stick-shaker" was provided. The technology of this system involved finding the best sensor and location to key on. The auto slat was implemented when it was found that the characteristics of a sealed slat, which had improved takeoff performance, had poor flying qualities at stall. The solution was to automatically open the slat gap as stall was approached.

Stall safety was greatly improved by use of the simulator. Early in demonstrating airplane stall and recovery techniques, it was found that the situation was too unforgiving to a student applying improper technique. Today, stall avoidance and recovery are carried out safely in the simulator. Stall characteristics are thoroughly investigated during flight tests involving, in many cases, more than 700 stalls. The designer is anxious to get as low a stall speed as possible for overall safety in addition to reduced structural weight. However, finding an ample stability and control configuration makes it difficult, but necessary, to achieve.

STRUCTURAL INTEGRITY

The importance of structural integrity to commercial aircraft safety is obvious. What wasn't so obvious was the fracture-mechanics problems encountered and the durability of the jet airplane that extended its life well beyond anything previous. The B-47 and the de Havilland Comet were the first large jets to become operational, and both encountered fatigue problems. The Comet encountered fuselage skin fatigue problems that led to a series of accidents. Subsequent investigation into these and the B-47 fatigue problems pushed the state of the art for aircraft structural design technology forward very rapidly. The B-47 and B-52 fatigue problems came to light after the U.S. Air Force started flying low-level radar avoidance missions.

The real challenge to commercial aircraft was that they would experience much higher flight-hours at lower stress levels. One hour of low-level flight was equivalent to 80 hours at cruise. These lessons learned were shared with all, and manufacturers have been benefiting from these improvements in aircraft safety ever since.

One of the material fracture-mechanics properties came to light. Laboratory data compared the tensile strength of three different aluminum alloys as affected by the length of a crack. In the past, it was general practice to design engineering structures such as aircraft, bridges, buildings, and pipelines to a required new, uncracked strength including a factor of safety. This factor of safety was intended to provide for degradation by corrosion, fatigue, damage, etc. It's clear today that engineering structures should be designed to have adequate strength after they have sustained fatigue, corrosion, and use damage to an inspectable level.

A lot of credit has to be given to the electron microscope. By electron microscope examination on failure surfaces of failed structures, it became clear that each time the structure was loaded, there would be crack growth. The initiation and growth of cracks could be identified with their prior loading. It was also clear that the fatigue and corrosion cracks of most aircraft materials start at the exposed surface of the material, permitting inspection for fatigue and corrosion cracks for most installations.

The attention of aircraft designers became focused on the rate at which cracks grow, the strength of cracked structures, and the variation of these factors for different materials. It was found that the rate of crack growth was a function of loading. It was then that considerable attention was given to understanding the loading cycles commercial aircraft would see in service.

STRUCTURAL SAFETY.   Criteria and procedures used in commercial airplane design over the last three decades have produced long-lived, damage-tolerant structures with excellent safety records. This has been achieved through diligent attention to detail design, manufacturing, maintenance, and inspection procedures. Structural safety has been an evolutionary accomplishment, with attention to detail being the key to this achievement. These design concepts, supported by testing, have worked well due to the system that is used to ensure that the fleets of commercial jet transports are kept flying safely throughout their service lives. This system has three major participants:

- The manufacturers that design, build, and support airplanes in service

- The airlines that operate, inspect, and maintain the airplanes

- The airworthiness authorities who establish rules and regulations, approve the design, and promote airline maintenance performance (Fig. 9-3)

Airplane structural safety depends on the diligent performance of all participants in this system. The responsibility for safety cannot be delegated to any single participant.

All jet transports are designed to be damage-tolerant, a concept that has evolved from the earlier fail-safe principle. On the whole, service experience with fail-safe designs has worked very well with thousands of cases where fatigue and other types of damage have been detected and repaired. The question being debated among experts in the industry is whether the fail-safe design practices used in the 1950s and 1960s are adequate as these airplanes approach or exceed their original economic life objectives. Note that there is no limit to the service life of damage-tolerant-designed airplane structure (see the section on aging aircraft next), provided the necessary inspections are carried out along with timely repair and replacement of damaged structure or preventive modifications for airplanes exceeding economic

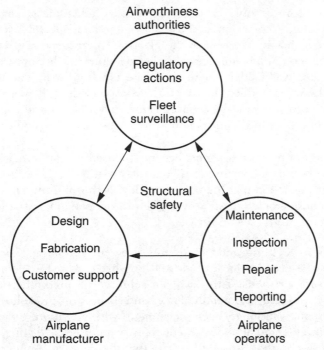

Airworthiness
authorities

Regulatory
actions

Fleet
surveillance

Structural
safety

Design

Fabrication

Customer support

Maintenance

Inspection

Repair

Reporting

Airplane
manufacturer

Airplane
operators

FIGURE 9-3   Structural safety system interaction.

design life objectives. Operational efficiency is affected by the cost and frequency of repair. Durability may, therefore, limit the productive life of the structure.

AGING AIRCRAFT.   One of the major problems facing the FAA and air carriers today is aging aircraft. By definition, aging aircraft are aircraft that are being operated near or beyond their originally projected design goals of calendar years, flight cycles, or flight hours. Nothing in the FARs pertains directly to the life of an airplane as measured in calendar years, but it is customary for designers to assume 20 years as the calendar life of an airplane. For the first generation of jet transports, some designers believed that the aircraft would be technically obsolete within 20 years. Using this measure, a number of the earlier models of the DC-9s, B-737s, B-727s, B-747s, and DC-10s still flying today would fall into this category.

The other two measures of age are the number of flight cycles and the number of flight hours accrued in service. Table 9-1 shows the design goals for U.S.-designed transports.

Industry designers have used the phrase *design goals*, but these goals have not represented the end of service life, nor are they contractually binding. Commercial aircraft were designed and certificated in the United States after World War II in the belief that with proper inspection, maintenance, and repair, the life of the airframe could be unlimited. The foundation for this premise was the adoption of the principle of *fail-safe design* by the FAA and the industry in the early 1950s. This

TABLE 9-1 Design Goals of U.S.-Designed Transport Jet Aircraft

AIRCRAFT	LIFE, HOURS	LIFE, FLIGHTS
DC-8	50,000	25,000
DC-9	30,000	40,000
DC-10	60,000	42,000
L-1011	60,000	36,000
707	60,000	30,000
727	60,000	60,000
737	45,000	75,000
747	60,000	20,000
757	50,000	50,000
767	50,000	50,000

SOURCE: *USA Today,* May 15, 1996.

rule required that a specified level of residual strength be maintained after *complete failure* or *obvious partial failure* of a *single principal structural element*. The early U.S. jet and propjet fleets were designed, tested, and FAA-certified to this rule without a specified life limit. The service experience acquired by this fleet by the middle to late 1970s had generally shown a satisfactory level of structural safety and provided many documented instances of the validity of the fail-safe concept.

However, in a few instances, failure to detect damage in a timely manner resulted in degradation of the residual strength to an unsatisfactory level. These instances and the recognition that economic factors combined with the rate of technological advancement would extend the desired airframe economic life beyond that anticipated at the time of design, testing, and certification prompted manufacturers, operators, and the FAA to agree that the fail-safe rule should be amended to require the use of fracture mechanics in defining the inspections and inspection intervals required for continued airworthiness. This agreement led to the current *damage tolerance* (*fail-safe*) rule of October 1978 and to the retroactive application of this rule to then currently certificated jet aircraft such as the DC-10 and the B-747. *Supplemental inspection documents* (*SIDs*), developed by the manufacturers, serve as the basis for the continued airworthiness of the first generation of jet transports.

Ten years later, the Aloha Airlines crash of 1988 focused public and congressional attention on aging aircraft. This accident and other structural failures stimulated a reexamination of the current approaches to the structural integrity of aging aircraft. Fatigue-initiated damage is the primary cause of concern about *aging aircraft*. When aircraft are properly inspected and maintained, corrosion and accidental damage should be understood and controlled long before the design life is reached. On the other hand, the problem of fatigue-initiated damage increases with time or, more properly, with use as measured by flight hours or flight cycles or both.

Fatigue damage to the fuselage is caused primarily by the repeated application of the pressure cycle that occurs during every flight. Fatigue damage to the wings is caused by the ground-air-ground cycle that occurs during every flight and by pilot-induced maneuvers and turbulence in the air. Thus, the design-life goal of flight hours is more important for the wings than it is for the fuselage. Fatigue-initiated damage is a random phenomenon in which the probability of existence at any specific point in the structure increases with time. Because detecting the damage is also probabilistic, the success of the damage tolerance process in preserving airworthiness lies in an acceptably low probability of the presence of damage and high probability of timely detection. A large transport airplane is a complex structure in which a large number of points are susceptible to fatigue cracking that could propagate to the point at which the residual strength would be less than the damage tolerance requirement. The number of points in the structure at which cracks will initiate increases with the age of the structure. As the number of initiation sites increases, the probability increases of not detecting at least one before the strength degrades to the limit. In other words, at some time the risk of not having limit-load capability at some point in the structure may be too great for the airframe to be considered airworthy. Thus, even an airframe designed to be damage-tolerant may reach a time in its life when additional inspections, maintenance, and repair will be required to maintain airworthiness. A common scenario for the fuselage includes the following:

- Cracks initiate at the edges of the fastener holes in the center of the panels along a splice.
- With repeated flight cycles, these small cracks link to form a patch spanning several holes
- The patch becomes sufficiently long that it is detected during scheduled checks before rapid growth occurs.

Rapid growth, if it occurs, will be arrested by *crack turning* at the tear straps or the frame or both, resulting in a safe depressurization that permits the pilot to safely land the airplane. Although this scenario has actually occurred, it is not the only possible scenario. At least two other scenarios can be envisioned in which cracking may not be arrested by the tear straps (e.g., if the fastener holes in the tear straps already contain small cracks or if there is widespread cracking in several adjacent bays). In these circumstances, the crack may continue to propagate rapidly along the splice and lead to uncontrolled depressurization, in which case neither the tear straps nor the frame is fulfilling the original fail-safe function. This, in fact, is what occurred during the Aloha Airlines incident—the uncontrolled crack propagation resulted in the loss of a large portion of the fuselage. The structural condition previously described, wherein widespread cracking in tear straps and adjacent bays occurs, is called *multiple-site damage* (*MSD*). Clearly, MSD is more likely to exist in heavily used aircraft. A key factor in maintaining the safety of aging aircraft is the determination of age in flight cycles, flight hours, or both, of the onset of MSD.

It is assumed in analyses of this phenomenon that the airframe was designed and manufactured as intended, such that aging is related to the *wear-out* phase of the well-known *bathtub curve*. Because neither design nor manufacturing control is perfect, cracks occur in airframes long before the design goal life is reached. This is not an aging phenomenon but should be considered as the population of locations in which the design or construction was deficient (i.e., local "hot spots"). Some are revealed by the airplane fatigue tests and others during the early service experience with the aircraft. As cracks are detected and corrective action is taken, the rate of new hot-spot cracking decreases with time; the net result is that the total population of crack locations at any given time would be small. Experience has shown that the risk of undetected cracks during this phase of the operational life can be controlled to a safe level by appropriate inspection and maintenance programs.

In contrast, during the wear-out, or aging, phase of the airframe, the frequency of cracking reaches the point at which the risk cannot be controlled solely by inspection and maintenance programs. In other words, the airframe has reached the age at which the probability of cracking in large areas of the structure is sufficient to produce an unacceptable risk of not detecting a critical level of damage. Current service with the post-World War II transport aircraft fleet designed by U.S. manufacturers to meet specified fail-safe requirements and desired design-life goals indicates that most of the aircraft have not yet entered this wear-out phase in which they are subject to widespread cracking of structural details of normal fatigue quality. That is, the onset of MSD is beyond the originally projected design-life goals for these aircraft, and therefore MSD is not an issue at this time.

The first 291 B-737 fuselages were manufactured using a cold-bonding process to adhesively join the fuselage skins at the lap splices. The lap splices were also mechanically fastened together with three rows of rivets. Service experience has shown that some of these bonds have failed, that is, the lap splices have disbonded. A disbonded lap splice has the potential for developing MSD at a relatively early age compared with the original life goal. Under the conservative assumption that the disbond existed at the time of the initial delivery of the aircraft, it has been determined that the onset of MSD in these first 291 aircraft may occur at 30,000 flights. Had they not disbonded, the splices clearly would not have been the sites for widespread cracking within the projected operational lifetime. The continuing airworthiness of the affected aircraft is being ensured through directives that mandate special inspections and a schedule for the replacement of several thousand fasteners in the lap splices with button-head fasteners of a larger diameter.

In summary, the following primary technical issues are posed by the aging aircraft fleet:

- *MSD,* the undesirable condition caused by widespread cracking of the structure, negates fail-safety and damage tolerance to discrete sources. Neither multiple-load-path nor crack-arrest fail-safe features of commercial transport aircraft can be depended on to protect the structural safety of the aircraft after the onset of MSD.

- *Corrosion,* a time-dependent process, decreases the size of structural members, leading to higher stresses and lower structural margins. Corrosion also has undesirable synergism with the factors that lead to cracking of the structure, factors that are not well quantified and are not considered in the damage tolerance and fail-safe design of the structure.

- *Nondestructive inspection* is the key to assessing the health of the aging aircraft fleet. It should not be relied on for ensuring the continuing airworthiness of an aircraft that may be approaching the onset of widespread cracking, that is, the threshold of MSD as measured in calendar years, flight cycles, or flight hours.

- *Structural repairs,* which are more prevalent in the aging aircraft fleet, are generally made to regain static strength and may not adequately fulfill damage tolerance and fail-safe requirements.

- *Terminating actions*—the FAA language that denotes the structural actions necessary to eliminate MSD—do not have the database in terms of component and full-scale testing comparable with that on which the original structures were certified. Thus, neither the design life of the terminating actions nor the inspection intervals for continuing airworthiness can be established without further testing and analysis.

Although risks to safety associated with aging aircraft are worthy of concern, no studies have been conducted that link the financial conditions of the airlines as a result of deregulation to the aging of the aircraft fleet; nor has any direct link been established between aging aircraft and increased risk.

The industry has responded to the issue of aging aircraft in two ways. An industrywide task force, the Airworthiness Assurance Task Force, recommended making mandatory many recommended maintenance procedures for older Boeing and Douglas aircraft and recommended requiring replacement of key components at certain points in an aircraft's service life. The FAA has adopted these recommendations. The cost of the task force's recommendations for Boeing aircraft alone is estimated at more than $1 billion over 10 years. In addition, the industry has placed massive orders for new aircraft.

Although the potential hazards of older aircraft should not be minimized, the life-cycle replacement and maintenance procedures now required by the FAA, if followed by the industry, will mitigate any additional risk associated with catastrophic failure of the airframe. The maintenance procedures to detect corrosion recommended by the industry task force, however, are quite extensive and have substantial cost implications for carriers and the FAA. Carriers are facing cost increases of roughly $2 million per aircraft. The FAA, already trying to expand its inspector workforce to meet existing standards, will be hard-pressed to ensure that the maintenance and replacement schedules are carried out.

CABIN SAFETY

Where is the safest place to sit in an airplane? The many people who posed that question in 1985 realized that surviving an airplane crash is often possible. But few of them also appreciated that survival and cabin safety depend on a great deal more than seating position, or that better emergency procedures and equipment, different cabin materials, and stronger seats can contribute to higher survival rates. In 1985, the NTSB addressed the issue of cabin safety on several fronts. It issued a study on emergency equipment and procedures relating to in-water air carrier crashes and found that equipment and procedures were either inadequate or designed for *ditchings* (emergency landings on water where there is time to prepare and that involve relatively little aircraft damage) rather than more common short- or no-warning in-water crashes. The NTSB recommended improvements in life preservers, passenger briefings, emergency-evacuation slides, flotation devices for infants, and crew post-crash survival training.

Later in the year, the NTSB completed a study on airline passenger safety briefings. The NTSB concluded that in the past, "The survival of passengers has been jeopardized" because they did not know enough about cabin safety and evacuation. The study also found wide variances and sometimes inaccuracies in oral briefings and in information on seatback-stored safety cards.

The NTSB reiterated a 1983 recommendation to the FAA to convene a government/industry task force to fully examine the issue of passenger education. In addition, 13 new safety recommendations were issued, including one that asked for new prelanding safety briefings to reinforce to passengers the pretakeoff briefing about seat belts, exit locations, and the location and operation of life preservers. Both studies touched on issues related to emergency evacuations, a subject of intense interest to the NTSB that gained public attention in 1984 when the FAA approved a request to deactivate the overwing exits on certain B-747 airplanes. The FAA said its regulations technically permitted the reduction in the number of exits. In testifying on the FAA's action, the NTSB said, "The exit reconfiguration of the Boeing 747 was ill-considered and not in the interest of safety of the traveling public." The testimony noted that in some emergencies, exits could not be opened or were unusable because evacuation slides failed to inflate, fire blocked access to them, or the position of the airplane when it came to rest made the exit unusable. The NTSB found that during postcrash fires, the number of exits available rarely approached the theoretical number mandated by regulations.

The NTSB's reaction to the FAA's approval of Boeing's request was part of widespread public opposition. Subsequently, the FAA indicated it would not permit modifications deactivating over-wing exits on the B-747 airplanes.

Effective evacuation of an aircraft also depends on flight attendants. In October 1985, the FAA issued a notice of proposed rulemaking to require protective breathing equipment for flightcrews and cabin attendants. The NTSB had recommended making this equipment available following its investigation of the fatal fire on the Air Canada DC-9 at Cincinnati International Airport in June 1983. The recommendation repeated one made a decade earlier after the NTSB participated in the investigation of a foreign accident involving a cabin fire. The NTSB believed that without protective breathing equipment, flight attendants could easily be incapacitated by fire and smoke, could not make effective use of fire extinguishers, and could be of no use during an evacuation. In 1985, the NTSB also supported a petition for rulemaking to set flight attendant flight-duty time limits.

SAFETY DESIGN FOR ATMOSPHERIC CONDITIONS

Long-duration, high-altitude, and high-speed flight can be significantly impacted by many diverse atmospheric phenomena. Four of the more important areas of concern include

- Turbulence
- Wind shear
- Volcanic ash
- Ice and precipitation

The first line of defense should be to detect and avoid these hazards. When this strategy is impractical, solution must address safe operations and maneuvers in the event of an encounter. Great progress has been made toward both.

TURBULENCE

It became apparent as jet airplanes started operating at high altitudes for extended periods that the structure of atmospheric turbulence was different from that at the lower altitudes. One advance that helped was the velocity-load factor, in *G*'s, altitude (VGH) recorder. Tests showed the need for further research, one avenue being the *High Altitude Clear Air Turbulence* (*HICAT*) program. Data collected in this program markedly changed the design criteria of commercial and military jet airplanes.

Early in the jet era, there were a number of what were referred to as "jet upsets" from severe air turbulence encounters. Study of these encounters led to a better understanding of how to safely fly jet aircraft in such an encounter. Results indicated that it was important to not change trim, to disengage the early autopilot's autotrim feature and airspeed/Mach hold, if necessary, and to fly attitude and let the airspeed and altitude vary somewhat.

Between 1983 and 1997, the NTSB investigated 99 turbulence accidents and incidents that resulted in two fatalities and 117 serious injuries. Most of these injuries—many of which involved fractures of the spine, skull, and extremities—were completely preventable if the occupants had been restrained by seatbelts.

Since 1972, the Board has made several safety recommendations to the FAA and the National Weather Service (NWS) and has been involved in numerous investigations dealing with the hazards of turbulence. For example, in March 1993, the engine of a B-747 cargo flight separated from the airplane in severe turbulence conditions while departing from Anchorage, Alaska. As a result of this investigation, the Board recommended that the NWS develop in greater detail turbulence forecasts using data from the National Weather Service's Doppler weather radars (WSR-88D). The NWS has implemented this recommendation.

The Board investigated another severe turbulence encounter involving a United Airlines B-747 that occurred in the western Pacific on December 27, 1997. This encounter resulted in one fatality and many injuries to passengers and flight attendants. The Board's investigation focused on turbulence forecasting, flightcrew training, dissemination of information on turbulence, and crew procedures in areas where turbulence is forecast.

The NWS Aviation Weather Center is also making progress in improving its clear air turbulence and mountain wave forecast products. Further, the FAA Aviation Weather Research program has a multidisciplinary team addressing the turbulence problem, and researchers from the National Center for Atmospheric Research (NCAR) are working on new algorithms for using data from the NWS Doppler weather radars to detect turbulence. The NCAR researchers are also developing software that may be able to turn airborne commercial aircraft into a real-time turbulence-sensing platform. The software will use onboard sensors and computers to measure and analyze turbulence as the aircraft flies through it. The data will be transmitted to the NWS, where they will be used to create accurate, real-time turbulence maps.

WIND SHEAR

With the increased frequency of flights, wind shear–related takeoff and landing accidents increased. The first part of the investigation into wind shear involved isolating the wind profile that got the airplane in trouble. Improved onboard data recorders on test airplanes showed the profiles were more severe than previously known. Additionally, the downburst phenomenon associated with certain weather conditions was discovered (Fig. 9-4). A three-pronged approach was initiated. The first prong involved training crews on avoidance of the phenomena, the second related to better detecting the conditions that can produce wind shear and alerting the crew, and the third prong dealt with getting maximum performance from the airplane if the crew inadvertently encountered wind shear.

Good-fidelity simulators were used for this analysis. In evaluating means of helping the crew get the greatest performance from the airplane, it was possible to use the latest technologies that were being incorporated in advanced flight decks. A very successful task force composed of individuals from the FAA, airplane manufacturers, and airlines made a great contribution to safety by producing a wind shear training aid for flightcrew avoidance and procedures for getting maximum performance in the presence of wind shear. Additionally, algorithms were developed and displays were modified to provide guidance and situational awareness, which enhanced flightcrew performance.

Another winds-aloft hazard is mountain waves. The loss of a 707 near Mount Fuji was the result of such an occurrence. We still have a lot to learn about wind in the vicinity of mountains, and the supercomputer's contribution to modeling this phenomenon gives hope for further improving the safe operation of aircraft in this environment. Also needed is further research on sensors capable of detecting these adverse winds.

VOLCANIC ASH

Volcanic activity has occurred throughout the world since the earliest of times. The first recorded impact on aviation was on March 22, 1944, when Mount Vesuvius

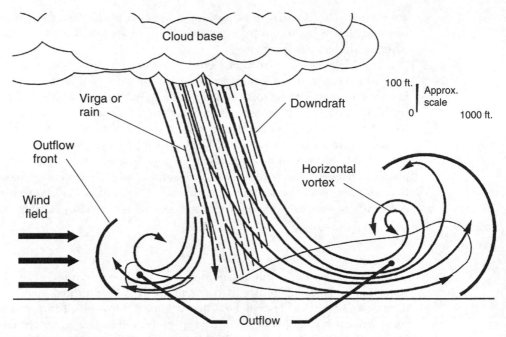

FIGURE 9-4   Basic features of a microburst.

did more damage to an airfield than any enemy activity. It was another 36 years before the next direct impact, and that was in 1980, when a civil Lockheed C-130 Hercules inadvertently penetrated a dense ash cloud from Mount St. Helens in western Washington following its second major eruption. The aircraft lost power on two of its four engines, but the crew was able to return to and land safely at McChord AFB. Since then, there have been five major eruptions worldwide.

Efforts were being made to avoid local areas where a known eruption had recently occurred. However, the extent of the problem was given new dimension when two 747s encountered volcanic debris from the eruption of Galunggung on Java in 1982 and experienced serious engine problems.

In all these occurrences, but especially those involving complete flameouts, credit goes to the crews for getting the problem quickly diagnosed and the engines relit. Global surveillance satellites have been very useful in tracking volcanic ash. Still more has to be understood as to hazard zones following eruptions. Dissemination of this information and the effect of volcanic ash on other aircraft systems critical to safe flight, such as the air data system, avionics, and crew windows, must be fully analyzed and communicated.

ICE AND PRECIPITATION

Ice and precipitation affect airplane operation in two areas. One is the effect they have on the airframe and engine, and the second is the effect on tire-ground contact.

Unlike the previous atmospheric topics, these elements are daily occurrences, depending on the season and location. Safe operations depend on a coordinated team effort by airline maintenance crews, flightcrews, and airport authorities, as well as ground and air traffic control. The technologies that help are many and depend on specific icing or precipitation conditions. Again, procedures play an important role. Flight simulator training continues to be an effective tool in developing correct takeoff and landing procedures for ice-contaminated runways to keep safe operation possible.

Developing all these procedures involves extensive wind-tunnel, environmental laboratory, and flight tests to fully understand the problems and show performance of the airplane under these conditions. The contribution to safety in some of these cases is focused on understanding the limitations of the airplane under adverse weather conditions. Only then is safety improved when those responsible for each aspect of the operation implement the procedures that come from these tests.

In-flight icing is one of the FAA's top weather research priorities. Improved operationally available, high-resolution, accurate forecasts of atmospheric icing conditions are needed. Several safety recommendations dealing with in-flight icing were issued as the result of the NTSB's investigation of the ATR-72 accident that occurred at Roselawn, Ind., on October 31, 1994. These recommendations and the findings from the investigation provided the major impetus to the icing research efforts of the FAA.

Then again on January 9, 1997, Comair Flight 3272, an Embraer 120, crashed near Monroe, Mich., destroying the airplane and killing all 29 people on board. There were reports of moderate icing in the area at the time of the accident. In May of the same year, the Board issued four urgent safety recommendations to the FAA regarding icing. Almost concurrently, the FAA issued a notice of proposed rulemaking to modify operating procedures in icing conditions. Comair modified its operating procedures based on the FAA's proposed rule, and the FAA issued a final rule. The Board also worked with the National Center for Atmospheric Research and the NASA Lewis Research Center regarding weather issues, and in January 1998, Board personnel traveled to Brazil, where the airplane is manufactured, to review all pertinent test data on icing and to perform studies in the engineering simulator.

Issues examined regarding this accident included flightcrew training, operations in icing conditions, and aircraft performance. This accident resulted in several significant research activities in icing. These activities include the quantification of the performance loss due to small amounts of surface roughness on a wing's leading edge and the performance penalties associated with pneumatic deice boot intercycle ice.

FLIGHT DECK HUMAN-MACHINE INTERFACE

The technology topics discussed to this point have dealt primarily with the design aspects of the airplane as a machine and its capability to operate safely in the atmospheric environment. This section takes a look at the technology that involves the human-machine interface. It involves the flightcrew and one or more other parties: maintenance, ground operations, weather advisers, air and ground traffic control, and others. It all comes together on the flight deck. The human-machine interface often becomes the determining factor in the event of an emergency, where correct, timely decisions and execution make the difference between life and death. As discussed earlier in this text, human factors is the science that deals with the human-machine interface in an attempt to maximize the potential for safe, efficient operation while eliminating hazardous conditions resulting from human error. This technology was incorporated only into certain details of the early jet airplanes. As the state of knowledge of airplane design and human-factor research and understanding has advanced, so has the jet transport flight deck improved.

Technology in flight decks has improved continuously since the early days of aviation. Notable advancements are radio communication, radio and inertial navigation, and approach systems. The jet engine greatly simplified cockpit controls and displays. The following are some of the flight-deck technology changes that have made a significant contribution to improving safety:

- Crew alerting and monitoring systems
- Simple system designs
- Redundant systems
- Automated systems (when essential)
- Map display
- Engine-indicating and crew-alerting system (EICAS)
- Displays with color enhancement
- Ground-proximity warning system (GPWS)
- The 757/767 flight deck
- Aircraft Communications Addressing and Reporting System (ACARS)
- Flight management system (FMS)

The crew-centered concepts in practice today are shown in Fig. 9-5.

The 707, 727, DC-8, and early 747, DC-10, and L-1011 flight decks used a standard arrangement of pilot, copilot, and flight engineer. Much like the previous aircraft, the airplane systems were designed for the flight engineer to be the systems operator. Large instrument panels were mounted behind and to the right of the two pilots. The flight engineer was expected to monitor and operate the vital hydraulic, electrical, fuel, air conditioning, and pressurization systems unsupervised. The design of the short-range 737 was changed radically to provide a flight deck to be operated by a two-person flightcrew.

The airplane systems were first simplified. The fuel system, for instance, has but three tanks: a right wing tank for the right engine, a left wing tank for the left engine, and a center tank to be used by both engines. The fuel boost pump capacity,

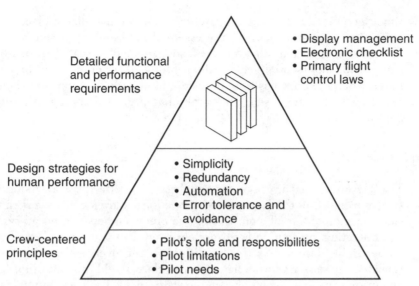

Detailed functional
and performance
requirements

- Display management
- Electronic checklist
- Primary flight
  control laws

Design strategies for
human performance

- Simplicity
- Redundancy
- Automation
- Error tolerance and
  avoidance

Crew-centered
principles

- Pilot's role and responsibilities
- Pilot limitations
- Pilot needs

FIGURE 9-5   Crew-centered concepts in practice.

line sizes, and fuel head were selected to permit fuel from the center tank to be used first, followed by wing fuel, without any crew action following the prestart checklist. When the center fuel was depleted, amber lights annunciated to the crew that the center tank pumps could be turned off. A crossfeed system was provided for nonnormal operations. The simplified system had other benefits besides crew workload reduction.

Multiple sources of power or supply were provided on all systems to have adequate system function when one or more elements failed. Multiple hydraulic pumps were provided, driven in different ways to provide a completely redundant system.

The control system on the Boeing 727 uses cables from the control column to hydraulic actuators for normal operations. However, if the two hydraulic systems fail, a system of cables to the flight controls provides a *manual reversion* method of controlling the aircraft. The 757 is a newer-generation aircraft and takes a different approach to redundancy. It has three hydraulic systems, engine and electrical pump, and a ram air turbine as a backup pump. This replaced the direct cables-to-controls connection. The 747 is similar, with four hydraulic systems and no direct cable-to-control connection. At least one hydraulic system and pump must be operating to move the flight controls. The Airbus A-320 also requires at least one hydraulic system and pump to move flight controls. The A-320's redundancies include three hydraulic systems, two engine-driven pumps, two electric pumps, a power transfer unit (uses one hydraulic system to pressurize another), and a ram air turbine in case ac electric power is lost.

Automatic operation of a system was provided for certain selected equipment failure cases to avoid the necessity of crew intervention at a critical time in flight. The 737 electrical system load-shedding feature is an example. In the event of a single generator failure on the two-generator nonparalleled electrical system, the remaining generator picks up the essential load and nonessential loads are shed. The galley power and other similar loads are shed so that the remaining generator can provide all essential loads without the need for crew attention or intervention. Later, when the crew has time to restore a second generator, the shed loads can be recalled. This same design philosophy has prevailed through the other two-crew designs on the 757, 767, and 747-400 aircraft.

CREW ALERTING SYSTEMS

The 737 introduced a very simple but elegant and effective crew monitor and alerting system. Almost all the airplane's system controls are located overhead and are outside the normal line of sight of the two crewmembers. When all systems are "on" and operating, no caution lights are observed. In the event of loss of equipment, an amber light annunciates the condition on the overhead panel. The loss is also repeated on master caution lights on the glare shield in front of each crewmember and in small caution panels also on the glare shield. The panel annunciation identifies the system affected and the location of the system controls overhead.

Since this was a new scheme at the time of certification, it got a lot of attention by the certification authorities and designers. To make sure that airplane system operation was not an inordinate burden or substantial workload for the flightcrew, many hours of flight in the simulator and during certification flight testing were devoted to measuring the time spent on airplane subsystem operation. Figure 9-6 shows the typical time a crew spends on various functions. Less than 1 percent of the crew's time from takeoff to landing is spent on system operation, including deliberate equipment malfunction simulation by the certifying agency. This careful attention to system detail design and improvements in monitoring capability has been used for virtually all jet models subsequent to the 737.

The *microprocessor,* which became available in the 1974–1975 time period, greatly improved the monitoring function. In today's modern cockpits and those on the drawing boards, all the functions in each of the airplane's systems are continuously monitored to ensure that the systems' parameters are within the proper operating range. When any parameter falls outside (either lower or higher than) the operating range, the crew is appropriately alerted and, when necessary, automatic functions such as described for the electrical system are performed.

GROUND-PROXIMITY WARNING SYSTEM.   Since the advent of powered flight, inadvertent ground or water contact has been a worldwide problem. While much early effort went into avoiding such accidents, no major advance occurred until introduction of the *ground-proximity warning system* (GPWS) in the early 1970s. Although there has been a marked reduction in controlled flight into terrain (CFIT) accidents since then, they still occur with distressing frequency and account for close to 75 percent of worldwide fatalities on commercial transports. For the air carrier fleet, the primary reasons are twofold: Either the airplane did not have GPWS (or it was inoperative), or the crew ignored the GPWS alert. The Flight Safety Foundation

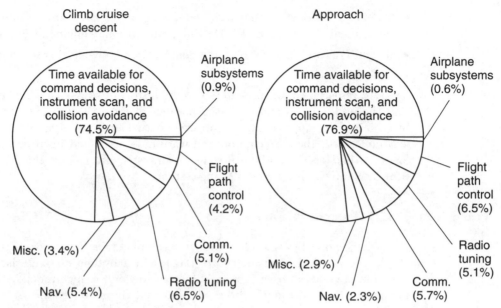

FIGURE 9-6 Typical jet workload distribution.

(FSF), along with others, has an aggressive program to essentially eliminate CFIT accidents.

ENGINE-INDICATING AND CREW-ALERTING SYSTEM.   The *engine-indicating and crew-alerting system* is a digital computer system using cathode-ray tube (CRT) displays to monitor and indicate propulsion and airplane subsystem information for the operation and maintenance of the airplane. EICAS interfaces with many airplane components and subsystems. Discrete inputs are implemented in a hierarchy that reflects operational and maintenance requirements:

• Flightcrew alert messages duplicate dedicated subsystem information (usually indicator lights) elsewhere in the flight deck.

• Status and maintenance messages provide lower-priority information on the condition of many subsystem components.

There were extensive research activities during the development of EICAS, some of which included

• The British Aerospace Corporation's Weybridge simulator efforts
• Reports published on Airbus's ECAM (electronic centralized aircraft monitor)
• Reports on McDonnell Douglas's EMADS (engine-monitoring and display system)

- A 1977–1978 Boeing study to integrate engine displays into a combined electro-mechanical display called IPEDS (integrated primary engine display system)
- Several FAA and industry research contracts on crew-alerting systems

FAA research contracts centered on the history of subsystem alerts and indicated that the proliferation and increased number of crew alerts could only be reversed by an integrated approach to a centralized crew-alerting system. At the time EICAS was conceived, the 767 caution and warning system had been developed on the basis of that research as the first major industry effort since the 737 to install a fully integrated central crew-alerting system.

Flight Deck—757/767 and B-747-400

Flight-deck noise levels are low enough to allow a true "headsets off" environment. While the forward windshields are flat for best optical characteristics, the side windows are curved to prevent turbulent airflow and reduce the associated aerodynamic noise. Wind-tunnel studies showed that the aerodynamic vortex created by the sharp angular change between the flat forward and flat number-two windows contributed to cockpit noise levels at cruise airspeeds. This source of noise has been eliminated in the 757/767 flight deck. The air-conditioning system is designed to further reduce flight-deck noise levels by means of ducting improvements and lower airflow velocities.

Vision characteristics are excellent inside and out the 757/767. Vision through the windshields exceeds Society of Automotive Engineers (SAE) recommendations, resulting in superior collision avoidance capabilities. An extra margin of safety on landing in adverse weather conditions is achieved by improving downward visibility and maintaining a low deck angle on approach.

In this "quiet dark" flight deck, few green or blue lights indicating normal system operation are used. Lighted pushbutton switches that combine the amber malfunction light with the shutoff switch are used to reduce the possibility of incorrect crew action.

The *integrated display system* (IDS) in the B-747-400 consists of six 8-inch-square screens. Although all are identical, each performs a different function depending on its location:

- Primary flight display (PFD)
- Navigation display (ND)
- Engine indication and crew-alerting system

The two outboard CRTs, directly in front of the pilots, function as the captain's and first officer's PFDs. Each pilot can use the PFD as the single source for all the primary flight instruments found on a traditional instrument panel. The tape formats used for altitude, airspeed, and heading/track indications were chosen for two reasons: First, they permit sufficient display resolution without disrupting the "basic T" instrument configuration; second, the tape formats more readily accommodate related supplemental information. This information, such as speed bugs and a trend vector, increases pilot situational awareness.

AIRCRAFT COMMUNICATIONS ADDRESSING AND REPORTING SYSTEM

The *Aircraft Communications Addressing and Reporting System* (*ACARS*) is a communication datalink system that sends messages, using digital technology, between an airplane and the airline ground base. The operational features of ACARS equipment and the ways in which ACARS is used in service vary widely from airline to airline.

There is nothing new about sending messages between the airplane and the ground. What makes ACARS unique is that messages can be sent, including fuel quantity, subsystem faults, and air traffic clearances, in a fraction of the time it takes using voice communications, in many cases without involving the flightcrew.

ACARS relieves the crew of having to send many of the routine voice radio messages by downlinking preformatted messages at specific times in the flight. These may include the time the airplane left the gate, liftoff time, touchdown time, and time of arrival at the gate. In addition, ACARS can be asked by the airline ground operations base to collect data from airplane systems and downlink the requested information to the ground.

Each ACARS message is compressed and takes about 1 second of air time to transmit. Because of the automatic reporting functions described above, the number of radio frequency changes that flightcrews must make is reduced on ACARS-equipped airplanes. Sending and receiving data over the ACARS network reduces the number of voice contacts required on any one flight, thereby reducing communication workload and costs. ACARS messages are limited to a length of 220 characters, which is adequate for routine messages.

The accurate reporting of event times, engine information, crew identification, and passenger requirements provides for a close control of any particular flight. Airplane system data, such as engine performance reports, can be sent to the ground on a preprogrammed schedule, or personnel on the ground may request data at any time during the flight. This allows ground personnel to observe the engines and systems and can alert them to problems to be investigated. More information on datalink communication systems is included in Chap. 8 on air traffic systems technologies.

FLIGHT MANAGEMENT SYSTEM

The *flight management system* (*FMS*) is an integration of four major systems: the flight management computer system (FMCS), the digital flight control system (DFCS), the autothrottle (A/T), and the inertial reference system (IRS). The basic functions of the FMS are

- Automatic flight control
- Performance management

- Precision navigation
- System monitor

The FMS is designed to allow crew access to the total range of its performance, navigation, and advisory data computation capability at any time and in any flight control mode. For example, when the airplane is under manual control, the pilots, at their option, can get flight optimization data from the flight management computer and appropriate "bugs."

The flight management computer is a major innovation in the FMS design. In addition to navigation, it performs real-time, fully automatic performance optimization and can control the airplane through the flight control system, including the auto throttle. While crews currently determine the most efficient speed and altitude by using the flight operations manuals and calculators, this function is usually performed only periodically. Further, because it is a digital system, the software and programmable databases in the FMS provide the adaptability and growth needed for present and future airline operations, particularly as navigation and air traffic control systems evolve.

MULTIPLE FLIGHT CONTROL COMPUTERS

On the newer-generation aircraft, two *elevator/aileron computers* (*ELACs*) have primary control of elevators and ailerons and perform the computations for spoilers and yaw damping. One ELAC is active while the other is a backup.

Three *spoiler/elevator computers* (*SECs*) provide primary control of spoilers (all three SECs are active). If both ELACs fail, the SECs also provide backup control of roll and pitch via spoilers and elevator.

Two *flight augmentation computers* (*FACs*) control rudder for turn coordination and yaw damping. They also compute limits for the flight envelope, wind shear, and speed information displayed on the PFD.

No single failure of an electrical, hydraulic, or flight control component will result in a reduction in operational capability.

CENTRAL MAINTENANCE COMPUTER SYSTEM

One technology that is able to assist in maintaining complex systems is the *central maintenance computer system* (*CMCS*). This system can collect, display, and provide reports of system fault information and test airplane systems. It provides maintenance personnel with a centralized location for both testing and access to maintenance data. The net result is a decrease in airplane turnaround time and an increase in dispatch reliability over previous-generation airplanes.

The CMCS was developed to monitor and troubleshoot the complex and integrated systems on the 747-400. It was also created to centralize ground testing, fault information storage, and real-time data monitoring. In previous-generation airplanes, testing, troubleshooting, and fault isolation are confined to a system-by-system approach, and faults that affect multiple systems are often difficult to isolate. Because so many systems on the 747-400 are interdependent, a fault in a single system can ripple through several other systems. Since the CMCS is linked to all major systems, it can provide simultaneous fault monitoring for related systems and can correlate multiple fault indications to a single fault.

Another major problem with previous-generation aircraft was the lack of a common test initiation procedure. Initiated tests went unused when they could have helped to isolate faults. The 747-400 CMCS provides a common design for testing interfacing systems.

FLIGHT PROCEDURES—TAKEOFF, APPROACH, AND LANDING

Procedures to ensure the flightcrew gets the best performance out of the airplane under all conditions are very important in realizing safety. All forms of analysis and simulation, from procedure trainers to all-up, full-function flight simulators, are used to ensure updated procedures. All conditions, functions, and routine and non-routine takeoff and landing scenarios can be safely tried and perfected.

TAKEOFF

Significant amounts of analysis and study have gone into understanding the physics involved in takeoff, rejected takeoff (RTO), and climbout. On modern jets, takeoff techniques involve accurately setting power, using proper rotation rates at V_r speed, and achieving and holding target climbout speed and attitude. Correct rotation techniques and procedures are highly stressed in pilot training and are essential to achieving consistent results.

Another problem encountered in early jets, which has improved but still needs further attention, is RTO overruns. Overruns may occur when a decision to stop is made at speeds above V_1. If the aircraft is near its field-length performance limit, initiation of the first stopping action must occur at or before V_1, and the maximum stopping procedure must be followed to prevent overrun. This procedure involves immediately retarding the throttles and applying full brakes, followed by actuating spoilers. Analysis of 67 rejected-takeoff accidents through 1990 found that two-thirds of the no-go decisions were made above V_1. Continuing the takeoff from these high speeds would have resulted in successful takeoff and climbout.

A significant amount of analysis and study has taken place over the years to more clearly understand and define takeoff decision speed and its correct use. Information has been developed and distributed to airlines and pilots through meetings and seminars, numerous written articles, and pilot training at manufacturers' schools.

Numerous improvements in aircraft and systems design have led to significantly improving safety during takeoff. The additions of autospoilers and autobrakes discussed earlier provide backup and assurance of achieving full braking performance in an RTO. Brakes and brake antiskid system design have greatly improved stopping capability on wet and dry runway surfaces. On many later models, non-critical warnings and sounds are inhibited during the takeoff roll and initial climb, allowing the crew to focus on the takeoff and reducing the possibility of unnecessary RTOs.

APPROACH AND LANDING

A review of worldwide jet transport accidents indicates that the approach and landing phase of the flight accounts for approximately 50 percent of all flight accidents. Most of these accidents involve failure of the flightcrew to stabilize the aircraft in speed and rate of descent soon enough before touchdown (at least by 500-foot altitude). This lack of stability takes the form of

- Excessive approach height over the threshold
- Floating down the runway
- Incorrect stopping techniques

Emphasis has been placed by the manufacturers on "flying by the numbers" in training aid publications. These numbers include approach speeds, procedures, approach path control, touchdown point, and flying technique. Aircraft design strives to build in approach-speed stability in the lift, drag, and speed relationships of the airplane. Improvements in engine design and engine acceleration characteristics provide better flight path control during descent and allow faster engine spin-up for go-around.

Many new and improved systems are now available at airports and on aircraft for glide slope control to the threshold and provide a consistent touchdown point 1000 to 1500 feet down the runway. Newer aircraft are fitted with autothrottles, automatic landing systems, autospoilers, and autobrakes that allow the aircraft to get on the runway quickly and attain full braking. Again, manufacturer training and education emphasizes consistency in approach and landing procedures to minimize the risk of an accident.

MODELING, DESIGN, AND TESTING TOOLS

Airplane design and technological progress seen this century could not have happened without parallel advancements in computing capability. This was no coincidence, for in many cases it was trying to solve airplane technology problems that gave stimulus for "there must be a better way." Two breakthroughs for the computer (and all avionics for that matter) were the vacuum tube and the transistor. The triode vacuum tube was invented by Lee De Forest in 1906, shortly after the Wright brothers' aeronautical success. It allowed development of an electronic computer.

The computer advancements made to date are what enabled not only all the onboard and support avionics systems but also all the analysis and testing that go into them. It is not possible to cover all this technological progress, as it has been explosive on all fronts. The important part of incorporating any technology is to first make sure it has been analyzed and tested thoroughly. Some of the technologies that have made contributions to safety are discussed next.

COMPUTATIONAL FLUID DYNAMICS

Computational fluid dynamics methods have been used extensively in the design of all new-generation aircraft. Advances in supercomputing technology over the years have allowed CFD to affect problems of greater relevance to aircraft design. Use of these methods allowed a more thorough aerodynamic design earlier in the development process, allowing greater concentration on related operational and safety-related features.

WING DESIGN.   The ability of CFD to do the inverse problem (i.e., given the desired resulting flow, what is the shape of the surface geometry?) has revolutionized the transport wing design process. Since the Wright brothers, wing design has been a "cut and try" operation, with the "try" taking place in the wind tunnel or in flight. The advent of sufficiently powerful computational methods allowed some of the try to be shifted to the computer. But the "cut" was still the designer shaping the wing based on experience and intuition. CFD allows a new approach in which the design engineer specifies to the computer the aerodynamic pressures desired on the wing, and the CFD code computes the geometric contouring of the wing surface that will produce those pressures. The engineer does all the design work and initial evaluation with CFD and then picks the best candidates, builds wind tunnel models, and tests them.

This inverse design technique was used in the development of the 777, allowing achievement of a level of aerodynamic performance not otherwise possible in a time-constrained development program. The CFD methods used for this application were based on nonlinear potential equations coupled with boundary-layer equations to account for viscous effects. Key to this use of CFD was the ability to model enough relevant physics and gain quick turnaround from initial geometry to completed solution. The resulting wing designs are thicker and lighter than their previous counterparts. These characteristics allow wings with greater span or less sweepback. Both features are conducive to better low-speed (landing and takeoff) performance, improved safety margins, and reduced community noise.

CAB DESIGN OF THE 757.   The 757, as originally conceived, was to share the cab of the 727. Only very late in aerodynamic development, when delivery dates were already being quoted to potential customers, was the decision made to develop a new cab. A new cab was desired for two separate but compelling reasons.

Improved downward vision from the 757 flight deck was desired to improve safety over that provided by the 727 configuration. The safety improvement comes from the additional rows of approach lights that can be seen during low-visibility landing. Additionally, it was desired to make the flight decks of the 757 (a standard body) and the 767 (a wide body) essentially identical, with all instruments and controls in the same place, which would make it feasible to obtain a common-type certificate that would allow a given crew to be certified to fly either type of aircraft. The 757 cab design had to accommodate the wide overhead panels and flight-deck instruments that had been designed earlier for the 767. The essential design challenge was to wrap a narrow low-drag aerodynamic shape about these wide components while lowering the number-one window frame to provide improved downward vision. A greater challenge was to produce this design in a very short time. Using computational methods, the 757 cab was carefully designed and tailored with complete success. The resulting elimination of areas of local supersonic flow over the cab at cruise conditions reduced both aerodynamic drag and noise compared to previous designs.

ENGINE/AIRFRAME INTEGRATION.   An early success was the improvement in understanding the interference drag of a pylon-mounted engine nacelle under the wing. CFD studies, along with specialized wind-tunnel testing, provided necessary insight into the flow mechanism responsible for the interference. This understanding led to development of design guidelines that allowed closer coupling of the nacelle to the wing. The Boeing 757, 767, 737-300/400/500 series and the KC-135R are examples of aircraft in which very closely coupled nacelle installations were achieved, allowing later-technology engines to be installed.

The 737 series of aircraft might not have remained in production beyond the mid-1980s if not for the installation of a modern, large-diameter, high-bypass-ratio turbofan engine. Installation had to be accomplished without incurring an excessive increase in interference drag, weight, or cost. The need, the opportunity, and the technology were available to provide a solution to this challenge. The resultant solution, which allowed a much larger-diameter engine to fit under the wing without increasing the main landing gear length, was made possible by using CFD techniques.

While the linear methods that were used for the above applications can simulate flows about complex configurations, they cannot accurately account for the nonlinear transonic flows that are more prevalent in installation of newer, more efficient, larger-diameter engines close to an aircraft's wing. During the aerodynamic design of a new airplane, risk of significant interference drag due to the engine exhaust was revealed through CFD analysis. Neither the earlier linear-based CFD methods nor conventional wind-tunnel testing techniques, which did not simulate the exhaust, would have detected this problem. Only a very expensive powered-nacelle testing technique would have assessed these interference effects.

Had this problem gone undetected until late in the aircraft's development when powered testing is usually done, any fixes would have been extremely expensive and time-consuming to implement. Again, the need, the opportunity, and the technology were available to provide a solution to this challenge.

Key to success of this application was the ability to model enough relevant physics and to provide solutions quickly enough to support the development schedule. CFD provided the ability to detect problems early in the development process, when fixing or avoiding them was least expensive.

Today, CFD is being used to better address the diverse requirements for nacelle design for cruise, engine-out second-segment climb, and engine-out ETOPS (extended range twin-engine operations). Nacelle design is being optimized to minimize drag at cruise conditions and provide acceptable engine-out drag for second-segment climb and ETOPS. Engine-out drag is not the only criterion of merit. If external flow separation occurs, the separation wake could impinge on other aircraft components, causing premature buffeting of the aircraft, perhaps indicating to the pilot that the aircraft is approaching stall conditions. If it is severe enough, the pilot might be inclined to reduce the angle of attack—a situation not desired during takeoff or climbout. CFD has become a very valuable tool for the designer to provide additional safety benefits that might not otherwise have been feasible.

WIND TUNNEL

Today, *wind-tunnel testing* complements the computational fluid dynamics tools, which focus efforts to a narrowed field of configuration options. During development of the 767 in the late 1970s, Boeing conducted tests in 15 tunnels in four countries for more than 20,000 hours, conducting 100,000 data runs, each consisting of recording an average of 65 independent parameters at approximately 25 different aircraft attitudes. In that 767 program, more than 100 variations of the wing design and untold numbers of different flap and slat positions were tested. Despite intensive use of CFD, the test program for the 777 involved about 20,000 hours of wind-tunnel testing up to the first flight.

The value of wind-tunnel testing goes beyond optimization of wing designs and performance. Safety involving loads determination for structural design, simulator database development, stability and control characteristics for both normal and failed configurations, and aircraft anti-icing control greatly benefits from the wind-tunnel-testing set of tools. The effective and efficient integration of propulsion systems onto the airframe is served, in part, by powered models that provide simulations of installed engines' flow fields.

Aeroelastic, or flutter, testing has long been a technique for understanding the flexible airplane. Boeing jets from the B-47 to the 777 have "flown" in the wind tunnel to allow designers to understand their dynamic characteristics. The envelope, not only of flight conditions of speed and altitude but also of loading variations due to payload options and fuel consumption as well, is explored and structural designs are validated.

PILOTED SIMULATION

The *piloted simulator* was recognized very early as a valuable tool in improving aircraft safety. It owes a lot to the pioneering efforts of Ed Link. Early in the initial transition of crews from props to jets, a number of training accidents occurred, particularly during engine-out training. It endangered not only those in the airplane but also those on the ground. The technology that allowed this hazardous training to occur in a simulated environment on the ground was a great benefit to flight safety.

Development of the simulator took two parallel paths, one for engineering evaluation and the other for flight training. Each benefited the other. The engineering simulation allowed the engineer and the test pilot to evaluate design options during design and to define the best operating procedures. It has also been used to evaluate accident and incident data to better understand what might have gone wrong and to find solutions.

At first, simulators were very crude. The engineering simulators were analog and could fly the initialized flight condition on instruments only. Later, a model board was constructed to represent a bird's-eye view of a large ground area using a miniature TV camera and closed-circuit TV that "flew" over the landscape model to provide an accurate view out the cockpit window. This feature greatly improved the evaluations that one could perform, but it had limitations. *Computer-generated imagery* (CGI) was the real breakthrough. It allowed weather and a variety of airports to be evaluated. This technology continues to advance and is providing more and more realism.

STRUCTURAL TESTS

Structural static tests are paramount to verifying the aircraft design's structural margin of safety prior to in-flight demonstration. Static tests to 150 percent of limit load were to show that the design margin was actually achieved. These tests were in effect before the jet transport era. The B-47 experience showed these structural static tests alone did not ensure full structural integrity. It was then that flight fatigue loads were required to be tested. This requirement produced a much more complicated and involved test but ensured a much safer product. These wing cyclic fatigue tests are performed on all commercial aircraft.

INTEGRATED AIRCRAFT SYSTEMS LABORATORY (IASL)

Flight-controls-system ground testing used to take place on the airplane prior to flight. As aircraft systems became more complex, there was a need to conduct this testing earlier, before the first aircraft was built, in a live functional mockup of the complete control system. This "iron bird" technology used for the 727 and later models has taken a quantum jump in Boeing's preparation for the 777. An all-new facility, the Integrated Aircraft Systems Laboratory, was constructed for testing of the fly-by-wire flight control system together with the other airplane systems. This facility allows Boeing to bring together and thoroughly test and validate the interactions and interrelationships of all 777 systems, computers, controls, and structures long before they would traditionally first come together in the number-one airplane.

The benefit of this approach is that integration and systems interface problems are found and resolved early. The laboratory is also useful to quickly resolve any problems that may be found in flight testing before the airplane is introduced into service.

FLIGHT TEST

The ultimate test of an airplane before it enters service comes when it goes through flight testing and is subjected to the real world environment. Flight testing has had to advance in four areas to keep pace with the jet airplane's advanced performance, expanded flight envelope, and the advanced technology incorporated into the airplane systems, such as autoland. These four areas include

1. New flight-test techniques
2. Improved accuracy of instrumentation
3. Improved data-recording systems
4. Real-time monitoring and data analysis

These advances have made it possible to conduct increasingly accurate and thorough tests to ensure that the airplane is ready for service. Typical of such tests is the takeoff minimum unstick speed V_{mu}. This test demonstrates the takeoff performance margin existing at unstick (takeoff), ensuring that the airplane will not encounter a hazardous condition such as loss of control or failure to accelerate. This is one of the key tests to ensure that engine thrust is maintained and the control and flap systems function as designed and provide the airplane with the low-speed, high-angle-of-attack controllability needed for safe operation. This test is also used to establish required operational parameters for takeoff field length. Flight tests such as this one have greatly contributed to the safety of the commercial airplane.

ACCIDENT/INCIDENT INVESTIGATION

The industry has always devoted a lot of effort to the technologies involved in finding the causes of incidents and accidents. The *flight data recorder* (*FDR*) and *cockpit voice recorder* (*CVR*) are indispensable tools in this task.

Substantial progress was made during 1987 in the NTSB's goal of persuading the FAA to require more and better cockpit voice recorders and flight data recorders aboard aircraft. Efforts by the NTSB to upgrade requirements for the use of FDRs and CVRs in keeping with the times have faced a long, uphill battle. Over the years, the NTSB had difficulty persuading the FAA to improve recorders on larger commercial jets and expand their use on commuter and some corporate aircraft, despite the proven value of recorders as a vital tool in accident investigations.

The FAA had issued a notice of proposed rulemaking to upgrade recorder requirements but failed to act on it for more than 2 years, despite NTSB prodding.

On March 25, 1987, however, the Secretary of Transportation announced a requirement for installation of CVRs on newly manufactured turbopropeller commuter aircraft carrying six passengers or more and having two pilots. The secretary also called for installation of digital flight data recorders on older jet aircraft, replacing the foil-type versions currently required. The announcement came after a commuter plane accident at Detroit Metro Airport on March 4, which dramatically helped focus congressional and public attention on the stalled flight recorder rulemaking. Just two weeks after the accident, a congressional panel approved report language directing the FAA to "take immediate steps to require the installation of CVR and FDR devices on all commuter aircraft in line with NTSB recommendations."

The number of FDR data channels and the crashworthiness of these vitally important records have increased significantly. The early generation of FDRs only recorded time, altitude, airspeed, vertical acceleration, heading, and radio transmission event marker. The number of parameters has increased so that today, with all-digital avionics systems onboard the new airplanes, a more detailed record can be made and stored for a longer time. This area is another way in which digital avionics contributes to aircraft safety.

Finally, aircraft manufacturers have dedicated safety staffs that not only support accident and incident investigations involving their aircraft but also oversee and observe others (as invitees) to minimize the opportunity for any recurrence. These preventive measures often involve industrywide participation.

FUTURE AIRCRAFT TECHNOLOGIES

The technological opportunities for continued improvement in flight safety are as good as ever. The analytical tools, computing capability, simulation, and testing capability all far exceed those of the past. And new technologies are continuing to unfold. These opportunities are challenging to all in the aviation community.

Of all the technologies available for improving safety while providing substantial gains for future growth, system capacity, and efficiency, electronics technology appears to have the most immediate promise. Accident statistics continue to bear out the fact that the manufacturers need to assist flightcrews in doing their tasks while improving their capacity. Chip-level redundancy combined with fault-tolerant design will provide far greater reliability levels for essential and critical functions. Airplane safety, capability, and total system capacity and efficiency will be enhanced.

Boeing is using a number of these features on the 777. This airplane has an advanced control system that makes greater use of advanced avionics to give flightcrews a greater degree of assistance in reducing workload and monitoring flight-critical conditions.

WEATHER DETECTION

Weather remains a major cause of delays and accidents. Real-time, accurate knowledge of the weather ahead is required. Aircraft must be able to detect weather conditions, alert the flightcrew to avoid dangerous situations, and effectively cope with turbulence, precipitation, and wind shear.

Today's airborne weather radar needs improvement. In the current air carrier fleet, we are limited to a plan view, and we need a quicker and in-depth understanding of the display. The future will likely bring advanced airborne weather radar with vertical as well as plan cross-sectional views. Ground and satellite weather detection improvements made available in the flight deck will combine to enhance flight planning, performance, and safety.

Onboard system enhancements to aid the pilot in unexpected wind shear encounters are being incorporated on current aircraft. These wind shear alert and guidance systems are being combined with crew training and improved ground-based alerting and weather radar systems. Emphasis is on detection and avoidance of hazards occurring along the flight path. If wind shear is unavoidable, the wind shear system quickly alerts the crew and provides pitch attitude guidance. Improvements are also planned to the current ground-based, low-level wind shear alert system. Terminal Doppler weather radars will be installed at critical airports. More information on weather automation systems is included in Chap. 8 on air traffic system technologies.

COMMUNICATION AND NAVIGATION SYSTEMS

As discussed at length in Chap. 8 on air traffic system technologies, aviation is transitioning toward a new communications, navigation, and surveillance/air traffic management (CNS/ATM) environment. This will inevitably change the way pilots communicate and navigate. For starters, most ATC messages would be sent by datalink to and from the airplane rather than by voice; navigation would be based more on satellites than on land-based navigation aids; and traffic surveillance would be achieved by networking tactical information among all aircraft. The benefits of these changes, particularly to the airlines, are potentially enormous. A global navigation and communications system will allow for higher traffic density and improved safety, while at the same time enabling airplanes to fly at optimum altitudes and on more direct routes.

As envisioned, the transition from today's airspace environment, in which routes are rigidly assigned by controllers on the ground, will move first to a system in which flight plans are originated in the airplane and approved by controllers on the ground, to the eventual goal of Free Flight, where flight plan changes will be initiated in the airplane in real time and simply monitored from the ground.

But this does not mean that the conventional terrestrial navigation systems will disappear overnight. In the United States, the FAA plans to keep very high frequency omnidirectional range (VORs) and distance-measuring equipment (DME) in operation for at least another decade. During this time, pilots' dependency on them will decrease as more flight operations use satellites to navigate. But the susceptibility of GPS to interference probably means VORs will be around for a long time to come, most observers believe.

Cockpits of the future will bring to the flight deck increased capabilities, some of which will include digital communications to accommodate high-speed data to and from the cockpit, WAAS/LAAS approaches and landings, GNSS-based en route navigation, and automatic dependent surveillance, allowing new sophistication in flight tracking and traffic avoidance. Also during this time, today's VHF voice communications will slowly give way to increased digital communications; and surveillance systems will become more sophisticated, moving away from today's Mode S transponders, air traffic control (ATC) surveillance radar, and Traffic and Collision Avoidance System (TCAS) to ADS-B (automatic dependent surveillance—broadcast). The result will be a better tactical picture of traffic and therefore far less need for voice callouts of other aircraft by air traffic controllers.

There is a tremendous need to provide data to the cockpit from the ground. In the future, digital communications are expected to make today's VHF communication radios obsolete, even those that are now undergoing updates for 8.22-kilohertz requirements in Europe. The new infrastructure will require radios that can transmit in both voice and high-speed digital formats. The VHF Datalink (VDL) Mode-2 system is already in Aeronautical Radio, Inc. (ARINC) operation, and the FAA began implementing it around 2002. What's coming are radios designed to eventually support VDL Modes 3 and 4 for digitized voice applications and ADS-B traffic surveillance. ADS-B avionics under development include a *cockpit display of traffic information (CDTI)* feature that enables the pilot to "see and avoid" other aircraft electronically. Using ADS-B technology, the aircraft emits signals that tell its position to other similarly equipped aircraft in the area. This information is depicted visually on a display in the aircraft cockpit and will greatly enhance a pilot's awareness of the position of other ADS-B equipped aircraft and lead to safer operations.

By reducing dependency on voice transmissions, pilot workload, presumably, will be greatly reduced. For en route applications, communications via datalink will be implemented first. Most observers agree it will probably be at least a decade before we see heavy use of digital communications in the terminal area, however. The transition from voice to digital communications will evolve fairly slowly, observers believe, in part because pilots need to become more comfortable with the idea of sending and receiving information in this format and in part because thousands of older airplanes will require new avionics to receive and send data via digital link.

Through VDL Mode-2, we will soon be able to send data to the cockpit at rates as high as 31.5 Kbps, which is fast enough for applications presently envisioned but may not be for future uses. This means avionics makers must develop new radios that can be updated through software changes rather than expensive hardware upgrades. If we have learned anything from the personal computer industry, however, it is that the new technology constantly being developed can quickly make yesterday's equipment obsolete. The challenge for aviation will lie in developing avionics that have reasonably long, useful lives.

The potential for change and innovation in the navigation and communications infrastructure is limitless; and when it finally evolves, the cockpit will be a markedly different place from what it is today.

DISPLAYS

Most observers did not imagine that the liquid-crystal display (LCD) would replace the cathode-ray tube as quickly as it did. Advances in commercial LCD production, however, helped bring costs down, while avionics makers enhanced off-the-shelf products, for unmatched levels of fidelity. Today, all new avionics suites for jets are being designed around large-format LCDs. Just how big the displays will grow depends largely on what happens in the market for commercial LCDs.

Synthetic vision and color flat-panel displays or three-dimensional holographic displays have the potential to eliminate pilot errors by providing virtual reality for the crew. As is the case in most of the commercial transport category airplanes and in virtually all general-aviation planes, the pilot must use mental gymnastics to locate him- or herself in space when flying in instrument meteorological conditions (IMC). Display technology coupled with GPS/GNSS accuracy can be the means for safety improvement in all weather conditions for all airspace users.

The future vision of avionics makers is to someday use displays that spread across the entire cockpit. Some are looking at new technologies such as gas-plasma displays or projection displays to meet this goal. For now, however, LCD is the most cost-effective way to bring the information needed for tomorrow's flight environment to the cockpit.

Using LCDs, avionics makers plan to integrate terrain, traffic, and weather information and display it on top of charts and approach plates, which will provide a seamless source for information, layered in such a way that pilots can better interpret the overall flight situation. Tomorrow's flight deck will be more dependent on sophisticated computer software that will greatly enhance the capabilities of the avionics, while at the same time adding to the complexity of systems. To help pilots better manage the technology, cursor-control devices, similar to a computer mouse, will be used, at first to scroll through simple menus and later to plan entire flights.

VOICE RECOGNITION

In just the last few years, voice recognition technologies have advanced tremendously. In fact, you probably use voice recognition in everyday life without even realizing it. For example, most telephone companies use voice recognition to look up and extract phone numbers from directory assistance databases. The quality of voice recognition today is so good that systems can understand different speech patterns with a high degree of reliability.

One way pilots may use voice recognition in the future is for radio tuning. A pilot would simply say, "Daytona Beach," and the radio would call up the correct frequency. Voice could also be used to configure aircraft systems when, after landing and turning off the active runway, the pilot commands the airplane to raise the flaps and perform other functions that, in today's cockpit, must be done manually.

Researchers are continually seeking ways to expand the range of computer-generated speech. At the Massachusetts Institute of Technology's Media Laboratory, experts in the field have been experimenting with synthetic voices since the late 1980s. They have designed computers that produce voices sounding remarkably human, which can even use varying inflections to appear happy, sad, excited, or angry. Soon, the classic robotic monotones of synthetic speech will be replaced by emphatic, easily understood voices, with subtleties in vocal stress used to convey different messages.

HEAD-UP DISPLAYS (HUDs)

HUDs will become much more commonplace in tomorrow's cockpits, experts believe, particularly if manufacturers can bring costs down. A number of HUD makers may shift their engineering focuses and begin looking at the use of smaller lenses and direct-projection HUDs, such as those used by the military in the C-17. This would eliminate the need for the large optics projectors and image collimators used today, making the systems far simpler to install and maintain.

HUD presents vital aircraft performance and navigation information to the pilot's forward field of view. The information on the HUD is collimated so that symbols appear to the human eye to be at infinity and overlaid on the actual outside scene. The use of HUD is gaining broad acceptance in aviation as flightcrews realize the tremendous benefits to situational awareness it can provide.

Also, *enhanced vision,* a system that allows pilots to see through low-visibility weather, could become a reality in the cockpit one day soon. An infrared (IR) camera, called the All-Weather Window, has been developed that is designed to extend Cat I approaches into visibilities as low as 700 RVR.

The All-Weather Window uses a cryogenically cooled (to 85 K, equivalent to −188°C) sensor that is specially sensitive to the IR radiation emitted by runway approach lights, enabling pilots to see the runway approach environment and the runway itself through rain, snow, fog, haze, and smoke.

PORTABLE COMPUTERS

Portable computers will begin to show up with greater regularity in the cockpit. A number of avionics makers have already introduced one or are developing such devices, which pilots would use to review charts and approach plates, perform weight and balance calculations, and plan flights, among other functions.

Eventually, with the inclusion of chart databases that can be viewed on the MFD and portable computers, paper may vanish from the cockpit. Of course, the transition from a paper chart that the pilot can hold in his or her hands to electronic-only format will be slow. Some companies are planning to introduce color printers for the cockpit that would be used in the event that the displays go blank or some other internal error prevents the charts from being displayed in electronic format. Printers in the cockpit will be an interim step for security purposes. Paper in the future will become almost bothersome. Paper means heads down, and once pilots realize it is not really necessary, there will be no need to have a hard copy.

FREE FLIGHT

As discussed in Chap. 8 on air traffic system technologies, in a Free Flight environment, pilots would be released from the rigid discipline of being spaced in nose-to-tail time blocks, along less-than-optimum routes, and often at inefficient altitudes.

Extended to its ultimate application, Free Flight would embrace the airplane's complete operation, from start-up at the originating airport to shutdown at the destination after having flown a direct, nondeviating course, using performance figures straight out of the airframe manufacturer's operating handbook.

To make this new era possible, operators will need to install new avionics, the most critical being based upon ADS-B. The current Mode-S transponders will eventually take on a much more sophisticated role as ADS-B is implemented, beginning around 2003. Also required for Free Flight will be VHF data/voice radios, which would be used to pass as much information as practical over controller-pilot datalinks. Voice mode will still be essential for emergencies and other nonstandard communication, but for the most part, tomorrow's skies will be much quieter.

Once the transition to a Free Flight–based traffic regime begins, observers believe airlines will begin updating the current fleet fairly rapidly. The technological challenges involved are huge, but not insurmountable, which means the future may be nearer than we realize. The goal for avionics makers, of course, is to ensure that the cockpit of the future is as intuitive to use—as pilot-friendly—as it is technologically capable and reliable.

KEY TERMS

Fly-by-wire

Extended-range twin-engine operations (ETOPS)

High-lift systems

Antiskid system

Automatic braking systems

Wing spoilers (speed brakes)

Thrust reversers

Aging aircraft

Fail-safe or damage tolerance design

Multiple-site damage (MSD)

Supplemental inspection documents (SIDs)

High-altitude clear air turbulence (HICAT)

Wind shear

Microprocessor

Ground-proximity warning system (GPWS)

Engine-indicating and crew-alerting system (EICAS)

Integrated display system (IDS)

Aircraft Communications Addressing and Reporting System (ACARS)

Flight management system (FMS)

Central maintenance computer system (CMCS)

Computational fluid dynamics (CFD)

Wind-tunnel testing

Piloted simulator

Computer-generated imagery (CGI)

Structural static tests

Flight data recorder (FDR)

Crew voice recorder (CVR)

Global Positioning System (GPS)

REVIEW QUESTIONS

1. Discuss some of the early developments in jet engine technology that were included in the Boeing B-47 and Dash 80 aircraft.

2. What was the purpose of pod-mounted engine installation?

3. What were some of the challenges to safety resulting from such radical airframe designs as highly swept wings, high wing loading, increased speeds, and long-duration flights at high altitudes?

4. What were some of the advances in high-lift systems as the industry transitioned from piston engine to jetliner operations?

5. Describe some technological improvements in the following stopping systems.

   a. Antiskid system

   b. Fuse plugs in the wheels

   c. Automatic braking systems

   d. Wing spoilers

   e. Thrust reversers

6. What is meant by a fail-safe design?

7. List and discuss the parameters that define the age of an aircraft.

8. Discuss factors that affect cabin safety and crash survival.

9. What are some of the technological advances that enhanced the flight-handling characteristics of today's generation of aircraft?

10. How did the early swept-wing designs affect stall characteristics?

11. Why was development of the electron microscope so important in determining structural integrity?

12. Who are the three major participants in the structural safety process? Describe their individual roles.

13. What were some of the early solutions to the problems of turbulence, winds, wind shear, ice and precipitation, and volcanic ash?

14. What approaches have been used in recent years to address the problem of wind shear?

15. How do ice and precipitation affect airplane operation? Describe several technologies designed to address this problem.

16. List and briefly describe at least five flight-deck technology changes that have made a significant contribution to improving safety.

17. What is the purpose of the engine-indicating and crew-alerting system (EICAS) and the Aircraft Communications Addressing and Reporting System (ACARS)?

18. What are the basic functions of the flight management system (FMS)?

19. What is computational fluid dynamics (CFD)?

20. Explain the "inverse design" technique as used in wing designs.

21. Why was a new cab design chosen for the B-757 and B-767 aircraft?

22. How does the wind tunnel complement CFD?

23. What is the purpose of the engineering simulation?

24. What are structural static tests? What advances have taken place in flight testing to provide more accurate and thorough tests?

25. What data are recorded on the FDR and CVR?

26. How have FDRs and CVRs been improved over the years?

27. Describe some of the problems and improvements needed to gather better weather information.

28. Describe at least three major features likely to be present in the flight deck of the future.

29. Discuss some of the technological advances in communications, navigation, and displays.

30. What is Free Flight? Is it worth it? Discuss.

REFERENCES

Boeing Airliner Magazine. Various dates. Selected articles. Published bimonthly. Seattle, Wash.

FAA/DOT. 2000. *Flight Standards 2000 Business Plan.* Washington, D.C.: U.S. Government Printing Office, February.

Godson, John. 1975. *The Rise and Fall of the DC-10.* New York: David McKay Company, Inc.

Green, William, Gordon Swanborough, and John Mowinski. 1987. *Modern Commercial Aircraft.* New York: Crown Publishers, Inc.

Gunston, Bill. 1980. *The Plane Makers.* Birmingham, England: Basinghall Books, Ltd.

Hall, John, and Ulf G. Goranson. 1992. "Structural Damage Tolerance of Commercial Jet Transports." FSF 45th IASS and IFA 22d International Conference. Long Beach, Calif.

Hennigs, N. E. 1990. "Aging Airplanes." *Boeing Airliner.* July/September: 17–20.

Henzler, K. O. 1991. "Aging Airplanes." *Boeing Airliner.* January/March: 21–24.

Ingells, Douglas J. 1970. *747: Story of the Boeing Super Jet.* Fallbrook, Calif.: Aero Publishers, Inc.

———. 1973. *L-1011 Tri Star and the Lockheed Story*. Fallbrook, Calif.: Aero Publishers, Inc.

National Research Council, Assembly of Engineering, Committee on Federal Aviation Administration Airworthiness Certification Procedures. 1980. *Improving Aircraft Safety: FAA Certification of Commercial Passenger Aircraft*. Washington, D.C.: National Academy of Sciences.

Norris, Guy. 1996. *Boeing 777*. Osceola, Wis.: Motor Books.

Schairer, George. 1992. "The Engineering Revolution Leading to the Boeing 707." FSF 45th IASS and IFA 22d International Conference. Long Beach, Calif.

Shaw, Robbie. 1996. *Boeing Jetliners*. Osceola, Wis: Motor Books.

Sinaiko, H. Wallace. 1961. *Selected Papers on Human Factors in the Design and Use of Control Systems*. New York: Dover Publications, Inc.

Soekkha, Hans M. (ed.). 1997. "Aviation Safety." Selected papers on flight operations. *Proceedings of the IASC-97*. Rotterdam, The Netherlands: VSP BV.

Steiner, John E. 1979. *Jet Aviation Development: One Company's Perspective*. Washington, D.C.: Smithsonian Institution Press.

Stekler, Herman O. 1965. *The Structure and Performance of the Aerospace Industry*. Berkeley, Calif.: University of California Press.

Taylor, Laurie. 1997. *Air Travel—How Safe Is It?* 2d ed. London, England: Blackwell Science, Ltd.

Taylor, Michael. 1982. *J. H. Boeing*. London, England: Janes Publishing Co., Ltd.

Taylor, Richard W. 1992. "Twin-Engine Transports—A Look at the Future." FSF 45th IASS and IFA 22d International Conference. Long Beach, Calif.

———. 1999. "Flight Safety: Management, Measurement and Margins." *Proceedings of the 11th annual EASS*. Alexandria, Va.: Flight Safety Foundation.

AIRPORT SAFETY

LEARNING OBJECTIVES

After completing this chapter, you should be able to

- Discuss the regulatory chronology of airport certification.

- Discuss the airport certification process.

- Identify and differentiate between the two types of airport certifications.

- Explain what the airport certification manual is and discuss its significance.

- Identify and locate the elements of an airport certification manual and discuss the requirements of each element.

- Locate and highlight relevant sections of Part 139 regulations and FAA Advisory Circulars.

- List and explain the required elements of an airport emergency plan.

- Identify the sources of safety hazards and the applicable regulations that govern the following:
  Airport terminal buildings
  Hangars and maintenance shops
  Ramp operations
  Aviation fuel handling
  Aircraft rescue and fire fighting (ARFF)
  Deicing

- Describe the airport surface operational environment.
- List the individual groups involved with airport surface operations.
- Define and discuss the following terms relating to runway safety: *surface incident, runway incursion, collision hazard, loss of separation, operational error, pilot deviation,* and *vehicle/pedestrian deviation.*
- Discuss runway incursions, trends, and statistics with respect to rates and severity.
- List and discuss strategies that are being proposed to minimize the occurrence of runway incursions.

INTRODUCTION

A commercial *airport,* by definition, is a tract of land (or water) that provides facilities for landing, takeoff, shelter, supply, and repair of aircraft and has a passenger terminal. This results in a diverse collection of complex operations each with its own set of safety issues. This chapter provides an overview of the certification process, the certification manual, and safety issues concerning airports. Extended discussions are conducted on some of the more hazardous operations such as fuel handling and runway incursions. Highlights of Part 139 and other relevant regulations are also reviewed.

AIRPORT CERTIFICATION

The Federal Aviation Act of 1958 was broadened in 1970 to authorize the FAA Administrator to issue operating certificates to certain categories of airports serving air carrier aircraft. This authority was established as a result of the Airport and Airway Development Act of 1970 that added a new section (612) to the 1958 Act. The 1970 Act also added to the 1958 Act a clause that barred any person from operating an airport or any air carrier from operating in an airport that did not possess a certificate if it was required to or if the airport was in violation of the terms of an issued certificate. The intent of the certification was to establish minimum safety standards for the operation of airports to ensure the safety of the flying public. To be certified by the FAA, airports are required to meet certain standards for airport design, construction, maintenance, and operations as well as fire fighting and rescue equipment, runway and taxiway guidance signs, control of vehicles, management of wildlife hazards, and record keeping. The FAA works with certificated airports to help them meet these standards through periodic consultations and site inspections. FAA regulations Part 139, adopted in 1972, govern the certification and operation of land airports, which serve any scheduled or unscheduled passenger operation of an air carrier that is conducted with an aircraft having a seating capacity of more than 30 passengers or that the FAA Administrator requires to have a certificate. This part does not apply to airports at which air carrier passenger operations are conducted only by reason of the airport being designated as an alternate airport. An airport that meets the grade is issued an Airport Operating Certificate. To obtain a certificate, an airport must maintain operational and safety standards under categories mentioned earlier. These requirements vary by the size of the airport and by the type of flights available. The FAA is granted the flexibility to issue certain exemptions to airports that serve few passengers annually and for requirements that might create economic hardships. The FAA provides guidance on meeting the requirements of Part 139 certification through documents called *Advisory Circulars* (*ACs*). The FAA may also issue *CertAlerts* through the Airport Safety and Operations Division of the Office of Airport Safety and Standards that serve as a quick means of providing both the public and FAA staff with important certification-related guidance.

The FAA issues two types of Airport Operating Certificates—an *Airport Operating Certificate* (*AOC*) and a *Limited Airport Operating Certificate* (*LAOC*). An Airport Operating Certificate is issued under Part 139 for an airport serving scheduled air carrier operations with aircraft having a seating capacity of more than 30 passengers. Fully certificated airports must maintain an *Airport Certification Manual* (*ACM*) that details operating procedures, facilities, and equipment and other appropriate information. A Limited Airport Operating Certificate is issued under Part 139 for an airport serving unscheduled air carrier operations with aircraft having a seating capacity of more than 30 passengers. Limited certificated airports must maintain an Airport Certification Specification (ACS) that details operating procedures, facilities, equipment, and other appropriate information. The ACS is similar in content to the ACM but has a lesser scope due to the nature of the limited certificate. In the United States, there are about 436 airports that hold Airport Operating Certificates, 127 airports that hold Limited Airport Operating Certificates, and 83 military airports that hold Airport Operating Certificates under an FAA exemption.

AIRPORT CERTIFICATION MANUAL

As mentioned above, every certificated airport that serves air carriers is required to have an Airport Certification Manual in accordance with Part 139. The ACM is a working document that outlines the means and procedures used to comply with the requirements of Part 139. While each airport has its own unique features and operational requirements, its Airport Certification Manual will contain the same basic elements, which include the following:

1. Airport accountability structure together with a listing of key personnel and their responsibilities
2. Airport grid map and a description of the airport layout, maps, and charts
3. Exemptions to Part 139 or any restrictions issued by the FAA
4. Identification system for runways, taxiways, aprons, and other movement areas

5. Runway and taxiway lighting and markings together with hold positions

6. Outlining aircraft movement areas

7. Procedures for using, storing, and disposing of hazardous materials

8. Allocating and maintaining safety areas

9. Methods for maintaining traffic and wind indicators

10. Procedures for maintaining lighting and lighting systems, and paved and unpaved areas

11. Procedures and coordination issues during construction projects

12. Self-inspection programs together with procedures to address corrective actions

13. Protection of NAVAIDS

14. Reporting of airport conditions

15. Location of obstructions to air navigation

16. Security and public protection

17. Crash, fire, rescue, and fire fighting procedures and equipment servicing

18. Control of ground vehicle traffic

19. Snow and ice removal

20. Bird and wildlife management plan

21. Airport Emergency Plan

The Airport Emergency Plan is an integral part of the ACM although it may be published and distributed separately to each user or airport tenant. The plan must be extremely detailed and provide sufficient guidance on response to all emergencies and abnormal conditions that the airport is likely to encounter. Emergencies include aircraft accidents; bomb threats; sabotage; hijackings; major fires; natural disasters such as floods, tornadoes, and earthquakes; and power failures. Emergency response plans should spell out procedures for coordinating internally with airport and fire fighting personnel and the control tower. In addition, procedures should be established for coordination with external agencies such as law enforcement, rescue squads, fire departments, ambulances and emergency transportation, and hospital and medical facilities. The Airport Emergency Plan must be reviewed at least annually, and there must be a full-scale exercise of the plan every 3 years.

COMPLIANCE WITH PART 139

If the FAA determines that an airport is not meeting its obligations, the FAA can impose a daily cumulative fine for each day the airport continues to violate any requirement of Part 139. In extreme cases, the FAA could revoke the airport's certificate or limit the areas of an airport where air carriers can land or take off.

OPERATIONAL SAFETY

As mentioned earlier, airport operations are complex and diverse, with hazards and their severity varying by the type of operation. For the purposes of our discussion, airport safety issues will be discussed under the following headings:

- Airport terminal buildings
- Hangars and maintenance shops
- Ramp operations
- Specialized services

  Aviation fuel handling

  Aircraft rescue and fire fighting (ARFF)

  Deicing
- Runway incursions

AIRPORT TERMINAL BUILDINGS

These facilities are laid out to minimize passenger delays and maximize throughput of travelers from the airport terminal entry point to the aircraft. Inadequate layout and facility size leads to overcrowding which can cause dangerous conditions that lead to injuries. Due consideration to safety should be given during construction and modification of the facilities so as to minimize hazards that cause falls and to protect people from harmful contact with machinery. Most of the design standards and requirements for building materials, fire protection, and building egress are covered by building codes (e.g., *Standard Building Code* and the *Uniform Building Code*); the *National Fire Protection Association* (NFPA); the *American National Standards Institute* (ANSI); and other local, state, and federal agencies. For operational issues, OSHA standards (which incorporate many of the standards from the above-mentioned organizations) should be adhered to. A summary of important safety considerations for airport terminal facilities follows:

- Emergency evacuation and egress routes, signs, and other marking should conform to the Life Safety Code, NFPA 101, and OSHA regulations 29 CFR Part 1910, Subpart E, "Exit Routes, Emergency Action Plans, and Fire Prevention Plans." Some additional notes are that
  - Exits should be well lighted and marked.
  - Panic hardware should be installed and properly working on the inside of exterior doors used for emergency exits.
  - Exits doors should be kept unobstructed and unlocked in the direction of egress.
  - Emergency evacuation plans should be posted in all buildings.
- All areas must be accessible to persons with disabilities (more information on this topic can be found in Advisory Circular 150/5360-14).

- All slip and trip hazards must be eliminated, and fall prevention strategies should be implemented on stairways, escalators, and moving walkways.

- Stairs, ladders, and ramps should be kept clean and free of obstacles or slippery substances and should be maintained in good condition. Both interior and exterior stairways should be sufficiently lighted so that all treads and landings are clearly visible. Prevention of falls, slips, and trips on any walking and working surface is regulated under 29 CFR Part 1910 Subpart D.

- There must be safe access to heating, ventilation, and air-conditioning systems for maintenance to permit safe and convenient replacement of heavy equipment (e.g., pumps and motors). In this connection a safe means for maintaining luminaries and windows must be provided.

- Skylight and roof openings must be provided with protective screens to prevent individuals from falling through.

- Doors should open automatically to eliminate inconvenience to people with baggage and to the handicapped. Also, doors should open away from the person going through.

- Escalators, elevators, and moving walks should comply with ANSI Standard A17.1-1996. Speeds for moving walkways and escalators should be set at the lower limit of the acceptable range.

- Moving walkways should have emergency stop buttons at both ends and, wherever feasible, every 50 meters apart. Attention-grabbing techniques should be installed to warn passengers that they are approaching the end of a moving walkway.

- Ticket counters should be ergonomically designed to permit safe body postures during baggage handling and data entry. Weight scales and postweigh-in baggage conveyors should be installed as close to floor level as practical to minimize lifting, and entrances to the scales should be left free so as to permit baggage sliding as opposed to lifting. OSHA regulates ergonomic hazards (or any other recognized hazard that does not have a specific regulation) under the *General Duty Clause* listed in Section 5(a) of the OSHA Act.

- All electrical systems and power service should be installed, maintained, and inspected according to the National Electrical Code, NFPA 70. OSHA regulates electrical practices under 29 CFR Part 1910 Subpart S.

- Fire extinguishers and sprinkler systems should be installed and maintained in accordance with NFPA 10, and portable fire extinguishers should conform in type and number to NFPA 13. OSHA addresses fire prevention under 29 CFR Part 1910 Subpart L.

- Maintenance and other operational issues:
  - Roofs should be inspected regularly, especially after violent storms.
  - Maximum allowable loading per square foot of floors should be prominently displayed in all storage areas to avoid damage or collapse.
  - Floors and stairways should not be waxed if it creates a slipping hazard.

- Planning and design guidance for airport terminals is included in Advisory Circular (AC) 150/5360-13 and AC 150/5360-9 for nonhub locations.

HANGARS AND MAINTENANCE SHOPS

Construction of hangars is covered by federal, state, and local building codes as well as many National Fire Protection Association standards. Operational issues are mostly covered by OSHA and EPA requirements. Some of the highlights include the following:

WALKWAYS AND WORKING SURFACES

- Adequate aisle space and access to fire extinguishers, hoses, sprinkler control valves, and fire alarm stations should be maintained.

- If work is performed at elevations in excess of 4 feet, the worker must be protected from falling by providing guardrails 42 inches high with an intermediate rail approximately halfway between the top rail and the standing surface. If guardrails are not feasible, an equivalent effective means of fall protection (safety harnesses, nets, etc.) must be provided. Regardless of height, if a work platform is adjacent to a hazard (e.g., dangerous equipment), then the worker must be protected from falling. Toe boards should be provided for the work platform on the open sides to prevent objects from falling on persons passing beneath or if there is moving machinery or if there is equipment with which falling material could create a hazard.

ELECTRICAL

- Large metal complexes such as aircraft docks should be permanently bonded and grounded and secured to a permanent structure such as a hangar floor. Primary sources of electric power to hangars should be approved for the location according to the National Electrical Code, NFPA 70, and aircraft hangars, NFPA 409.

- All other electrical operational issues such as grounding, wiring, receptacles, and circuit breakers are covered by 29 CFR Part 1910 Subpart S.

COMPRESSED GAS, FLAMMABLES, AND HAZARDOUS AND TOXIC SUBSTANCES. Storing and handling of compressed gases, and flammable and combustible liquids are addressed under 29 CFR 1910 Subpart H, Hazardous Chemicals. Many of these standards reference or incorporate those developed by the NFPA and the Compressed Gas Association. Toxic and hazardous substances are addressed under 29 CFR Part 1910 Subpart Z.

- 29 CFR 1910.101(b) addresses the in-facility handling, storage, and utilization of all compressed gas cylinders by referencing procedures specified in the Compressed Gas Association Pamphlet P-1-1965. Compressed gas cylinders should be secured and stored upright when not in use and be stored in carts designed for that purpose when in use. Gas cylinders should have their caps on except when in use. Cylinders of compatible gases should be stored together, and oxygen cylinders should be stored separately from flammable gas cylinders and flammable or combustible substances.

- General-purpose compressed air systems should not exceed 30 pounds per square inch unless specifically required (e.g., for tire inflation and special-purpose pneumatic tools). The practice of using compressed air to clean personal clothing should be avoided.

- All flammable and combustible materials (paints, hydraulic fluids, greases, solvents, cleaning agents, etc.) should be stored and used in accordance with NFPA 30. Information on the use and handling of specific materials and processes can be found in 29 CFR Part 1910 Subpart H and 29 CFR Part 1910 Subpart Z, Toxic and Hazardous Substances.

- Storage of flammables should be limited to quantities that are permitted for immediate short-duration uses in the shop areas. These chemicals should be stored in labeled cabinets designed and approved for flammable storage. Bulk storage of flammables should be in separate buildings designed for that propose that include design features such as explosion-proof electrical systems, spill containment, and automatic fire protection. Storage quantity requirements for shop and bulk storage can be found in 29 CFR 1910.106.

- Painting and stripping are hazardous activities that use toxic chemicals and therefore must be performed in well-ventilated areas that are separate from other maintenance activities and are designed for that purpose (e.g., paint booths). In addition, painting and stripping generate toxic wastes that must be contained and either recycled or disposed of appropriately. Therefore, these operations are governed by both OSHA and EPA regulations.

- Miscellaneous on-site use of all chemicals is covered under OSHA's Hazard Communication Standard, 29 CFR Part 1910 Subpart 1200. This standard comprehensively addresses the issues of chemical hazard evaluation, employee hazard communications, and employee personal protective equipment and procedures. In summary, this subpart requires:
  - Developing and maintaining a written hazard communication program for the workplace, including lists of hazardous chemicals present on site
  - Labeling of containers of chemicals in the workplace and containers of chemicals being shipped to other workplaces
  - Preparing and distributing material safety data sheets (MSDS) to employees
  - Developing and implementing employee training programs that detail the hazards of chemicals on site and what employees must do to protect themselves

MISCELLANEOUS

- The control of hazardous energy (lockout and tag-out) standard (29 CFR 1910.147) covers the servicing and maintenance of machines and equipment in which the unexpected energization or start-up of machines or equipment, or release of stored energy (electrical, hydraulic, pneumatic, or gravity), could cause injury to employees. This standard requires maintenance personnel, through written equipment-specific procedures, to deenergize stored energy and positively block off all potential energy sources to systems that are being worked on by using personalized locks and tags.

- Material handling equipment such as overhead and gantry cranes that are used to move large aircraft sections are regulated under 29 CFR 1910.179. Slings, chains, and hoists used to move or lift smaller aircraft parts and sections (e.g., aircraft jacking) are regulated under 29 CFR 1910.184.

- The use of powered material handling equipment, such as fork trucks and hand trucks, and the use and maintenance of batteries used to power these and other equipment are regulated under 29 CFR 1901.178.

- Hoist units such as those used to reach large aircraft tails should comply with safety requirements listed under 29 CFR Part 1910 Subpart F, Powered Platforms, Manlifts, and Vehicle-Mounted Work Platforms.

- Heating and ventilation systems in hangars should be in accordance with NFPA 409 or the equivalent.

- Welding, cutting, brazing, and other hot work that has the potential to start fires is regulated under 29 CFR Part 1910 Subpart Q.

- Noise evaluation and hearing protection for processes such as riveting, sand blasting, grinding, polishing, and other noisy operations that exceed 85 decibels (time-weighted A scale for 8 hours) are regulated under 29 CFR 1910.95.

- The safe use of pedestal grinders and other grinding wheels is regulated under 29 CFR 1910.215, Abrasive Wheel Machinery.

RAMP OPERATIONS

The ramp area is generally designed for aircraft, not the vehicles that service and/or operate in the proximity of the aircraft. Most of the signs and markings are for aircraft. The ramp area sees a diverse collection of high-paced activities that involve aircraft, vehicles, and individuals working in close proximity to one another. Some of these activities include

- Aircraft ground handling that may include taxiing, towing, chocking, parking, or tie-down
- Aircraft refueling
- Aircraft servicing—catering, cleaning, food service, etc.
- Baggage and cargo handling
- Power supply
- Routine checks and maintenance

Individuals involved with the above activities are exposed to several of the occupational hazards (and hence regulated by similar regulations) mentioned earlier, including

- Cuts (from antennas, pitot tubes, static discharge wicks)
- Slips, trips, and falls (on the ground and from elevations)
- Strains and sprains (from baggage handling)
- Exposure to hazardous materials (fuel)
- Contact with moving parts (propellers) and bumps (undersurface of fuselage)
- Electrical hazards (tools, motor, generators)—electrical wiring should conform to National Electric Code, NFPA 70, for class I, Group D, Division I hazardous locations
- Biohazards (bloodborne exposures during cleaning, covered under 29 CFR 1910.1030)
- High-pressure air and other fluid exposures from pressurized systems
- Noise from engines and other equipment

In addition, there is a potential for significant injury from exposure to jet blast, moving aircraft, and moving vehicles. Since most of the operations take place outdoors, weather conditions (heat, cold, snow, rain, ice, and wind) can increase the risk of injury. Hazards due to aircraft refueling are discussed in the next section.

SPECIALIZED SERVICES

These comprise

- Aviation fuel handling
- Aircraft rescue and fire fighting (ARFF)
- Deicing

AVIATION FUEL HANDLING.   Fuel handling is an important safety issue not only to the fuelers, but also to other airport personnel, the traveling public, and the operation of the aircraft. Failure to adhere to safe operating procedures when fueling aircraft and/or transporting fuel from one location to another on the airport can result in major disasters. This potential for disaster is well recognized as millions of gallons of aircraft fuel are handled each year without major incidents for the most part. The basics types of aviation fuel are aviation gasoline (avgas), jet A or jet B (commercial jet uses) fuels, and JP series (military jet uses) fuels. Guidance on fuel handling in an airport environment is covered under 14 CFR Part 139.321(b) and the FAA advisory circulars in the 150 series (e.g., AC 150/5230-4), all of which reflect NFPA standards (Aircraft Fuel Servicing, NFPA 407, 1997, and Aircraft Fueling

Ramp Drainage, NFPA 415, 1997). It is also necessary that fueling operations be conducted in accordance with procurement document specifications and all other applicable local, state, and federal regulations. Special attention must be paid to federal, state, and local hazardous materials regulations and to agency-specific fuel spill avoidance requirements. The major concerns with fuel handling are

- Health hazards to fuelers
- Fuel contamination
- Explosions and fires during fueling or fuel transfer
- Explosions and fires during fuel tank repair
- Hazards from spills

HEALTH HAZARDS TO FUELERS. Repeated contact with aviation fuels can cause skin irritation and dermatitis. Skin area exposures to fuels should be washed immediately with soap and water. Eyewash stations should be conveniently located for immediate flushing in case the eyes or face comes into contact with fuel. Fuels such as avgas contain additives such as benzene. Hence, fuels can be toxic if inhaled or swallowed. Fuels can also affect the central nervous system by causing narcosis, which leaves an individual in a state of stupor or unconsciousness. Threshold limits for fuel vapors in the breathing zones are established by the *American Conference of Governmental Industrial Hygienists (ACGIH)* and 29 CFR 1910 Subpart Z (which reflects ACGIH standards).

FUEL CONTAMINATION. Fuel contamination is a major safety issue as it can affect the operation of the aircraft. Contamination can occur by refueling an aircraft with contaminated fuel or the wrong fuel grade or type. Fuel is considered contaminated if it contains any material other than what is called for in the specifications. The two most common contaminants found in fuels are rust and water (free or emulsified). Other contaminants found include microbial growths, paint, metal, rubber, and lint. Commingling, or the mixing of different fuel grades and types, is also a serious issue. For example, mixing avgas with jet fuels can reduce the volatility and antiknock features of fuels needed for reciprocating engines, which can result in engine failure. This could also lead to lead deposits in turbine engines. To prevent commingling, Advisory Circular 150/5230-4 requires the use of separate pumps, lines, standard couplings, and color codes for each type of fuel. Also, all fuel-dispensing systems should be clearly marked and labeled to indicate the grade and type of fuel they contain.

EXPLOSIONS AND FIRES DURING FUELING OR FUEL TRANSFER. There are three essential requirements for a fuel to burn. First, the fuel must be in its vapor form; second, it must develop the right fuel-to-air ratio with the surrounding atmosphere; and third, there must be a source of ignition to start the fire. In fuel handling, the first two conditions invariably exist, and there is very little one can do to completely eliminate these hazardous states. Hence, the primary objective during fuel handling is to control or eliminate ignition sources. Several ignition sources and their control are discussed below.

Static electricity.   Static charge discharge is one of the most important sources of ignition and is a constant threat to safe fueling. When two dissimilar materials come in contact with each other, a static charge of electricity develops by the exchange of positive and negative charges across the contact surfaces. Static charges build up when fuel gets pumped through the fuel lines. The quantity of static electricity produced is directly proportional to the fuel flow rate. Also, due to their lower volatility, jet fuels are generally more susceptible to ignition from static discharges than are aviation fuels. Static electricity also gets generated when fuel is allowed to free-fall through the air, as when a tank is filled through a spout. An electrically charged atmosphere can build up charges on an aircraft, as can rain, snow, or dust blowing across the aircraft.

While completely eliminating static charge is impractical, several steps can be taken to minimize its buildup. Some of these steps are as follows:

- Follow correct *grounding* and *bonding* procedures, as listed in AC 00-34, "Aircraft Ground Handling and Servicing," 1972, and reproduced below.
  Fueling from a Tanker
  - Connect grounding cable from truck to ground.
  - Connect grounding cable from ground to aircraft.
  - Bond vehicle to aircraft, either with "Y" or separate bonding cable.
  - Bond fuel nozzle to aircraft before opening aircraft filler cap.
  - Disconnect grounding cables in reverse order on completion.
  Fueling from Hydrant
  - Connect grounding cable from dispenser to aircraft.
  - For over-wing refueling, bond grounding cable from hose nozzle to aircraft before removing aircraft tank cap.
  - For under-wing fueling, discharge electricity by touching nozzle to aircraft prior to connecting. Ground wire is not necessary.
- Avoid pumping contaminated fuels since dirt and water particles in the fuel will add to the static charge accumulation.
- Reduce fuel flow rates to decrease turbulence, where feasible. Reduced flow rates allow longer periods for the static charge to dissipate.
- Prevent fuel from free falling through the air to the bottom of the tank.

Sparks.   Arcing may occur when electrical connections are made or from faulty equipment and improperly maintained ramp vehicles. To avoid sparks, the following precautions should be taken:

- Aircraft batteries should not be installed or removed, nor should battery chargers be operated during fueling operations.
- Aircraft ground power units should be located as far away as possible from the fueling points.

- Battery-powered tools as opposed to electrically powered tools should be used near aircraft during fueling.

- Aircraft radios and radars should be shut off. Also, fueling should not be performed within 100 feet of energized airborne radar.

- Flashings used near the fueling area should be UL-approved for use in such locations.

- Workers near the fueling area should be advised against wearing footwear with metal nails or cleats or metal plates on the heels.

Miscellaneous.   Other issues to be considered for minimizing ignition sources include these:

- No-smoking rules should be strictly enforced.

- Employees should not be allowed to carry matches, lighters, and other smoking paraphernalia when engaged in fueling operations.

- Welding, cutting, or other hot work should not be conducted within at least 35 feet (preferably 50 feet) of fueling. Similar distance restrictions should be maintained for open-flame lights and exposed-flame heaters.

- Fuel pits should be located at least 50 feet away from a terminal building or concourse.

EXPLOSIONS AND FIRES DURING FUEL TANK REPAIR.   Maintenance work on aircraft fuel cells and tanks, and bulk fuel storage tanks present similar inhalation and fire hazards as described above. It is extremely important that cells and tanks be drained and thoroughly vented before any maintenance work is attempted, especially hot work such as welding, brazing, and cutting. If tanks are large enough for an employee to enter and perform work, then confined-space testing, entry, and rescue procedures should be adhered to as required by 29 CFR 1910.146. Positive pressure should be employed for venting and should be undertaken with an approved blower (with non-sparking blades and an isolated nonsparking blower motor) using a grounded hose.

HAZARDS FROM SPILLS.   Fuel spills present varying degrees of safety issues depending on the size of the spill and the kind of response required. While simple flushing may be appropriate for small spills, larger spills may require the use of absorbents, adsorbents, and chemical fixants. Leaks from underground storage tanks (USTs) present a different set of safety issues. Spills and USTs are regulated by EPA (as mentioned in earlier chapters) and local and state jurisdictions. Also, depending on the response planned by the facility, OSHA regulations 29 CFR 1910.120 could apply. Some of the issues to consider in connection with fuel spills include

- Leak/spill prevention
- Emergency response procedures

- Reporting and notifications
- Spill control/containment
- Cleanup procedures
- Employee training

AIRCRAFT RESCUE AND FIRE FIGHTING (**ARFF**).   Responding to aircraft crashes is a high-risk job leaving individuals exposed to a wide variety of safety hazards such as dealing with fire, smoke, chemicals, blood, and other body fluids and lifting in awkward postures. AARF training, therefore, should include varying levels of OSHA training depending on the level of responder involvement. To develop an understanding of the safety issues involved with aircraft rescue and fire fighting, a review of ARRF is in order.

As mentioned earlier, all certificated airports must have a plan to respond to all conceivable emergencies. The airport emergency plan is an integral part of the Airport Certification Manual. A major emergency that an airport has to plan for is aircraft crashes, which are handled under the aircraft rescue and fire fighting plan. This plan is based on the largest aircraft that is likely to be serviced by the airport. Detailed requirements and procedures for ARFF equipment and agents are contained in

- Part 139.315 (Aircraft rescue and fire fighting: Index determination)
- Part 139.317 (Aircraft rescue and fire fighting: Equipment and agents)
- Part 139.319 (Aircraft rescue and fire fighting: Operational requirements)
- FAA Advisory Circulars in the 150 series

Aircraft rescue services and fire fighting equipment are based on an index as determined by the length of the air carrier aircraft and the frequency of its departure. If there are five or more average daily departures of air carrier aircraft in a single index group serving a given airport, the longest index group with an average of five or more daily departures will be the established index for the airport. If there are less than five average daily departures of air carrier aircraft in a single index group serving that airport, the next-lower index from the longest index group with air carrier aircraft in it will be the index established for the airport. The minimum designated index for a certificated airport is index A. Highlights of index classifications and other ARFF requirements are summarized below.

INDEX CLASSIFICATIONS

INDEX A

Aircraft length: Less than 90 feet

Requirements: One vehicle carrying at least

1. 500 pounds of sodium-based dry chemical or halon 1211; OR

2. 450 pounds of potassium-based dry chemical and water with a commensurate quantity of aqueous film-forming foam (AFFF) to total 100 gallons, for simultaneous dry chemical and AFFF foam application

INDEX B

Aircraft length: At least 90 feet but less than 126 feet

Requirements: Either of the following:

1. One vehicle carrying at least 500 pounds of sodium-based dry chemical or halon 1211, and 1500 gallons of water, and the commensurate quantity of AFFF for foam production; OR
2. Two vehicles:
    a. One vehicle carrying the extinguishing agents as listed above in Index A(1) or Index A(2); AND
    b. One vehicle carrying an amount of water and the commensurate quantity of AFFF so that the total quantity of water for foam production carried by both vehicles is at least 1500 gallons.

INDEX C

Aircraft length: At least 126 feet but less than 159 feet

Requirements: Either of the following:

1. Three vehicles:
    a. One vehicle carrying the extinguishing agents as listed above in Index A(1) or A(2); AND
    b. Two vehicles carrying an amount of water and the commensurate quantity of AFFF so that the total quantity of water for foam production carried by all three vehicles is at least 3000 gallons; OR
2. Two vehicles:
    a. One vehicle carrying the extinguishing agents as listed above in Index B(1); AND
    b. One vehicle carrying water and the commensurate quantity of AFFF so that the total quantity of water for foam production carried by both vehicles is at least 3000 gallons.

INDEX D

Aircraft length: At least 159 feet but less than 200 feet

Requirements: Three vehicles:

1. One vehicle carrying the extinguishing agents as listed above in Index A(1) or Index A(2); AND

2. Two vehicles carrying an amount of water and the commensurate quantity of AFFF so that the total quantity of water for foam production carried by all three vehicles is at least 4000 gallons.

INDEX E

Aircraft length: At least 200 feet

Requirements: Three vehicles:

1. One vehicle carrying the extinguishing agents as listed above in Index A(1) or Index A(2); AND
2. Two vehicles carrying an amount of water and the commensurate quantity of AFFF so that the total quantity of water for foam production carried by all three vehicles is at least 6000 gallons.

OTHER ARFF REQUIREMENTS.   Some of the operations requirements listed in Part 139.319 include

- Foam discharge capacities
- Equipment requirements and capability
- Extinguishing agent substitutions
- Change of status in classification indexes
- Vehicle communications
- Vehicle marking and lighting
- Vehicle readiness
- Response and drill requirements
- Personnel selection and qualification requirements
- Training requirements

DEICING.   Ice and snow on control, airfoil, and sensor surfaces can have serious repercussions on the operation of the aircraft. Hence, when freezing or near-freezing conditions are likely to be present, the aircraft should not be allowed to take off before being sprayed with deicing fluid. Deicing presents five distinct hazards:

- Damage to the aircraft by the deicing equipment
- Application of deicing fluid to areas where it should not be applied, such as static ports, pitot heads, angle-of-attack sensors, the engine, and other inlets
- Hazards (inhalation and ingestion) to the deicing crew and to the environment, since deicing fluids are considered hazardous materials and are moderately toxic

- Hazards to the crew and passengers if the cabin air intakes are not shut off during deicing
- Falls, slips, and other safety hazards during the application process

Deicing crews should have training as required by OSHA's Hazard Communication standard (29 CFR 1910.1200). Also, Material Safety Data Sheets (MSDS), which list health hazards and protection information on the deicing fluids being used, should be made readily available. Vapors or aerosols of deicing fluids can cause nose and throat irritation, headaches, vomiting, and dizziness. For high-volume applications it is recommended that personnel be fitted with respirators in accordance with 29 CFR 1910.134. Ingestion of deicing fluids can affect the kidneys and the central nervous system. However, these fluids are so highly diluted with water that large quantities would need to be consumed to be lethal. Application crews usually work on elevated platforms (baskets at the end of booms) and would therefore have to be adequately protected from falling through safety harnesses and guardrails. Finally, deicing spent fluid is a regulated waste and therefore should be used, collected, and recovered/disposed of in accordance with EPA and other federal and local regulations.

RUNWAY INCURSIONS

BACKGROUND. The prospect of two planes colliding is horrifying. It conjures an image of twisted, flaming wreckage falling from the sky. And yet such catastrophic collisions can easily occur on the ground when the safe, coordinated use of runways breaks down. The worst aviation disaster on record was, in fact, the result of two 747s colliding on a runway at Tenerife Airport in the Spanish Canary Islands. It was a foggy night in 1977 when the captain of one of the 747s failed to obey a "hold in position" instruction from air traffic control and started his takeoff roll before the other 747 cleared the runway. This catastrophic incident resulted in the loss of 583 lives (Landsberg, 1995).

Over a three-decade period spanning 1970 to 2000, there have been 14 catastrophic runway incidents at towered airports that have resulted in at least one fatality (Wenham, 1996; Flight International, 2000). One study (Barnett et al., 2001) predicts that over the next 20 years, runway collisions at towered U.S. airports could kill as many as 700 to 800 people, resulting in more U.S. domestic jet deaths than all other causes combined. Serious injuries were predicted at about 200. With runway incursions averaging about one a day, the opportunities for a catastrophic incident are high, which makes this issue of preventing runway incursions one of the FAA's top priorities. Runway incursion prevention has also been on the National Transportation Safety Board's list of "Most Wanted Transportation Safety Improvements" for the last decade.

The U.S. National Airspace System (NAS), through its approximately 460 towered airports, handles more than 180,000 takeoffs and landings each day. This volume makes it the largest and busiest aviation transportation system in the world. Safe and efficient handling of this system requires the dedicated effort of more than 600,000 certificated pilots, 15,000 air traffic controllers, and hundreds of thousands of airport personnel. Despite its complexity, this system of interacting humans, procedures, and technologies is also the safest system in the world, mainly through the extreme priority placed upon safety by the FAA, the aviation community, and other regulatory bodies. Despite its enviable safety record, there is increasing evidence that safety may be compromised by events on the airport surface known as runway incursions. To provide a better understanding of the problem of runway incursions, terminology and definitions as used by the FAA to describe aspects of runway safety are now reviewed.

DEFINITIONS

A *surface incident* is an event during which authorized or unauthorized/unapproved movement occurs within the movement area or an occurrence in the movement area associated with the operation of an aircraft that affects or could affect the safety of flight.

A *runway incursion* is any occurrence at an airport involving an aircraft, vehicle, person, or object on the ground that creates a collision hazard or results in a loss of separation with an aircraft that is taking off, intending to take off, landing, or intending to land.

A *collision hazard* is any condition, event, or circumstance that could induce an occurrence of a collision or surface accident or incident (e.g., a pilot takes an unplanned or evasive action to avoid an aircraft, vehicle, object, or person on the runway).

A *loss of separation* is an occurrence or operation that results in less than the prescribed separation between aircraft, or between an aircraft and a vehicle, pedestrian, or object.

All runway incursions are surface incidents, but not all surface incidents are runway incursions. To qualify as a runway incursion, an aircraft that is taking off, intending to take off, landing, or intending to land must encounter both of the following conditions:

1. At least one aircraft, vehicle, pedestrian, or object must be on the runway.
2. A collision hazard or a loss of separation must occur.

Runway incursions are classified into the following four categories.

1. *Operational error* (OE). This is a failure of the air traffic control system that results in a loss of separation as defined in Federal Aviation Administration Order 7210.56A.
2. *Pilot deviation* (PD). This is the action of a pilot that results in violation of Title 14 of the Code of Federal Regulations or a Federal Aviation Regulation.

3. *Vehicle/pedestrian deviation* (*VPD*). This is any entry or movement on the movement area by a vehicle (including aircraft operated by nonpilots) or pedestrian that has not been authorized by air traffic control.

4. *Miscellaneous.* This covers a situation that occurs and cannot be attributed to any of the three categories above (e.g., equipment failure).

AIRPORT SURFACE ENVIRONMENT.   The typical airport surface environment is a complex system of markings, lighting, and signage coupled with layouts that vary by airport. Under adverse weather conditions, the complexity of the environment increases as visibility diminishes and visual cues get concealed. Cryptic signs and surface markings together with inexperienced pilots pose additional difficulties as does nighttime flying where airport lights blend with background city lights. It is within this environment that large numbers of individuals with varying levels of experience, training, and language proficiency must concentrate on performing their tasks against a backdrop of intense radio communications for coordinated actions and procedures that are needed for smooth and safe operations—all this while interfacing with large numbers of airplanes and ground vehicles of differing makes and models that are in close proximity to one another. Given the complexity and intensity of the operation, it is easy to understand why even well-trained, highly conscientious individuals remain vulnerable to error, especially when faced with unusual or unexpected situations. Runway incursions are invariably the result of human error. With expected growth in air traffic over the next several years likely to exceed planned capacity requirements, even if the incursion rate is reduced or held steady, the absolute number of these events could increase over time.

Since runway incursions result from human error, it would be beneficial to review some of the jobs and tasks involved with conducting airport operations so that one can get a better understanding of which task elements have the greatest potential to cause errors. Airport ground operations are managed by air traffic controllers, pilots, and airport personnel. These groups jointly manage aircraft and vehicular traffic, and monitor and maintain conditions of runways, taxiways, aprons, signs, markings, and lighting. Controllers serve as the hub for the entire operation. Issuing clearances, instructions, and advisories to pilots and other ground vehicle operators, controllers guide all traffic that operates on the aprons, taxiways, and runways. Controllers monitor the surface conditions together with the identity, location, and movement of aircraft and other vehicles by looking out through the tower windows. Pilots for their part use airport layout maps, and surface taxiway/runway signs, markings, and lighting that they observe from the cockpit window, to guide them from the runways to the gates and back. Pilots and controllers rely primarily on visual feedback to maintain situational awareness and separation. Vehicle operators manage baggage, fuel, catering, and miscellaneous ground handling functions. They are required to follow controllers' instructions to maneuver, especially if they have to cross runways. A sampling of vehicles required at most airports includes

• Fuel trucks for aircraft refueling

- FAA vehicles for monitoring and maintaining navigation equipment, radars, visibility-measuring equipment, and certain lighting systems
- Airport authority vehicles to check for foreign objects and monitor runway conditions
- Maintenance vehicles for maintenance tasks
- Snowplows and mowers for maintaining runways and surrounding areas
- Baggage, catering, and utility trucks for flight service
- Emergency vehicles for crash and fire rescue
- Construction vehicles for maintaining airport surfaces or new construction

As can be seen from the preceding discussion, there is ample opportunity for error given the significant combination of tasks, procedures, and equipment that must be coordinated to permit the operation of the system.

INCURSION TOTALS AND RATES. Table 10-1 lists the number of runway incursions by various categories together with the overall runway incursion rate per 100,000 airport operations or flights from 1992 to 2001 (FAA, 2001). Over this 10-year period, runway incursions have increased by 164 incidents, or 75 percent. If one were to factor in the number of airport operations, this represents an increase of 69 percent in the runway incursion rate. While there appeared to be a decline in incursions in the early 1990s, the incursion rate began to gradually increase from 1995 with a surge of 34 percent in 2000 over that of 1999. The rate dropped, however, by 8 percent in 2001, and the total number of incursions for the first 6 months in 2002 is about 20 percent lower than that for the same period in 2001. It appears (as of this writing) that the initiatives being put in place (see control strategies later) to control runway incursions are working.

If one were to consider incursions attributed to error type, pilot deviations surged 132 percent over the 10-year period, more than any other type of incursion. *Vehicle/pedestrian deviations* recorded a 119 percent increase, and operational errors, associated with air traffic controllers, decreased by 2 percent during this same period. Of the 2872 runway incursions recorded over the 10-year period, the 1483 pilot deviations accounted for 51 percent, operational errors accounted for 28 percent, and vehicle/pedestrian deviations accounted for about 20 percent of the total incursions.

INCURSION SEVERITY. The FAA commissioned a study to analyze runway incursions based on severity. The study covered a 4-year period from 1997 to 2000. Four severity categories were selected for classifying runway incursions. Category A represented the highest risk of collision or accidents, category B represented a major or significant potential for collision, category C represented a minor potential for collision, and category D represented little or no chance for collision. Classifications of incursions into these categories were based on five operational dimensions of human reaction time, the need for corrective actions to avoid collision, environmental conditions, aircraft and/or vehicle speed and proximity (distance) between aircraft and vehicles. The study

TABLE 10-1 Runway Incursion Types, Totals, and Rates with Airport Operations

Calendar Year	Operational Error (OE)	Pilot Deviation (PD)	Vehicle/ Pedestrian Deviation (VPD)	Total Runway Incursions	Towered Airport Operations	Incursion Rate for Total Incursions
1992	90	92	37	219	63,018,680	0.35*
1993	74	84	28	186	61,946,482	0.30
1994	83	66	51	200	62,452,572	0.32
1995	65	125	50	240	62,074,306	0.39
1996	69	146	60	275	61,817,425	0.44
1997	87	132	73	292	64,440,947	0.45
1998	91	183	51	325	66,211,734	0.49
1999	78	182	61	321	68,672,240	0.47
2000	87	259	85	431	67,480,097	0.64
2001	88	214	81	383	65,456,366	0.59
Total	812	1483	577	2872		

*Sample calculation: $(291 \times 100,000)/(63,018,680) = 0.35$.

revealed that 80 percent of the incursions were in the lower-severity C and D categories. However, this represented a 48 percent increase in the total number of incursions over the study period. While the high-severity incursions (category A) remained relatively constant, category B (major-severity) incursions increased by 20 percent. This study also revealed that the number of runway incursions among types of aircraft operations was in proportion to their use of the U.S. airspace. General aviation represented 58 percent of the air traffic volume and had 60 percent of the overall incursions. Commercial carriers controlled 38 percent of the traffic volume and had 38 percent of the overall incursions. Military operations accounted for just 2 percent of the incursions. Interaction between types of aircraft operations revealed that runway incursions mainly involved two general-aviation aircraft that were predominantly in the minor C and D categories. There was no correlation between the overall runway incursion rate and airport traffic (operations) volume. However, the rate of category A and B incursions at the busiest airports was about twice that of the nonbusy airports.

CONTROL STRATEGIES. The FAA's office of runway safety has outlined a strategic plan to achieve a safer runway environment in a document entitled *The National Blueprint for Runway Safety* (FAA, 2000). This document sets goals, establishes accountability, and outlines a timetable of key activities that are designed to measurably reduce the risks of runway incursions and surface incidents. This document groups recommendations under seven major categories:

1. Training
2. Technology
3. Communications

4. Procedures

5. Airport signs, markings, and lighting

6. Data, analysis, and metrics

7. Local solutions

TRAINING.   There are 17 specific initiatives listed under this category. Training initiatives are intended to increase the knowledge, skills, and overall performance of pilots, controllers, vehicle operators, and other individuals who interact on the airport surface. Some of these initiatives include the following:

1. Reviewing written exams and practical test standards administered during certification of flight instructors and pilots to determine if greater emphasis needs to be placed on airport signs, markings, and lighting, and requiring crew resource management training for cockpit crew.

2. Enhancing existing training for air traffic controllers to emphasize
   - Situational awareness
   - Task and resource management
   - Understanding of individual and team roles and responsibilities
   - Decision making
   - Coordination, communications, and team building
   - Retraining and refresher training
   - Incursion examples that have occurred at the airports the controllers operate in

3. Procuring hardware such as state-of-the-art visual simulators for training controllers

4. Developing awareness and educational material for the aviation community

5. Reviewing Federal Aviation Regulations Parts 61, 121, 129, 135, and 139 to determine if changes are needed to enhance the safety of runway operations

6. Requiring airports to factor in runway incursion prevention plans during airport construction and planning projects

7. Establishing national standards to guide airports in developing programs and standard operating procedures to regulate operations of vehicles and aircraft on the airport surface

In this connection, the Aircraft Owners and Pilots Association's Air Safety Foundation (AOPA, 2001) provides an interactive program designed to teach pilots about runway incursion avoidance. Successful completion of this course satisfies the ground school requirement of the FAA "Wings" Pilot Proficiency Award Program. For the control tower, the FAA has developed the Training Proficiency Unit (TPU) and a four-part computer-based instruction course to educate air traffic controllers on runway incursion prevention.

TECHNOLOGY.   The primary use of technology is to improve human performance by enhancing human operator skills while compensating for human shortcomings. Some promising technologies are now reviewed.

1. *Airport surface detection equipment (ASDE)—X series.*   This is a traffic management system that provides radar surveillance of surface movement of aircraft and ground vehicles; and it is especially useful during periods of low visibility such as rain, fog, and night operations. The system uses the Airport Target Identification System (ATIDS) to provide data tags on arriving and departing aircraft. Cooperating vehicles have data tags as well. The system uses a combination of surface movement radar and transponder multilateration sensors to display aircraft position labeled with flight call-signs on ATC tower displays. *Multilateration* is the process of determining a transponder's location in two or three dimensions by using a surveillance sensor. ASDE-X will utilize a fusion platform, allowing it to accept input from multiple sensors (i.e., ADS-B, ATIDS, and radar). Automatic dependent surveillance—broadcast (ADS-B) equipped aircraft and vehicles automatically transmit important information such as latitude and longitude, velocity, altitude, heading, and identification, using the Global Navigation Satellite System (GNSS) and aviation datalinks. The integration of these sensors provides data with an accuracy, update rate, and reliability suitable for improving airport safety in all weather conditions. The FAA has identified several dozen airports in the United States as candidates for ASDE-X systems. The ASDE-X system also enables decision support tools that provide

- Positive correlation of flight plan information with aircraft position on controller displays
- Seamless surveillance coverage of the airport from arrival through departure
- Elimination of blind spots and coverage gaps
- Surveillance data with an accuracy and update rate suitable for

  Situational awareness in all weather conditions
  Departure and arrival queuing
  Conflict detection and resolution
  Taxi route conformance monitoring

2. *Loops technology (LOT)*   Loops is a lower-cost alternative to the ASDE-X series system and is used at airports where there is a localized problem not requiring the full capabilities of an ASDE-X system.

3. *Airport movement area safety system (AMASS).*   AMASS is a software enhancement to the ASDE-X system that provides controllers with aural and visual alerts to potential collisions on the runway. It processes data from ASDE-X and ARTS systems to predict aircraft movement for arriving and departing aircraft as well as aircraft and/or vehicles that infringe on the runway surface. It does so by correlating speed, time, and distance algorithms to assess potential collision situations. AMASS is being developed to have commonality in displays, keyboards, and the visual specification with ASDE-X.

4. *Precision runway monitoring (PRM).*   PRM is a nonrotating, nonmechanical, circular phased array antenna, which is electronically scanned to increase airport capacity, reduce delays, and increase the safety of parallel operations. It scans four to five times faster than existing surveillance radars, and it tracks and displays each aircraft and updates its position and velocity every second. Using PRM, controllers are able to monitor the progress of each aircraft in real time and issue directions as required to maintain safe aircraft separation. Two production systems are in use at the Minneapolis–St. Paul and Lambert–St. Louis international airports. Additional systems have been contemplated for the John F. Kennedy, Philadelphia, San Francisco, and Atlanta/Hartsfield international airports.

5. *Orincon Corp.'s ground safety tracking and reporting system (GSTRS).*   This system allows controllers to monitor traffic flow on runways and taxiways and to identify potential incursions (Aerospace Engineering, 2000). GSTRS is weather-independent and does not have the blind spot problems of existing ground radar. Marketed as a joint effort of ARINC and Orincon, GSTRS is based on advanced signal processing and neural network technologies. Its technology incorporates off-the-shelf hardware that is more cost-effective than existing FAA ground tracking radar.

6. *Three-dimensional runway surface markings (MITRE-CAASD).*   MITRE Corporation's Center for Advanced Aviation System Development (CAASD) has been conducting research for the FAA on an innovative runway marking project that it created to help pilots and others operating vehicles on airport surfaces see critical "hold short" markings better. Painted on airport taxiways, these visual markings indicate areas on airport surfaces that cannot be crossed without prior permission from air traffic control. Airport surface markings and signage help aircraft and vehicle operators navigate the airport surface and identify their locations to air traffic controllers. The CAASD project utilizes three-dimensional artistic techniques to make airport surface markings more conspicuous to pilots and vehicle operators on the airport surface. The technique, based on anamorphic projection, is the one that artists have used since the Renaissance. Basically, this technique creates the illusion that a two-dimensional object is three-dimensional. The concept of three-dimensional runway markings is the brainchild of MITRE engineer James Krantz after he viewed a figure painted on the Community Bridge, an arts project in Frederick, Md. The artist, William M. Cochran, had used anamorphic projection to give the figure of an angel a strikingly realistic, three-dimensional appearance when viewed from an angle. Krantz noted that when viewed head-on, the figure appears distorted, but when viewed from above, it seems to emerge from the bridge surface. He envisioned runway surface markings that would appear to be actual barriers when viewed from the cockpit of an aircraft.

7. *Runway status lights (RWSL).* RWSL is a radar-based safety system intended to improve on-airfield situational awareness by providing pilots with an unambiguous visual advisory of runway status through a combination of computer-controlled surface lighting systems.

Communications.   Under the communications thrust, communications between all individuals involved with airport surface operations will be simplified and standardized. Some of the proposed recommendations include

- Requiring all ground and aircraft radios that operate in the NAS to be retrofitted with antiblocking and anti–stuck microphone radio technology
- Standardizing taxi routes at busy airports to reduce congestion
- Teaming up with the *International Civil Aviation Organization* (ICAO) to develop English language proficiency standards and harmonize ICAO/FAA phraseology
- Requiring predetermined surface movement instructions to be read back by pilots and ground personnel

Procedures.   There are 11 initiatives proposed under this category of recommendations which are intended to manage and control the actions of pilots, controllers, and vehicle operators. Some of these initiatives are as follows:

- Segregate ground vehicles from airport area vehicles whenever possible.
- Study the amendment of 14 CFR 91.129(i) to require specific instructions to cross runways.
- Revise 14 CFR Parts 121 and 135 to ensure that training adequately addresses surface operations.
- Encourage Part 121 and Part 135 operators to develop SOPs for airport surface operations.
- Develop and implement short-term memory aids for controllers (to permit remembering aircraft hold positions).
- Develop national standards for operating practices for controllers, emphasizing *best practices* for pilot operating procedures and controller memory management.
- Scrutinize the practice of clearing multiple aircraft to land in sequence on the same runway.
- Enforce procedures for ground vehicular traffic and establish accountability for all ground operations personnel.

Airport Signs, Markings, and Lighting (ASML).   This strategy emphasizes improving the overall airport environment to increase visibility, improve safe and efficient movement of aircraft, and assist pilots in recognizing and understanding airport signs, markings, and lighting. Recommendations include these:

1. Comprehensively review and revise all ASML requirements and standards, and develop objective standards for airport maintenance staff to determine when signs need repainting and surfaces need refinishing.

2. Put out safety advisories at all towered airports on operating and communicating procedures together with regularly updated taxi diagrams, airport signs, markings, and lighting issues. Note that the FAA has teamed up with the AOPA to provide airport taxi diagrams for more than 455 of the busiest airports.

3. Encourage the use of Airport Improvement Program funds for installation and maintenance of security fencing and ASML.

4. Train pilots to identify conditions that adversely affect runway safety, and require them to report these conditions to the appropriate airport's regional office.

DATA, ANALYSIS, AND METRICS.   This initiative is designed to understand the causes of runway accidents and to measure the effectiveness of controls. Some specifics under this category include

- Developing standardized methods for analyzing and identifying human-factors aspects of accidents, establishing and maintaining a national database of incursions, and making these data available to researchers to perform trends and comparative analyses between airports.
- Developing software tools to code and analyze data.
- Establishing a nonpunitive, confidential incursion data collection program.
- Conducting comparative analysis between airports that have frequent incursions with those that do not have as many incursions. Analysis parameters should include, but are not limited to, traffic density, peak traffic activity, congestion, and controller work load.
- Investigating runway incursions by an independent organization.

LOCAL SOLUTIONS.   This thrust encourages development of runway safety initiatives at the local level to solve problems that are unique to a particular airport. In addition, participation in information sharing at the regional, national, and global levels is encouraged so that successes and lessons learned in one locale can be applied with changes, where appropriate, to other regions. Local airports and regions are encouraged to maintain and update a central information source of newly implemented corrective actions and their effectiveness in preventing runway incursions. To promote safer ground operations, the FAA offers all airports the ability to create their own web sites (FAA, 2000) on which they can post airport-specific information and advisories. Informative booklets on airport basics, operations at towered and nontowered airports, phraseology, and emergency procedures are also available from the FAA.

CONCLUSIONS.   Despite causing civil aviation's worst disaster, runway incursions are very infrequent events, and collisions that result in deaths are even more infrequent. Barnett's 10-year review of the literature (Barnett et al., 2001) revealed the occurrence of only 4 catastrophic collisions in the United States from 1989 to 1998. The FAA's 2001 report on incursion severity for the period of 1997 to 2000 documents a total of 1369 runway incursions spread over 266 million hours of operation. This approximates 5 runway incursions per million operations. Of these 1369 incursions, 81 percent were minor in nature and only 3 runway incursions (0.2 percent) resulted in collisions. While this is a good record by any measure, there is serious concern that an increasing number of airplanes, vehicles, and people are mistakenly coming dangerously close to each other. With predicted increases in air traffic volume in a system that is already constrained, the number of incursions would most certainly increase. What's more, even if the incursion rates stay the same, there will be an increase in the absolute number of incursions if nothing is done about it, which of course would increase the probability of catastrophic incursions. It is for this reason that the FAA places a very high priority on controlling runway incursions. Despite the fact that pilot errors result in more runway incursions than any other cause, minimizing runway incursions requires a joint effort from all groups involved in the management of the airport operations system, i.e., the pilots, controllers, vehicle operators, and other ground staff.

KEY TERMS

Airport

Airport Operating Certificate

Limited Airport Operating Certificate

FAA Advisory Circulars

CertAlerts

Airport Certification Manual

Airport Emergency Plan

Standard Building Code

Uniform Building Code

National Fire Protection Association (NFPA)

American National Standards Institute (ANSI)

(OSHA Act) General Duty Clause

American Conference of Governmental Industrial Hygienists (ACGIH)

Bonding

Grounding

Aircraft rescue and fire fighting (ARFF)

Runway incursion

Surface incident

Collision hazard

Loss of separation

Operational error

Pilot deviation

Vehicle/pedestrian deviation

Runway incursion incident rates

Runway incursion severity categories (A, B, C, and D)

Airport surface detection equipment (ASDE)—X series

Multilateration

Loops technology (LOT)

Airport movement area safety system (AMASS)

Precision runway monitoring (PRM)

Runway status lights (RWSL)

Airport signs, markings, and lighting (ASML)

Three-dimensional runway surface markings (MITRE-CAASD)

Orincon Corp.'s Ground Safety Tracking and Reporting System (GSTRS)

International Civil Aviation Organization (ICAO)

REVIEW QUESTIONS

1. Which act was responsible for initiating the airport certification process and what were its salient features?

2. What part number regulates airport certification? What are its highlights, and when was it first promulgated?

3. Discuss the highlights of the airport certification process and their significance.

4. What are the two types of airport certifications and why are they issued?

5. What is the Airport Certification Manual and why is it so significant?

6. List at least 10 elements required in the Airport Certification Manual.

7. Where would one locate the requirements of the Airport Certification Manual?

8. List and explain the required elements of an airport emergency plan.

9. List five to seven safety hazards associated with each of the following, and cite the regulatory violation associated with the hazard:

   a. Airport terminal buildings

   b. Hangars and maintenance shops

c. Ramp operations
d. Aviation fuel handling
e. Aircraft rescue and fire fighting (ARFF)
f. Deicing

10. What is a runway incursion?

11. List and explain the different categories of runway incursions.

12. Describe the airport surface operational environment.

13. List individual groups together with their functions and the equipment they use for airport surface operations.

14. Define and discuss the following terms: *surface incident, runway incursion, collision hazard,* and *loss of separation.*

15. Discuss runway incursion trends with respect to rates and severity.

16. List and explain seven strategies to minimize the occurrence of runway incursions.

17. List and explain the technologies being proposed to eliminate runway incursions.

18. What is being done in training, ASML, procedures, and communications to prevent runway incursions?

REFERENCES

Barnett, A., G. Paul, and J. Ladelucs. 2001. *Fatal US Runway Collisions Over the Next Twenty Years,* FAA Grant Report.
FAA. 2000. *National Blueprint for Runway Safety.* FAA Office of Runway Safety.
———. 2001. *Runway Safety Report—Runway Incursion Severity Trends at Towered Airports in the United States, 1997–2000.* FAA Office of Runway Safety.
Flight International. 2000. *Airline Safety Review of the 1990's.* January 25 issue. Reed Business Information.
Gesell, L. E. 1988. *The Administration of Public Airports,* 2d ed. Chandler, Ariz.: Coast Air Publications.
Landsberg, B. 1995. "Stop Look and Listen." *AOPA Pilot,* vol. 38, no. 8, pp. 156–159.
National Safety Council. 2000. *Aviation Ground Operation—Safety Handbook.* 5th ed. Itasca, Ill.: National Safety Council Press.
Rodrigues, C. C. 2001. "Safety Analysis of Runway Incursions." *Proceedings of the International Conference on Industrial Engineering—Theory, Applications and Practice.* November 18–20. San Francisco, Calif.
Wenham, T. P. 1996. *Airliner Crashes.* Stephens Limited, Sparkford, Somerset, UK.
Wood, R. H. 1997. *Aviation Safety Programs—A Management Handbook.* 2d ed. Englewood, Colo.: Jeppesen Sanderson, Inc.

WEB REFERENCES

Access to Airports for Individuals with Disabilities. ADA Advisory Circular. http://www1.faa.gov/arp/pdf/5360-14.pdf
Advisory Circulars: http://www2.faa.gov/arp/150acs.cfm?ARPnav5acs
Aerospace Engineering. July 2000. http://www.sae.org/aeromag/prodshow_7-00/05.htm
AOPA. 2001. *Air Safety Foundation Runway Safety Program.* http://www.aopa.org/asf/runway_safety/
AOPA Home. 2001. *Air Safety Foundation Airport Taxi Diagrams.* http://www.aopa.org/asf/taxi/
CertAlerts: http://www2.faa.gov/arp/certification/certalert.cfm?ARPnav5cert
FAA. 2000. *Runway Safety Program.* http://www.faa.gov/runwaysafety
———. 2001. *Runway Safety Program—Statistics and Data.* http://www.faarsp.com/h2.html

Guidelines for submitting a supervisory fueling safety program: http://www2.faa.gov/arp/safety/fsguide.cfm?ARPnav5safety

Part 139: http://www.access.gpo.gov/nara/cfr/cfrhtml_00/Title_14/14cfr139_00.html

Part 139 certification: http://www1.faa.gov/arp/certification/overview.cfm?ARPnav5cert

Planning and design guidance for airport terminals: http://www1.faa.gov/arp/pdf/5360-13.pdf

AVIATION SECURITY

LEARNING OBJECTIVES

After completing this chapter, you should be able to

- Distinguish between security and safety.
- List the categories of attacks on civil aviation.

- Discuss the salient features of
  - Airport security, air carrier security, and indirect air carrier security
  - The Anti-hijacking or Air Transportation Security Act of 1974
  - The Air Carrier Standard Security Program (ACSSP)
  - The Air Marshal Program
  - The Aviation Security Improvement Act of 1990
  - The Antiterrorism Act of 1996
  - The White House Commission on Aviation Safety and Security, 1996
  - The Aviation and Transportation Security Act of 2001
- Discuss the role of intelligence gathering and analysis concerning security.
- Describe how international influences can affect the effectiveness of security measures.
- Discuss the theory and concepts behind the following security strategies:

  Imaging technologies
  Trace detection technology
  Explosive detection systems (EDS)
  Metal detectors
  Biometrics
  Strengthening of aircraft and baggage containers
  Cockpit door reinforcement
  Computer-assisted passenger screening

INTRODUCTION

Security and safety are often considered synonymous topics, and the discussion of one invariably invokes references to the other. However, it is important to realize that there is a fundamental difference between the two. *Safety* usually refers to measures taken against the threat of an accident, whereas *security* refers to protection from threats motivated by hostility or malice. While the end objective of safety and security is to minimize risk by preventing injuries and loss of lives and property, at their core safety practices are designed to prevent *unintentional* acts whereas security practices are designed to avert *intentional* acts. The intentional acts addressed in this chapter are the attacks against commercial aviation. While attacks on civil aviation can be carried out against airliners, airports, and airline offices, airliners have been the most common target for attacks. The spectrum of attacks on civil aviation includes

- Bombings, shooting, and miscellaneous attacks at airports
- Bombings, shooting, hijackings, and miscellaneous attacks (e.g., commandeering) on aircraft on the ground or in flight

- General-aviation and charter aviation aircraft attacks
- Off-airport facility attacks
- Shootings (from the ground) at aircraft during takeoff and landing

This chapter reviews the history of attacks against civil aviation together with the regulations and security measures that evolved to minimize the occurrence and severity of these attacks.

REVIEW OF ATTACKS ON CIVIL AVIATION

The first attack on civil aviation dates back to 1930 when Peruvian revolutionaries seized a Fokker F-7 aircraft in South America. The incident received little attention and never resulted in any international effort to combat potential threats to international aviation. Therefore, until the mid-1960s, airlines and airports gave security matters little attention. Low-technology security applications, such as airport fences, were intended as a safety measure to separate aircraft from wildlife rather than terrorists. However, attacks against civil aviation rose rapidly in the 1967–1976 decade. While there were only 32 worldwide hijackings from 1961 through 1967, there were 290 hijacking attempts (successful and unsuccessful hijackings) worldwide during the following 4 years after 1968. In the United States, from 1930 to 1967, only 12 commercial aircraft hijackings were attempted, of which only 7 were semisuccessful (Dorey, 1983). However, in 1968 there was a surge of hijackings in the United States in which several flights traveling to the Caribbean were repeatedly detoured to Havana, Cuba. Of the 22 hijackings in 1968, 19 of these flights were redirected to Cuba. The climax occurred in 1969 when 33 regularly scheduled U.S. airliners were hijacked (7 of these attempts failed). By 1969, the number of U.S. passengers and crewmembers who were being detoured to Cuba totaled 1359 (Moore, 1991). This led to a great deal of money being spent on security by the air carriers and the FAA on x-ray systems, magnetometers, training programs for screening personnel, and air marshals, after which the relative level of security immediately rose. Consequently, the threat of hijacking in the United States as well as the actual incidences of hijackings diminished over time, giving way to more frequent occurrences of both threats and actual assaults against U.S. flag carriers in Europe, Asia, and Africa, peaking with the bombing of Pan Am Flight 103 over Lockerbie, Scotland, in December 1988. The decades that followed saw a decline in attacks against aviation. As can be seen from Fig. 11-1, all three major forms of attack on airliners display almost the same trend over time, rising steeply in the 1967–1976 decade and then declining in the following two decades. Also, note the 1969 peak in the 40-year worldwide hijacking trend in Fig. 11-2.

From 1947 to 1996 there were 1098 total incidents of attacks on airliners, compared with 129 attacks on airports and 249 attacks on airline offices. Unlike

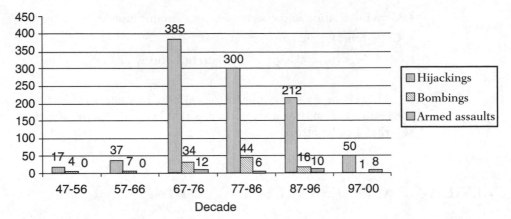

FIGURE 11.1 Frequency of main forms of attack on airlines. (Source: http://www .gwu.edu/~~cms/aviation/track_ii/merari.html)

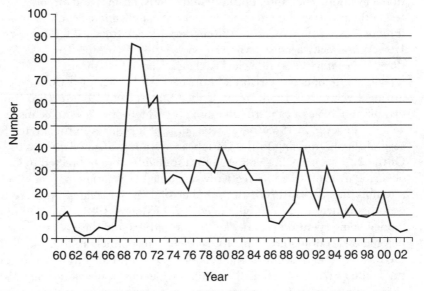

FIGURE 11.2 Worldwide air carrier hijackings.

attacks on airliners, attacks on airports (bombing and armed assaults) peaked in the 1977–1986 decade. With regard to the modes of attack, hijackings were by far the most common form of attack on commercial aviation. Hijackings constituted 87 percent (959 of 1098) of all incidents on airliners during this 50-year period. The other significant forms of attack on airliners were bombings (on the ground or in midflight) and armed assault (shooting at aircraft on the ground or in flight, throwing hand grenades, etc.).

While the number of attacks was on the decline, the severity of these attacks demonstrated a different trend. The decade of the 1980s was a disastrous one for aviation. This period confirmed the existence of a dangerous trend toward greater violence against air transportation. Overall, 25 planes were sabotaged by explosives, causing 1237 casualties. Some of the more infamous incidents included the Air India bombings in June 1985 in which 329 people perished off the coast of Ireland, the Korean Air disaster in 1987 in which 115 lost their lives, the midair explosion in 1989 of the French UTA flight over the Niger desert in which 171 were killed, and the bombing of Colombian Avianca in 1989 near Bogotá in which 107 people died. In addition to these sabotages there were 17 other documented failed attempts between 1982 and 1987, as noted in the President's 1989 Commission Report on Aviation Security and Terrorism. By comparison, there were 650 deaths in the 1970s and 286 deaths in the 1960s. Things cooled off in the 1990s only to give way to the events of September 11, 2001, when 3247 (including onboard passengers, crew, and people on the ground) were killed in four separate terrorists acts on the same day.

The increases in severity of aviation attacks were the results of terrorists changing their tactics and philosophies and making use of new technologies. They gained access to more sophisticated and lethal technologies, such as automatic weapons and deadly plastic explosives. They attacked airports by using pistols and bazookas, and they developed numerous ingenious ways to turn innocuous-looking suitcases and radios into lethal bombs. The character of airline hijackings also changed from the lone hijacker of the early 1960s who was making a personal or political point to its use by the 1970s as an organized terrorist tactic. In the era leading up to September 11, 2001, hijacking for the most part did not explicitly threaten the lives of hostages. The operating philosophy was to negotiate with the hijackers until an amicable settlement was reached, thereby minimizing casualties. However, after the September 2001 attacks, where hijackers rammed airplanes into buildings, murdering passengers and people on ground and killing themselves, there is every likelihood that future hijackings will incur some casualties.

THE REGULATORY MOVEMENT

Prior to the formation of the Transportation Security Administration (TSA) in 2001, the responsibility for protecting the users of commercial air transportation against terrorist and other criminal acts belonged to the FAA. On November 19, 2001, the security responsibilities of the FAA were transferred to the TSA, which was placed within the Department of Transportation (DOT). The regulatory evolution of aviation security is now discussed.

AIRPORT SECURITY

Airport security issues were initially governed by subparts under Part 139. Subpart 335 (Public Protection) and Subpart 337 (Wildlife Hazard Management) were intended to prevent inadvertent entry of unauthorized persons or vehicles to the aircraft movement area and to prevent damaging collisions with wildlife other than birds through fencing or other means. Subpart 325 (Airport Emergency Plan) required each certificate holder to develop and maintain an airport emergency plan designed to minimize the possibility and severity of personal injury and property damage on the airport in an emergency. This concentrated more on managing the emergency rather than on preventing it. Part 139 was mainly concerned with deterring *mistaken* entry of humans and animals into air operations areas. However, to stem the increase of hijackings during the 1970s, there was a need to specifically deter access to air operations areas by individuals and ground vehicles that were *unauthorized* to do so. Part 107 was therefore issued in 1972, which required the airport operator to immediately adopt and put into use facilities and procedures designed to prevent persons and vehicles from unauthorized access to air operations areas. Under Part 107, airports were required to

- Improve or establish protection against unauthorized access to air operations areas
- Establish authorized access to air operations areas through a suitable identification system
- Identify vehicles operating in air operations areas

As an example, FAR 107.14 mandated the upgrading of access control identification and security systems at all major commercial airports in the United States. Such systems were required to ensure that only authorized personnel could enter restricted areas. Additionally, passenger screenings as well as carry-on item checks were required for high-density flights. By the end of 1972 several hijacking acts, including the unenviable distinction of a California man being hijacked twice in a single trip, led the government to issue two emergency regulations. One gave carriers 30 days in which to institute a 100 percent search of all passengers and carry-on items, and the other gave airport operators 60 days in which to station at least one law enforcement officer at each passenger checkpoint during boarding and reboarding. These resolutions caused several problems for carriers and airport operators. At most airports, carriers had been sharing a single metal detector (magnetometer). This had been sufficient because they were required to screen only "selected" individuals. Now every passenger was required to be screened, and there were not enough magnetometers in existence to meet the new requirements. Carriers did not have enough time to recruit, hire, and train the required number of additional employees.

Airports' funding problems were equally severe. Funds were not budgeted for full-time law enforcement officer coverage, and the airports and their city or joint authority were hard pressed to find the necessary additional funds in such a short time. Despite all these problems on January 5, 1973, the 100 percent screening of air passengers and their carry-on items was instituted. The results, as expected, were long lines of travelers who were quite tolerant of these procedures that were in place to stop hijackings. The significant difference between this rule and previous antihijacking measures was the universality of the new regulation. Previously, FAA had required air carriers to conduct a weapons scan of only those passengers who fitted a hijacker profile, about 1 percent of the 500,000 passengers boarding airliners daily. As a result of this program, not a single airliner was hijacked in the United States in 1973.

AIR CARRIER SECURITY

On January 31, 1972, Part 121.538 was issued. Within this rule, air carriers were required to adopt and implement a screening system that would detect weapons and explosives in carry-on baggage or on the person of passengers. An order was issued requiring each carrier to submit its screening program to the FAA Administrator no later than June 5, 1972. The carrier's security program was required to

- Prevent or deter unauthorized access to its aircraft
- Ensure that a responsible agent or representative of the certificate holder would check in baggage
- Prevent cargo and checked baggage from being loaded aboard its aircraft unless they were handled in accordance with the certificate holder's security procedures

In response to a Presidential statement ordering renewed emphasis on air security, Part 121.538 was amended again on March 9, 1972. The screening requirement was to be put into immediate effect, and the new deadline became May 8, 1972. Further refinements took place on September 11, 1981 when Part 121.538 was rewritten as Part 108 of the FARs. Both parts were first amended as a result of the President's 1972 Commission Report on Aviation Security and Terrorism.

Part 108 based security requirements upon aircraft complexity rather than upon certification. It categorized airplanes into three groups according to configured seating capacities: more than 60 seats, 31 to 60 seats, and less than 31 seats. The rationale used by the FAA was that larger airplanes would be more attractive to the hijacker. Part 108 required the adoption of a comprehensive security program for operations with 31 through 60 seats. The program was to be comparable to that required for operations with airplanes having more than 60 seats. The smaller operator, however, would only have to implement those portions of the program that required the following:

- Procedures for contacting a law enforcement agency and arranging for a response to an incident when needed.

- Instructions for all crewmembers and internal employees in the appropriate procedures. Each operator must also be prepared to implement its full security program upon notification of specific threats by the FAA.

For operation of smaller aircraft of 1 to 30 seats, no security program was required unless passengers had uncontrolled access to a sterile area. In those cases where passengers had uncontrolled access, or where passengers were discharged into a sterile area, provisions were required to properly screen the passengers. Carriers were required to control access to the sterile area through surveillance and escort procedures or through screening procedures of another carrier.

In summary, by virtue of federal regulations in Part 107, airports were responsible for preventing unauthorized access to the air operations area and for providing law enforcement support at passenger-screening stations. Part 108 required airlines to institute programs that would prevent firearms and explosives aboard their aircraft through passenger and baggage screening and other methods. While Parts 107 and 108 addressed criminal acts and aircraft sabotage, Part 139 addressed operational safety issues of separating the public from air operations.

ANTI-HIJACKING OR AIR TRANSPORTATION SECURITY ACT OF 1974

This Act (Public Law 93-366), promulgated in August 5, 1974, contained two titles:

- Title I, Anti-hijacking Act (punitive provisions for hijackers and security standards for foreign air transportation and services)
- Title II, Air Transportation Security Act (security regulations for U.S. airports and carriers)

This law provided the statutory basis for the December 5, 1972, rules requiring carriers to institute 100 percent screening of passenger and carry-on items, and for airport operators to station at least one law enforcement officer at each passenger checkpoint during boarding and preboarding. Additionally, as part of its obligation under this Act, the FAA began a research and development program that emphasized the development of devices to protect air travelers against acts of criminal violence and aircraft piracy.

AIR CARRIER STANDARD SECURITY PROGRAM (ACSSP)

Application of the above-mentioned regulations varied greatly across the airlines. Regulatory requirements were not implemented consistently across the industry. Airlines had complete discretion on how they implemented the requirements. They also had the right to express their concerns and suggest changes based on trial and error. This resulted in inconsistent application of regulations across the airline industry, and in early 1975 the Air Transportation Association sought to work out a single standard security program. Their effort produced the FAA's Air Carrier Standard Security Program (ACSSP), which attempted to bring some structure to the diverse interpretations of the new rules. In 1976, all but a few of the carriers accepted the program, which is mandatory today.

INDIRECT AIR CARRIER SECURITY

FAR 109 regulations that were established in 1979 govern indirect air carrier security and provide additional protection against criminal activity. This part prescribes aviation security rules governing each air carrier, including air freight forwarders, food service, and cooperative shipping associations engaged indirectly in air transportation of goods. Each indirect air carrier is required to have a security program designed to prevent or deter the unauthorized introduction of explosives or incendiary devices into any package cargo intended for transportation by air.

AIR MARSHAL PROGRAM

The increase in the rate of hijackings during the late 1960s and early 1970s caused the public to exert pressure on the U.S. government to implement security procedures at airports and to mandate security requirements for U.S. air carriers. This resulted in the establishment of the Anti-Hijacking Program of the Federal Aviation Administration. One element of this program was the federal Air Marshal Program, which began as the *Sky Marshal Program* in 1968 and continued through the 1970s and part of the 1980s as a program that was initially instituted to stop hijackings to and from Cuba. The Air Marshal Program was criticized by airlines and passengers who considered it more an airborne risk than a deterrent for hijackers. This program gained in importance, however, after the hijacking of TWA 847 in June 1985. During this incident, two Lebanese Shiite Moslems hijacked a Boeing 727 departing from Athens and diverted it to Beirut, where they were joined by additional hijackers. During a two-week confrontation, the hijackers demanded the release of Shiite prisoners held by Israel and murdered Robert Stethem, a U.S. Navy diver who was a passenger on board the plane. In response to this hostage nightmare and the rapid surge in Middle East terrorism, then-President Ronald Reagan directed the Secretary of Transportation, in cooperation with the Secretary of State and the Attorney General, to explore immediately an expansion of the FAA's armed Federal Air Marshal Program aboard international flights for U.S. air carriers. On August 8, 1985, Congress enacted *Public Law 99-83*, the *International Security and Development Cooperation Act*, which established the explicit statutory basis for the FAA-Federal Air Marshal Program and allowed for assessment of security at foreign airports and approval of foreign air carrier security programs. This statute authorized Federal Air Marshals to carry firearms on board and to make arrests without warrant for any offense against the United States committed in their presence, if they had reasonable basis to believe that the person to be arrested had committed or was committing a felony. Three weeks after the TWA 847 hijacking, the FAA imposed new regulations in the form of *FAR 108.14* (*14 CFR 108*), requiring that all scheduled carriers and public charter operators carry Federal Air Marshals on a priority basis, without charge, and that they provide seating selected by the marshals, even though it might mean bumping full-fare passengers.

AVIATION SECURITY IMPROVEMENT ACT OF 1990

On December 21, 1988, Pan Am flight 103 exploded over the village of Lockerbie, Scotland, killing the 259 people aboard and 11 people on the ground as well as damaging several residential buildings. Investigations by a flight commission found that the crash was due to an explosion of a bomb placed in the luggage compartment of the aircraft. It was asserted that the luggage was coming from direct passengers boarding in Frankfurt, as well as from some possible suitcases transferred from Air Malta flight 180 to Pan Am 103 at the Frankfurt Airport.

Prior to the bombing of Pan Am 103, the U.S. civil aviation security system was not sophisticated enough to detect the level of device used to destroy the plane. The system, created in the early 1970s, was designed to prevent hijackings. In the mid-1980s, it was converted to an antisabotage system at high-threat international locations. Accordingly, the 1988 Pan Am 103 tragedy can be regarded as an aberration because the U.S. antisabotage security measures in the FAA's Air Carrier Standard Security Program (ACSSP) were designed to prevent such a tragedy. U.S. government officials noted that Pan Am violated ACSSP security requirements at Frankfurt, Germany, and Heathrow, United Kingdom, in allowing the suitcase with the explosive device to get on board its aircraft.

The major issue in the Pan Am 103 incident is the reconciliation of baggage and passengers. U.S. carriers were required to conduct a positive baggage-passenger reconciliation in 1988 at designated international locations. According to investigation findings, the Pan Am tragedy occurred because the airline was x-raying all interline bags at certain international high-threat locations instead of conducting a reconciliation process and physically searching all unaccompanied bags. It had discontinued the ACSSP, started in 1984, that required the reconciliation process for interline bags and consequently failed to reconcile the number of bags previously checked by interline passengers with the number of passengers who actually boarded the plane at Frankfurt. Indeed, the ACSSP required any unaccompanied bag from a high-threat location to be physically searched before it could be carried on the plane. Because Pan Am did not identify and physically search all the unaccompanied interline bags, flight 103 left Frankfurt with several extra bags, one of which contained a bomb.

This failure was duly noted in the 1990 report of the Presidential Commission on Aviation Security and Terrorism, established in the aftermath of the disaster. The commission's mandate was to comprehensively study and evaluate the practices and policy options with respect to preventing terrorist acts involving aviation. The commission made a number of recommendations to prevent the recurrence of such a tragedy. *Aviation Security Improvement Act of 1990* (Public Law 101-604), passed on November 16, 1990, implemented many of the recommendations of the commission. This Act has been described in the 1992 FAA Annual Report to Congress as perhaps the most comprehensive, far-reaching legislative initiative designed to improve all aspects of aviation security. It mandates many regulatory actions affecting several agencies, requires new reports, creates new organizations and staffing requirements, and empowers the FAA to promote and strengthen aviation security through an expedited, more focused research and development (R&D) program.

The Aviation Security Improvement Act of 1990 generated many changes, especially within the FAA. It created a new high-level position of assistant administrator for civil aviation security to head the office of Civil Aviation Security, which had been established in 1962. Originally designed to deal with crimes, its mandate was expanded and its staff, which had grown from 126 to 684 in 1990, now dealt with all aspects of security, including drug and narcotics interdiction. In addition, a new office of Intelligence and Security was established within the U.S. DOT to coordinate transportation security activities. It is headed by a director of intelligence and security, who reports directly to the Secretary of Transportation.

The 1990 Act also charged the FAA with enhancing its oversight of airport security and appointed FAA security managers to major airports. The enhanced oversight responsibilities of security officials working for these managers have expanded well beyond checking passengers as they board to safeguarding the entire airport. The FAA now carries out periodic threat and vulnerability assessments, laying out guidelines and regulations for

- Airport security design
- Public notification of threats
- Security personnel investigation and training
- Cargo and mail screening
- Research and development activities
- International security negotiations

ANTITERRORISM ACT OF 1996

An amendment to the *Antiterrorism Act of 1996* required the FAA to ensure that the same security measures (not merely similar ones, as was the case) used by U.S. carriers on routes into or from the United States would be implemented by non-U.S. air carriers on those routes. It generated a controversy due to the argument that many of those carriers were not considered as targets by terrorists (at least not targeted to the same degree, as determined by intelligence assessments) and therefore should not be required to apply the same level of costly security measures demanded of those that were. Because of its international status as a major world power, the United States and, by extension, U.S. air carriers were more likely targets of international terrorists. Of course, the United States could have ensured identical security measures by simply reducing its own level of security to that used by other air carriers on the same routes. Given the perceived levels of threat and the political imperative in the United States not to decrease security measures, this was not considered a viable option. Nor would it have been particularly wise.

WHITE HOUSE COMMISSION ON AVIATION SAFETY AND SECURITY, 1996

The late 1990s witnessed significant changes in the direction and emphasis of aviation security in the United States. The principal triggering event for this new emphasis and importance was the catastrophic loss of TWA 800 off Long Island, New York, in July 1996. The early model Boeing 747 was carrying 230 passengers and crew when it exploded minutes after departing John F. Kennedy International Airport, bound for Paris. The terrific force of the explosion had torn the aircraft apart, and the disturbing recovery images, along with vivid eyewitness accounts, riveted the attention of a shocked U.S. public for many weeks. And the irony was that the FBI eventually ruled that TWA 800 was not the result of a terrorist act. It was an all too familiar scene. Only two months earlier, a McDonnell-Douglas DC-9 operated by ValuJet Airlines had crashed into the Florida Everglades, killing 110 people. The aircraft was on fire and losing control, and the crew struggled to land the crippled airliner. The crash scene was particularly gruesome.

These back-to-back crashes shook the foundation of the aviation community. The traveling public was frightened, and the media questioned the perceived safety and security of domestic airline operations. Within weeks, President Clinton announced the creation of the *White House Commission on Aviation Safety and Security*. Chaired by Vice President Gore, the commission set an aggressive agenda for reviewing the safety of the air transportation system and issued initial recommendations within two months. The final report, issued five months later, outlined sweeping changes calling for regulatory reform and additional research directed toward new, safer technologies. Special attention was given to an action plan to deploy new high-technology machines to detect the most sophisticated explosives. The report included 57 recommendations, 31 dealing with improvements in security for the traveling public. The commission recommended that the federal government consider aviation security a national security concern. It created within the FAA a new high-level position of assistant administration for civil aviation security to head the office of Civil Aviation Security.

How was it that a possibly irrelevant event produced so much change in the way aviation security was regarded in the United States? For comparison, following the bombing of Pan Am flight 103 over Lockerbie, Scotland, in December 1988, although major changes were accomplished in the way the FAA handled security and organized itself—placing local security managers at major airports, raising the status of security in the agency hierarchy, and conducting background checks of certain airport and air carrier employees—few major changes in the set of baseline security measures within the United States were evident to the traveler. Further, there was no consensus on applying further significant measures, such as the installation of trace equipment for checking some carry-on baggage, as was done in a number of other countries. In general, it was felt by many skeptics that the threat of a major terrorist action against aviation security in the United States was virtually nonexistent.

Several other events had come together at about the same time to enable this significant change to occur. First, several major terrorist events within the United States, beginning in 1993, made it clear to the U.S. public that the existence of two large oceans no longer guaranteed the absence of major international terrorist acts on our territory. These attacks were well known at that time and included the World Trade Center bombing and the murders at the headquarters of the Central Intelligence Agency near Washington, D.C. Since those events, there were continuing indications of terrorist activity, including a plot to bomb several major targets in the New York City area. Even though aviation had not yet been specifically attacked at home, the history of terrorist attacks on civil aviation (including U.S. targets) overseas makes it clear that such a possibility existed and must be guarded against. Further, the revelation in early 1995 of the plot by Ramzi Ahmed Youssef to destroy a large number of U.S. civil aircraft in Asia demonstrated that U.S. aircraft could still be targets of international terrorists. Concerns were heightened by the realization that Youssef had previously been in the United States.

The existence of a serious terrorist threat within the United States impelled the FAA to convene an outside advisory panel composed of representatives from other government agencies, air carriers and airport authorities, and various citizens and professional groups with the purpose of recommending improvements in baseline aviation security measures. In fact, by coincidence, this baseline working group was formed only hours before the crash of TWA 800. The working group's recommendations were passed on to the *White House Commission on Aviation Safety and Security*, formed shortly after the crash, and had a major impact on its first and final reports. Many of the White House Commission's recommendations (over 30 dealing with security issues) were given the force of law and financing by ensuing congressional action.

A second piece of the groundwork for the major change in aviation security was the emergence of successful new security technology, both in explosives detection and in other areas such as human factors and aircraft container hardening. The existence of an approved explosives detection system, the CTX 5000, manufactured by InVision and certified by the FAA in 1994, made it conceivable that effective technical measures could be taken to block the introduction of explosives aboard aircraft. Further, the rapid development and improvement of trace explosives detectors raised the possibility of redundant technical measures to check baggage for explosives, based on a totally different technical approach.

Combined with the apparently successful bombing of a U.S. aircraft, practically within sight of New York City, the situation in July 1996 made the social and political pressure to institute significant improvements in baseline security measures irresistible. Within three months, congressional legislation appropriated federal funds for a large-scale purchase of expensive security equipment. This was a first in the United States, with one minor exception about 25 years ago. Air carriers, not the federal government, have traditionally had the responsibility for such purchases.

Legislation authorized other security enhancements, such as background checks on security screeners, vulnerability assessments at airports, and the increased use of dogs for detecting explosives. While security implementation strategies continued throughout the late 1990s, they lacked the urgency and thoroughness that would be needed to thwart the events of September 11, 2001. Improvements in aviation security developed complications because government and industry often found themselves at odds, unable to resolve disputes over financing, effectiveness, technology, and potential impacts on operations and passengers. To make a point of this, the inspector generals (IGs) of the FAA conducted approximately 173 tests at 8 U.S. airports between December 1998 and May 1999. Much to their dismay, the inspector generals were able to gain a total of 117 unauthorized accesses to restricted areas during the testing period. The IGs were able to gain access by following authorized employees into restricted areas, riding unguarded elevators, going through unlocked gates, and walking through cargo areas. The investigators concluded that problems that plagued airport operators in the past still existed, and that airport operators and air carriers had major deficiencies in the areas of unauthorized access control procedures, employee training, and FAA oversight programs (Anderson, 2000).

AVIATION AND TRANSPORTATION SECURITY ACT OF 2001

The terrorist events of 2001 changed the face of aviation forever while fundamentally modifying the thinking and approach to security in the United States. This catastrophic event called for marked new trends and measures in security not only in aviation but for all modes of transportation. On September 11, 2001, four passenger aircraft were hijacked and crashed by terrorists. The first airplane was an American Airlines 767 (flight 11) flying from Boston to Los Angeles when it was hijacked and flown into one of the World Trade Center towers. All 11 crewmembers, 76 passengers, and 5 hijackers were killed. The second jet that was crashed into the second World Trade Center tower was a United Airlines 767 (flight 175). All 9 crewmembers, 51 passengers, and 5 hijackers were killed. The impacts against the towers, together with the heat generated from the explosion of the aircrafts, caused both towers to collapse. The third aircraft was an American Airlines 757 (flight 77) on a flight from Dulles to Los Angeles. It was hijacked and flown into the Pentagon, collapsing part of the structure. All 6 crewmembers, 53 passengers, and 5 hijackers were killed. The fourth aircraft was a United Airlines 757 (flight 93) on a flight from Newark to San Francisco. It was hijacked and crashed into a field near Pittsburgh, Pa. All 7 crewmembers, 34 passengers, and 4 hijackers were killed. The total death toll from this event was 3274.

On November 19, 2001, Congress enacted the *Aviation and Transportation Security Act (ATSA)*, Public Law 107-71, 115 Stat. 597, which established the Transportation Security Administration (TSA) as an operating administration within the Department of Transportation (DOT), headed by the Under Secretary of Transportation for Security (Under Secretary). The Under Secretary took office on January 7, 2002. Prior to September 11, civil aviation security functions were the responsibility of the FAA and aircraft carriers. However, after the attacks, legislators mandated that the federal government take the lead to secure aviation. Also, to ensure uniform and nationwide implementation of these new measures, the responsibility for airport security operations was placed under the jurisdiction of the new *Transportation Security Administration (TSA)* that was housed in Department of Transportation. The new federal aviation security law included a key provision that would launch a nationwide pilot program in 20 or more U.S. airports to test new and emerging security technologies. The measure, which was also part of freestanding legislation introduced in October 2001, was expected to dramatically improve airport security by promoting the development and use of cutting-edge technologies, such as biometric authentication, Global Positioning System applications, enhanced communication systems, and database integration protocols.

Pursuant to ATSA, the TSA is responsible for security in all modes of transportation, including civil aviation under Chapter 449 of Title 49, United States Code, related research and development activities, and other transportation security functions exercised by DOT. The TSA is specifically responsible for the day-to-day security screening operations for passenger air transportation and intrastate air transportation under 49 U.S.C. §§ 44901 and 44935. This responsibility includes hiring, training, testing, and deploying or arranging for federal security screeners, federal security personnel, federal law enforcement officers, and federal security managers at U.S. airports. The TSA must also research, develop, and deploy security equipment and programs at U.S. airports; coordinate transportation security intelligence information; coordinate transportation security efforts with federal and state agencies; and deal with threats to transportation. To summarize, the TSA is required to

- Enter into contracts with private security screening companies for U.S. airports and directly supervise the companies' performance of screening services. Air carriers used to contract with private security screening companies to perform these services.

- Pay the costs deemed necessary by the TSA of screening services where the air carriers currently perform their own security screening of passengers and property and reimburse air carriers for their costs for providing ground security coordinators to assist TSA in overseeing the work performed by private security screening companies at U.S. airports.

- Assume functions related to aviation security from the FAA including personnel, property, records, unexpended fund balances, and related authority. Existing FAA orders, rules, regulations, grants, and proceedings would remain in effect until modified, terminated, superseded, set aside, or revoked.

- Deploy security personnel at airport secure areas, and consider the physical security needs of air traffic control facilities, parked aircraft, servicing equipment, fuel and supplies, adjacent automobile parking, and other ground or water transportation mode access areas.

- Provide to Congress its plan to screen all checked bags at U.S. airports and on ATSA's optional enhanced security measures, including installation and operational dates and schedules.

- Establish acceptable performance levels for aviation security, including screening and access control, and provide Congress with an action plan containing measurable goals and milestones.

- Assess and recommend to airport operators:
  - Commercially available means to prevent unauthorized access to secure areas
  - Effective biometric systems
  - Access point surveillance
  - Card/key-based access systems
  - Emergency exit monitoring
  - The elimination of "piggybacking" into secure areas by unauthorized persons

- Replace contract security screeners with its own federally hired, trained, and tested employees on a graduated basis until TSA has deployed sufficient federal security screeners, as well as federal security managers and federal law enforcement officers, to screen all passengers and property at required U.S. airports. It will provide for alcohol and controlled-substance testing, as well as background and criminal history checks, for contract security screeners and TSA employees.

- Assess the existence of dual-use items that seem harmless, but may be used as a weapon; maintain a list of such items; and include it in screener training.

- Establish a pilot program for private airport security screening companies.

- Deploy sufficient explosive detection systems (EDSs) to screen all checked baggage at U.S. airports, or screen all checked baggage by alternative means.

- Work with airport operators to strengthen area access control points and consider biometrics/technology deployment to identify individuals.

- Impose a uniform fee, the September 11 Security Fee, on certain passenger enplanements to offset the costs of providing civil aviation security services.

TRANSPORTATION SECURITY ADMINISTRATION REGULATIONS

On February 22, 2002, five days after assuming responsibility for inspecting persons and property, the Under Secretary of Transportation for Security issued rulemaking transferring the Federal Aviation Administration rules to Title 49 of the

Code of Federal Regulations. This created the Transportation Security Regulations (TSR). The TSRs appear in Title 49, Code of Federal Regulations (CFR), Chapter XII, which includes Parts 1500 through 1699. This is broken out into subchapters. Subchapter A contains administrative and procedural rules. Subchapter B contains rules that apply to many modes of transportation. Rules for civil aviation security are contained in Subchapter C.

Subchapter A, 49 CFR 1500, outlines to whom this rule applies and defines the terms used in the TSRs.

- *Part 1520* addresses the protection of sensitive security information, information that was found in 14 CFR Part 191 prior to the transition. This part outlines the type of information that may not be released under the Freedom of Information Act. Contained in this section is the duty to protect any information that is given to a person in performance of his or her duties and the responsibility to report to DOT when he or she becomes aware that sensitive security information has been released to unauthorized individuals.

- *Part 1540* outlines the delegation of the Under Secretary of Transportation for Security's authority. Part 1540 also contains prohibitions regarding making fraudulent or intentionally false statements or entry in compliance reports or applying for an access or identification medium that will be used in the aviation system. Also prohibited by Part 1540 is the interference with screening personnel while they are performing their duties and the carriage of weapons, explosives, or incendiaries by individuals into specified areas at airports. Individual responsibility outlined in Part 1540 includes
  - The security responsibilities of employees and persons who access the airport
  - The responsibilities of persons who wish to enter any area that requires screening
  - The responsibility of airmen to present certain certifications to TSA for inspection when so requested. Prior to the transition this information was contained in 14 CFR Parts 107 and 108.

- *Part 1542* contains information previously found in 14 CFR Part 107 prior to the transition. These airport requirements for the most part were transferred unaltered from 14 CFR Part 107 and primarily address access control and law enforcement support. However, there are some changes in section 1542.209 that addresses fingerprint-based criminal-history records checks.

- *Part 1544* addresses requirements for aircraft operators in the United States. These requirements were contained in 14 CFR Part 108 prior to the transition. Part 1544 continues to require that the aircraft operator not permit persons to have unauthorized explosives, incendiaries, or weapons when on board an aircraft. Although TSA will conduct most inspections, if the aircraft operator becomes aware that a person has an unauthorized explosive, incendiary, or weapon, the aircraft operator must not permit it on board.

- *Part 1546* provides the rules for foreign air carriers that operate within the United States. The security sections that were found in 14 CFR Part 129 have been incorporated and reorganized for ease of use. Also, several administrative requirements were updated, for example, the procedure for adopting and amending a security program.

- *Part 1548* provides the rules for indirect air carriers that operate within the United States. Requirements that were previously found in 14 CFR Part 109 have been incorporated and reorganized for ease of use. In addition, like Part 1546, the administrative procedure for adopting and amending a security program was updated.

- *Part 1550* was created to require security programs for both passenger and all-cargo operations using aircraft with a maximum certified takeoff weight of 12,500 pounds or more.

ROLE OF INTELLIGENCE

Since the late 1980s, the issue of intelligence gathering and analysis concerning transportation has been taken more seriously. Efforts have been made to reduce turf wars that had become commonplace among the Federal Bureau of Investigation (FBI), Central Intelligence Agency (CIA), Defense Intelligence Agency (DIA), and other intelligence collection and analysis institutions, as well as such intelligence-consuming organizations as the U.S. DOT and FAA. The U.S. DOT office of Intelligence and Research was established to make policy and to coordinate the efforts of the different DOT agencies. It does not, however, have the power to enforce its decisions or to ensure that its requirements are met.

Neither the FAA nor the U.S. DOT office of Intelligence and Research actually collects or, in the strictest sense, analyzes intelligence data. Both groups remain dependent on the intelligence agencies for their intelligence products. Responsibilities for terrorism are divided among the CIA, State Department, National Security Council (NSC), DIA, and FBI. An interagency coordinating mechanism was originally set up in 1986 to eliminate turf battles, but it has instead become merely another actor in the process, concerned with defending its own prerogatives. When the White House did not assume responsibility for transportation security in the late 1980s, the State Department began playing a greater role in coordinating and facilitating policy integration, especially in relation to non-U.S. agencies.

Much remains to be done. The national intelligence community remains somewhat insular and parochial, with definite hierarchies raising both organizational and security concerns. National transportation agencies, as well as state or local entities, are frequently excluded from full and open disclosure of intelligence obtained and processed at the national level. Overcoming bureaucratic and security hurdles and developing a mechanism to create a continuous supply of timely intelligence to all potentially affected levels and actors holding appropriate security clearances remain a priority. Consistent with the need to protect sources, consumers want such intelligence to include as much background information, timely updates, and current operational information or intelligence as possible. The process of "sanitizing" classified information to officials with lower or no security clearances is a real problem. Although senior officials may well have all the available details, many others need to know more than just "there is a serious threat."

INTERNATIONAL INFLUENCES

The cost-benefit ratio of current U.S. security efforts is not insular. If U.S. security measures become radically more effective than those of Germany, France, Switzerland, or South Africa, the terrorist motivated to harm U.S. interests is likely to seek U.S. targets in one of these countries. From the perspective of the international aircraft operator, strengthening a single link of the security chain (e.g., the United States) has a limited effect on the strength of the entire chain. There is little or no overall economic benefit to improving aviation security inconsistently.

There are implicit assumptions about the behavior of the threat and of the effectiveness of security measures. For instance, most current security measures are based on the assumption that the terrorist wants to detonate an explosive device onboard an aircraft in flight or near it on the ground. Present measures are, to a large degree, effective against such a threat.

However, if the terrorist were to acquire shoulder-launched missiles (a perfectly realistic scenario, especially given that much terrorism is state-sponsored), present security measures would become pointless. In fact, there has been one failed attempt as of this writing in which a shoulder-launched missile was targeted at an aircraft. If aviation security is made nearly perfect, the terrorist pursuing an optimal strategy is likely to target other forms of transportation, other forms of industrial activity, or any other valuable and sufficiently vulnerable asset. A positive benefit may be achieved by the aviation industry, but that has the effect of displacing the threat to some other target, which then experiences a "negative benefit" in the form of an increased threat. When benefits are analyzed on a national scale, the interactive effect described here attaches greater uncertainty to the result and suggests that overall benefits are less than they may seem when narrowly evaluated only in an aviation environment.

Aviation illustrates the multinational aspect of the transportation security problem well. U.S. airlines are heavily involved in international travel and are becoming more and more intertwined with foreign airlines through code sharing and other arrangements. Hence, the issue of U.S. transportation security challenges matters such as illegal immigration and the smuggling of drugs, hazardous wastes, and terrorist materials.

National and international security concerns arise because such economic and technological integration is not accompanied by political integration. There is no central authority in world politics, much less one with power to enforce its mandates. It is a self-help international system in which states, international organizations, corporations, other nonstate actors, groups, and individuals try to pursue opportunities and avoid dangers.

For example, in the case of Pan Am flight 103, the bomb was placed on the plane in Frankfurt through an intramodal movement from a foreign airline originating in still another country by a terrorist group supported by yet another country's intelligence service. The tragedy involved activities in five countries.

Responsibility for aviation security is primarily a government function at the national level. In many cases, the Ministry of the Interior designs, operates, and oversees airport security; and a state-level agency usually recruits, employs, trains, and supervises security employees. National systems, however, vary widely in terms of organization, orientation, policies, and procedures.

In Germany, for instance, state and local government agencies are responsible for screening airline passengers and their carry-on baggage. In France, the central government is responsible for these activities. In the United Kingdom, however, the private British Airport Authority manages such functions. To varying degrees, authorities in these countries and the United States conduct a passenger-bag reconciliation process with screening of checked international baggage. The FAA also recently initiated passenger-baggage reconciliation in the United States at selected airports on a trial basis.

At selected high-threat international locations, the United States requires U.S. airlines to review passengers in relation to established profiles, selecting some (and their checked and carry-on bags) for additional security screening. Only recently has the United Kingdom required the screening of all checked international baggage. France, by contrast, has tended to place relatively less emphasis on such security measures.

When foreign air carriers are state enterprises or are partially owned by the government, they often play a more active security role, especially outside their country's borders. The United States is the only country in which all air carriers are for-profit corporate enterprises. In other countries the carriers are government-owned or heavily subsidized. For U.S. carriers, an increase in security costs—like most other government mandates, such as noise abatement, safety, and handicapped accessibility—reduces corporate profitability. Especially in European countries, such costs are borne by the government and have little or no adverse effect on the competitive posture of the carrier.

This is part of the "level playing field" issue in which non-U.S. carriers are held to somewhat different standards by their own governments and are also subsidized to meet the requirements that the FAA imposes on them to operate within our borders. U.S. flag carriers must charge higher fares on internationally competitive routes to recover security costs that are often not borne by competing foreign flag carriers, and, generally, higher fares mean fewer passengers.

During the first Gulf War, the FAA required a series of heightened security measures for all U.S. airlines and airports, making no differentiation between large, international airports and small rural, domestic airports. The security measures produced two obvious economic effects. The first was increased costs, which arise from such procedures as increased inspections, expanded screening measures, increased patrol, overtime payrolls, and enforcing new parking restrictions. The second effect is loss of revenue. Allowing only ticketed passengers past screening points reduces concession revenues, and parking lot closures reduce cash flow from the airport's second-largest source of revenue. There are also huge overtime payrolls for increased patrols and inevitably a general—albeit temporary—decline in traffic because of exaggerated public perceptions of the domestic hazards. The estimated total costs plus lost revenue ran into the millions per week at the peak of activity. Hindsight reveals that the domestic threat to civil aviation at the time was either minimal or nonexistent. Critics point out that it would be better to spend the money on intelligence to get an accurate fix on the real threat. Hindsight also points out that an incremental expenditure on security, like safety beyond an established baseline, is extremely high.

Israel's El-Al is particularly active in security and is widely regarded as a model to be emulated. The airline's small size enables it to carry out security activities such as extensive passenger and baggage screening and searches that are difficult to implement in larger-scale operations. By contrast, U.S. airlines generally have opposed implementing the degree of stringency in Israeli-style security systems and in recent years have been moving away from such trends. Partly as a result of U.S. government requirements, however, many have been willing to adopt such measures at most high-threat locations. American Airlines led the way in this regard by hiring an Israeli security company in 1986 to help it initiate more stringent security measures.

U.S. carriers frequently rely on the host government to provide security at least to the degree that they conform to procedures outlined in International Civil Aviation Organization (ICAO) Annex 17. The U.S. government recognizes that many host governments, with the exception of Israel, are either incapable of providing or unwilling to provide the level of security necessary to counter terrorist actions against U.S. aviation at high-threat locations.

The FAA's ACSSP compensated for this deficiency by mandating high-level security system requirements at high-threat locations. The ACSSP requires individual U.S. carriers to provide additional protection in these high-threat locations, especially when the host government is not required under ICAO Annex 17 to provide the level of security necessary to protect U.S. international aviation from terrorists.

The U.S. government does ask host governments to help implement these additional security measures. The costs of these measures are borne by the individual carriers. The *International Security and Development Cooperation Act of 1985* gives the FAA authority to assess security measures at foreign airports in accordance with ICAO Annex 17 standards. If the FAA finds that an airport fails to administer and maintain effective security measures, the Secretary of Transportation is required to take appropriate actions up to or including suspension of service. The U.S. DOT also issues public warnings about the level of security at foreign airports if it falls below international standards.

Considerable efforts have also been made to achieve international coordination. The ICAO, with some 184 member states, has attempted to deal with the issue of security by establishing overall standards and practices through Annex 17. Although the annex includes important measures and represents an accepted international standard subject to implementation by state authorities, it is not adequate to detect sophisticated bombs. International organizations that base policy on consensus of their members often tend to accept the lowest common denominator, which, however well intentioned, may be inadequate to the challenge. Ensuring effective implementation by all members is a further problem.

REVIEW OF SECURITY TECHNOLOGIES

Commercial aviation can be protected from the threat of explosives in two ways: by preventing explosives from reaching the aircraft (e.g., by using explosives detection technologies) or by mitigating the effects of an explosive by protecting the aircraft from an onboard explosion (e.g., via aircraft hardening and hardened containers). While each of these approaches will be discussed individually, a combination of these two approaches may provide the best protection of commercial aviation.

The first stage of the deployment of new, advanced security equipment began in 1997 when Congress appropriated $144 million to this end. It included the purchase of 54 additional explosives detection systems (beyond the 3 that were already being tested in an airport demonstration project), some 20 other units of advanced bulk detection equipment, and nearly 500 trace detection devices. This instrumentation was to be placed in major U.S. airports, beginning with Chicago O'Hare and JFK International Airport in New York. The first round of deployment was completed in 1999. After the events of September 11, 2001, the implementation plan was sped up, and all airports were required to install baggage and passenger screening equipment by December 2002. Some of the technologies used in these scanning and detection systems will now be reviewed.

IMAGING TECHNOLOGIES

Imaging technologies work either by sensing the natural radiation emitted by the human body (passive imaging) or by exposing subjects to a specific type of radiation and then measuring the radiation reflected by the body (active imaging). These systems can detect metallic weapons or plastic explosives by sensing the differences in reflected radiation between the human body and the weapons or explosives. The screening systems then generate televisionlike digital images. Operators are trained to identify potential threat objects in these images, often with the assistance of image-enhancing software that highlights unusual features.

X-Ray Imaging.   X-ray units used for inspecting carry-on luggage and people use low-dosage, low-energy radiation. Higher-dosage units are used for checked baggage. Older-generation units produced images of low quality and poor contrast. Many of the older analog systems have given way to modern digital video amplification and processing systems that produce clearer images and increased contrast ratios. Current technologies use either x-ray or millimeter wavelength electromagnetic radiation together with digital image transfer and processing. Regulations that regulate the use of x-ray systems are listed under TSR Parts 1546.209 and 1544.211.

Current x-ray devices are equipped with a lot of image evaluation and enhancement features to assist the operator. Newer devices display images in different colors according to the level of energy radiation, thickness, density, or level of threat to help the operator differentiate among materials and display a combined image that is color-coded by whether an object is metallic, organic, and so on. Additional features include the ability to automatically select the best contrast for every spot within the x-ray image. The advantage of a multicolor scale becomes evident when one is examining overlapping materials.

Some x-ray equipment uses pattern recognition software to outline areas of a bag for closer attention by the operator. Some use threat-image projection (TIP) to insert a stored image of a threat object into an actual image of a bag, in order to help the operator stay alert (and to monitor performance).

Passive Millimeter-Wave Imaging.   This technology is based on the principle that any object not at absolute zero temperature will emit electromagnetic energy at all wavelengths. These systems operate in the near-100-gigahertz range. The energy released by an object is sensed by a receiver that is designed for the purpose and is converted to an image. An important feature of this technology is its ability to accomplish imaging by gathering the radiation emitted naturally from the human body without introducing artificial radiation. Thus, no health risks are associated with this technology. This technology is still in development.

Active Millimeter-Wave Imaging.   *Active millimeter-wave imaging* combines a source of millimeter-wave energy with a detector and works much as radar does, "illuminating" the area to be searched with millimeter waves and then imaging the reflected waves. This technology uses low-energy, low-intensity reflected x-rays to scan an object to generate an image. Active imaging techniques require radiation, which raises some health concerns.

As with all imaging technology, the operator must be trained to develop a level of competency to interpret the image in order to determine whether a metallic or nonmetallic weapon is present. Also, imaging techniques can display images that are somewhat revealing of the human anatomy. This is likely to cause privacy concerns as passengers are likely to object to their images being displayed to the operator (and possibly others around). Sample imaging systems that are currently being marketed are reviewed below.

American Science and Engineering (AS&E).   The following three product lines are offered by AS&E: (http://www.as-e.com/):

- The BodySearch Personnel Inspection System is a personnel inspection system for screening individuals for the presence of concealed weapons, drugs, and illegal contraband. BodySearch is a less intrusive alternative to strip searches when law enforcement officers are authorized to perform them.
- The AS&E's 101ZZ system allows inspection of an object from two sides and is useful for scanning large numbers of both small and oversized objects.
- The PalletSearch Cargo Inspection System can inspect large, dense, palletized mixed cargo and is especially effective in detecting organic contraband in confused or cluttered environments.

Heimann Systems.   HI-SCAN x-ray inspection units combine the extensive benefits of the scanning process with state-of-the-art image processing. A special generator, developed and manufactured by Smiths Heimann (http://www.heimannsystems.com/), produces the x-rays needed for scanning. Data gathered by ultrasensitive semiconductor detectors are digitally stored, and displays are of high-contrast and in color. A special feature of Smiths Heimann x-ray technology is the extremely low radiation dosage.

Secure 1000 Nicolet Imaging Systems/Teledyne.   The Secure 1000 (http://www.dspguide.com/secure.htm) is an electronic imaging system used to detect weapons, explosives, drugs, and other dangerous objects concealed under a person's clothing. The person being examined stands in front of the system for a 3-second front scan and then turns around for a 3-second rear scan. A digital image is immediately displayed on the security operator's computer screen showing the shape and location of objects hidden under the clothing. The Secure 1000 operates by scanning the subject with a narrow beam of x-rays. Some of these x-rays will penetrate a few millimeters into the body, backscatter (reflect) in the opposite direction, and be gathered by sensitive x-ray detectors. Advanced image processing algorithms use this information to generate the images.

Portable X-Ray Imaging System.   The RTR-4 (http://www.americanaimpex .com/x-ray.htm) is the latest and only fully digital portable x-ray equipment available to explosive ordnance disposal (EOD) professionals. It provides the ability to quickly and efficiently search for weapons, drugs, and contraband in areas too difficult or time-consuming to search by hand. The RTR-4 is an extremely popular x-ray system for EOD applications.

TRACE DETECTION TECHNOLOGY

Trace detection technologies are based on the direct chemical identification of either particles of explosive material or vapor-containing explosive material. Thus, the presence of a threat object or bomb is inferred from the presence of particulate matter or vapor. The main difference between trace detection and electromagnetic imaging is that in trace detection, a sample of the explosive material must be transported to the instrument in concentrations that exceed the detection limit. Trace detection technologies cannot be used to detect the presence of metallic weapons. All trace detection equipment is regulated by TSR Part 1544.213. The two distinct steps in trace detection are sample collection and chemical identification. Trace detection practices are more commonly used for baggage screening (as opposed to people screening) in aviation security. Some basics of this technology are discussed next.

SAMPLE COLLECTION.   Explosive substances can be detected by instrumentation in their vapor or solid form. Initial efforts in the development of trace detection technology were focused on collecting vapor around the person or baggage. However, because many modern explosives do not readily give off vapor at room temperature, the focus has expanded to include detection of particulates of explosive materials on the skin and other surfaces.

If traces of explosive material are to be detected, they must be concentrated from an air sample (vapor technologies) or dislodged from a substrate (particulate technologies). In vapor detection, large amounts of air must be collected, from which small amounts of the substances of interest are extracted. In particle detection, pieces of explosive material must be removed from the surface to which they are adhering. Both trace detection approaches have strengths and weaknesses depending on the type of explosive material being sought. Vapor technologies are more effective for detecting explosive materials with high vapor pressures, while particulate technologies are more appropriate for detecting explosive materials with low vapor pressure such as military plastic explosives.

Samples can be taken either by having the subject walk through a portal or by passing a hand-wand device over the subject or piece of baggage. Either method may be implemented as a *contact* or *noncontact* technique. In contact portal sampling, passengers walk through a portal by pushing open a door or by rubbing against paddles or brushes. In a noncontact system, an airstream passes over the passengers as they walk through the portal. Hand-wand devices may be used to sample air around the person or to make physical contact. In general, contact methods focus on gathering particulates of explosive material from the hands or clothing of the subject. Noncontact methods may use the airstream to dislodge particles, or they may distill a sample of explosive vapor from the airstream.

Although using a hand-wand device is a potentially efficient sample collection technique, it is more labor-intensive and more time-consuming than collecting samples by using an automated portal. The optimum solution may be to attach a hand-wand device to a portal-based trace detection system as a higher-level surveillance accessory. This is a common technique used with metal detection portals. In trace detection, a single chemical-identification instrument could be served by both the portal and the hand-wand device sample collection mechanisms.

A problem in all trace detection approaches is clearing vapors or particles of explosive materials from the sample collection mechanism so that subsequent readings are not influenced by previous traces of explosive materials. The baseline readings must be monitored to alert operators to elevated levels of contamination before the contamination results in the shutdown of the equipment.

IDENTIFYING EXPLOSIVE MATERIALS.   After a sample is collected, a variety of commercially available chemical-identification technologies may be used to determine if the sample contains any explosive materials. The detection limit of most technologies under consideration by the FAA is generally sufficient to identify explosive materials in a sample. Even average-performance mass spectrometers are capable of measuring and identifying ultra-trace quantities of relevant chemicals, but more highly trained operators may be required to maintain a high level of detection capability. Much of the work sponsored by the FAA in trace detection technologies concentrates on integrating a particular chemical-identification technique with a sampling technique.

The chemical-identification part of the trace detection instrument is likely to be smaller than the portal or sample collection portion of the system. Therefore, for portal-type systems, airport accommodation is generally not dependent on the size of the chemical-identification component of the system but will more likely be affected by how the sample collection is implemented.

Some chemical-identification technologies may be small enough to be incorporated into handheld instruments and thus have potential use in hand-wand devices. Technologies under consideration include pyroluminescence (for detecting solid particles of explosives, e.g., gun powder), chemical sensors, and ion-drift spectroscopy. As discussed above, these technologies must be combined with a sample collection technique. The need to move large amounts of air in collecting a sample most likely means that available sample collection techniques may limit the application of these technologies to continuous surveillance methods relying exclusively on the use of hand-wand devices. Sample trace detection systems that are currently being marketed include the following:

- *Thermedics EGIS 3000* (http://www.aerospace.nasa.gov/library/chicago/checkpt.htm). The EGIS system is a freestanding analytical unit that uses high-speed gas chromatography with chemiluminescence detection. Samples are collected by wiping a surface with a specially designed filter. The filter is then placed into the analytical unit, and the EGIS begins an 18-second analysis to determine the presence of any kind of explosives in the sample.

- *Barringer Ionscan 400* (http://www.tslbrasil.com.br/Ionscan.pdf).   Ionscan is a detection identification device that uses ion mobility spectrometry to screen and search for trace quantities of explosives that contaminate baggage, electronics, and cargo. Samples are collected either by wiping a surface with a cotton cloth in a swab holder or by using a battery-operated handheld "vacuum cleaner" that uses a filter card. The sample is then placed onto a sample tray and slid into an analyzer which automatically confirms and identifies explosives within 5 seconds.

- *Ion Track Itemiser* (http://www.iontrack.com/applications/facilities/). The Itemiser also uses ion mobility spectrometry to screen and search for trace quantities of explosives that contaminate various objects. Samples are collected either by wiping a surface with a paper filter or by using a battery-operated handheld "vacuum cleaner" that uses a paper filter. The sample is then dropped into an analyzer which confirms and identifies explosives within 5 seconds.

- *CTX-5000SP* (http://www.invision-tech.com/products/ctx9000.htm). Employing transmission x-ray data, the CTX-5000SP system (from InVision Technologies) acquires an overall map of objects in luggage. It then uses strategic computer tomography to identify objects that may be explosives. This system is automated, using red and green lights to indicate the presence or absence of a threat. When it detects a potential threat, the operator is alerted and then uses the instrument's threat resolution features to validate that threat.

EXPLOSIVE DETECTION SYSTEMS (EDSs)

Bulk *explosive detection systems* includes any device or system that remotely senses a physical or chemical property (or combination thereof) of an object in an attempt to detect the presence of an explosive concealed in a container (e.g., passenger baggage). Since scrutinizing checked baggage is considered by the FAA to be the most critical component of the security program, bulk explosives detection equipment has become the foundation of TAAS (total architecture for aviation security). The critical performance metrics for explosives detection equipment include the probability of detection P_d, the probability of false alarm P_{fa}, and the throughput rate. This equipment must function without causing unreasonable delays. A TSA-certified EDS that passes the FAA's certification test must meet set detection rates and false-alarm rates (exact values are classified) and have a baggage throughput rate of at least 450 bags per hour. EDS replaces the old flawed thermal neutron activation (TNA) technology.

A number of nuclear and x-ray-based techniques have been investigated, but only three systems [all based on x-ray computed tomography (CT)] have been certified. The use of EDS is governed by the TSA under Part 1544.213. There are only a few TSA-FAA certified EDSs currently on the market. Nevertheless, some non-certified explosives detection devices have been selected for deployment.

Most of the bulk explosives detection equipment that have been deployed, or are being considered for deployment, are x-ray-based, with the exception being the InVision device, which is based on a nuclear quadrupole resonance measurement technique. The x-ray-based technologies that have been deployed include transmission x-ray, dual-energy x-ray, and CT. Of these, only the CT-based technology has passed the FAA bulk explosives detection certification test. Equipment using other technologies are being selected because they have large enough apertures to handle oversized bags, and operational data on this equipment would be useful. The performance baseline for the noncertified explosives detection devices is set by the

TSA. The sole electromagnetic instrument scheduled for deployment is a nonimaging technique based on nuclear quadrupole resonance (NQR). Currently, Boeing-Siemmens possesses the contract to deploy 1100 EDSs on the 438 airports around the country. Sample EDSs that are currently being marketed are reviewed below.

CTX 9000 INVISION TECHNOLOGIES (http://www.invision-tech.com/products/ctx9000.htm). The CTX 9000 DSi system is the world's fastest FAA-certified explosives detection system. FAA-certified at 542 bags per hour, it features alternate operational modes yielding even higher throughputs. The CTX 9000 DSi uses technology derived from medical computed tomography (CT). As the conveyor moves each bag through the machine, the system creates a scan projection x-ray image. From this image, the powerful onboard computer determines which areas need "slice" images, taken by the rotating x-ray source. Using sophisticated computer algorithms, the CTX 9000 DSi analyzes these images and compares their CT properties with those of known explosives. If a match is found, the system alarms and displays the object on the screen. The operator views the screen image to determine whether a real threat exists and then follows established protocols for threat resolution.

VIVID CT30 (http://www.dsxray.com/Vivid.asp?ProductCode5VividCT). This third-generation EDS combines dual-energy x-ray technology with a compact, high-resolution CT system, to produce a product that has the highest-performing FAA/TSA-certified EDS in its category. The VCT30 system uses a patented dual-energy x-ray technology to determine the effective atomic number of all objects in the bag. The CT subsystem provides highly accurate density information of selected objects. First, the bag is passed through the dual-energy x-ray subsystem, and objects of interest are identified. Next, the bag is moved into the CT subsystem, where tomographic slices are acquired for each object of interest. The VCT30's advanced algorithms automatically analyze the dual-energy x-ray and CT images to confirm if the object of interest has properties consistent with those of explosives. If the automated detection algorithm identifies a suspicious object, then the operator is presented with a high-resolution image of the complete bag and the suspect objects are highlighted in the operator interface. The operator can use advanced imaging tools to quickly determine if an object in question is a real threat.

EXAMINER 3DX 6000 (http://www.dsxray.com/Linescan.asp?ProductCode5eXaminer). This system utilizes an advanced computer tomography image of the entire bag or parcel, which is automatically analyzed and displayed in either a two-dimensional or three-dimensional image. The eXaminer 3DX 6000 is designed for ease of installation and support. The CT subsystem is the smallest and lightest system of its class. It has standard power requirements and is internally climate-controlled. The highly reliable design provides assured availability in any security environment.

METAL DETECTORS

Screening procedures currently used in U.S. airports, at least during routine operations, involve metal detection portals for screening passengers and x-ray imaging systems for screening hand-carried baggage. Metal detection devices impose a time-varying magnetic field in the space within the portal that induces eddy currents in metallic or ferromagnetic objects passing through that space. Various methods are used to detect these eddy currents; and when they exceed a preset level, an audible or visual alarm goes off, and an operator intervenes to ascertain the presence or absence of a dangerous object or weapon. The effectiveness of this security screening system depends not only on the performance of the equipment, but also on the performance of the personnel operating the equipment and resolving the alarms.

Metal detectors vary from portals to handheld systems depending on the application. Most passengers are required to pass through a portal-type metal detector, but a more scrutinized search could be conducted with a handheld or portable metal detector. Metal detectors are not as effective as other systems, and their deficiency lies in their inability to catch all forms of dangerous weapons. Their greatest weakness is that they do not detect metals incapable of being magnetized. Newer-generation metal detectors can search for ferrous or nonferrous objects. Metal detectors are used broadly around all the airports and are one of the most important sources of security.

The newest advance in metal detectors is the multizone concept. Through a multizone approach for portal-type metal detectors, sensitivity can be assigned to specific areas of the body zones that the metal detector would scan. Each zone has its own sensitivity adjustment. This permits the detection response of each zone to be independently adjusted. The horizontal zone corresponding to the foot region of the body can be made more or less sensitive than the waist region or visa versa. Use of metal detectors in airport security is regulated by TSA under its Part 1544.209. Sample metal detection systems that are currently being marketed are reviewed below.

B.O.S.S. (http://www.omni-security.com/product2/boss2.html). *Body orifice security scanner (B.O.S.S.)* is a fast, nonintrusive, inexpensive, high-sensitivity detector designed to detect metal objects hidden in body cavities. It is commonly used in corrections facilities to scan inmates for weapons and contraband objects hidden in anal, vaginal, oral, and nasal cavities. B.O.S.S. minimizes the need for unpleasant intrusive manual searches and increases the safety of officers and subjects. Because the measurement method is nonintrusive, it eliminates the liability and safety issues associated with manual searches. It also saves the time and expense involved in x-raying a person suspected of concealing contraband metal. The person being screened simply positions his or her chin near the oral sensor and then sits momentarily in the chair. The entire procedure takes just a few seconds. Magnetic field sensors housed in the seat of the chair and the oral sensor assembly automatically scan for the presence of metal. Audio and visual alarms are activated when metal is carried into the magnetic field. An alarm remains activated for the duration that an object is within the low-intensity magnetic field. The B.O.S.S. measurement system detects ferrous and nonferrous objects that are either stationary or moving.

SENTRIE OMNI (http://www.dsxray.com/Sentrie_Omni.asp?ProductCode5Omni). The Sentrie omni is a rugged, handheld metal detector designed to detect metallic weapons and objects using noncontact, physical screening procedures. The Sentrie

Omni incorporates high-quality construction with a sophisticated electronic design for reliability. The Piezo Audio System offers extended battery life due to extremely low power consumption. This is particularly valuable for screening the general public in high-volume applications that are typically encountered in airports, correctional facilities, and courthouses.

PM200HD (PORTAL TYPE) (http://www.dsxray.com/PM.asp?ProductCode5PM200HD). The PM200HD High Discrimination Detector is designed especially for high traffic screening applications. It detects weapons efficiently and discriminates against harmless personal possessions. Therefore traffic flow can be maintained at an optimum rate in locations where fast throughput is essential.

BIOMETRICS

The term *biometric* is used to refer to the emerging field of technology devoted to identifying individuals by using biological traits such as those based on retinal or iris scanning, fingerprints, or face recognition. *Biometric technology* is defined as the automated use of physiological or behavioral characteristics to determine or verify identity. This technology has traditionally been used in place of passwords in top-secret government areas but is increasingly finding uses in a wide range of security applications. One such application is to prevent unauthorized employees from gaining access to certain areas and assets. Another is to quickly identify low-risk users, such as prescreened airport passengers, so that security personnel can focus on the much smaller category of "high-risk" passengers.

Biometric systems are basically of two types: verification and recognition. By measuring a physical feature or repeatable action of the individual, we establish a biometric identification. Verification systems require that the individual seeking access have some sort of identification, such as a card, that is then matched with some physical characteristic of that person to make the verification. A reference measurement of the biometric is obtained when the individual is programmed into the device. Biometric measurement variables include

- Fingerprint (optical, silicon, ultrasound, touchless)
- Facial recognition (optical and thermal)
- Voice recognition (not to be confused with speech recognition)
- Iris scan
- Retina scan
- Hand geometry
- Signature scan
- Keystroke scan
- Palm scan (forensic use only)

The FAA has been instrumental in advancing the applications of biometrics in the aviation industry. Late in 2002, the FAA established the Aviation Security Biometrics Working Group (ASBWG), a multiagency working group, to accelerate its study of biometrics into existing airport security systems. This group is co-chaired by the FAA and the Department of Defense's Counter Drug Technology Development Program office. Likely applications include employee identity verification and access authorization, protection of public areas in and around airports using surveillance, passenger protection and identity verification, and aircrew identity verification (ground and en route). Sample biometric technologies that are currently being marketed are reviewed next.

FACEIT IDENTIX CORPORATION (http://www.identix.com/). FaceIT ARGUS is a state-of-the-art, automated, full-time facial surveillance and identification engine that allows each camera of a surveillance system to serve as a diligent observation point even when the video is not being actively observed. This system makes it possible to turn any passive closed-circuit TV system into an active, real-time identification and observation point. The system can handle an unlimited number of camera inputs and, through the Internet, be linked to any size database. The system is fast, accurate, and automatic.

HANDKEY RECOGNITION SYSTEM (http://www.handreader.com/products/hk/pi_handkey.htm). The HandKey uses the uniqueness of three-dimensional hand geometry to control access to restricted areas. The HandKey can operate as a complete stand-alone access control station. Door lock operation and alarm monitoring of the door status are provided. An external alarm circuit is included for operating an audio or visual alarm. HandKeys can be networked together through various communication methods: Ethernet, Modem, RS-232/422/485 tied in with HandHet for Windows for a complete access control solution for multiple openings.

STRENGTHENING AIRCRAFT AND BAGGAGE CONTAINERS

Another approach to aviation security is to try to strengthen aircraft frames and to plan redundancies in vital systems such as controls, electrical systems, and hydraulics, to mitigate the effects of bomb blasts in flight. A further alternative is to use hardened baggage containers that can control the effects of bomb blasts in checked baggage.

Retrofitting is difficult. It is easier and more practical to incorporate such design measures during the design stage of the aircraft. The FAA has engaged in extensive studies with military experts and airframe manufacturers to learn how aircraft fail due to explosions in flight and to discover measures to increase chances of aircraft survival. Explosives tests have been carried out to check calculations, both in the United States and in the United Kingdom. The best known of these efforts was the explosives testing on a Boeing 747 in Bruntingthorpe in England in May 1997. Several simultaneous experiments were run with four independent bombs.

The experiments tested the effects of various protective measures to different parts of the interior cargo hold of the aircraft and also tested a model of a hardened baggage container.

The aircraft hardening experiments in this case were run by experts from the United Kingdom, not the United States. They appeared to indicate some promise for the future, in which the application of material of relatively small weight may contribute significantly to the resistance of aircraft to bombs at certain locations in their cargo holds. However, due to the size of the current U.S. fleet, hardening an aircraft is neither economically feasible nor reachable. Therefore, hardened containers are being investigated by the FAA and other international airworthiness authorities as an alternative near-term solution.

Using HULDs (hardened unit-loading devices) to protect aircraft from explosive attacks is not a new concept. Airlines that operate in high-risk areas of the world have been using custom-built containers. Because these containers are much too heavy for general use, only one or two are used per aircraft for carrying select items. However, new advances in technology and the need for increased security after the terrorist attacks of 2001 will trigger the greater use of these containers.

Regarding hardened baggage containers, the FAA has focused until now on ULD-3 versions, suitable for wide-body aircraft. However, the industry is currently engaged in designing containers that are substantially smaller and could fit a narrow-body aircraft. In 1994, a hardened ULD-3 container developed by JAYCOR was tested successfully, using bombs that were comparable in size to those used in past terrorist events. This container was not a great deal heavier than many currently used aluminum ULD-3s. Further development was needed, however, to incorporate doors into the container that would make its use by air carriers operationally feasible. The FAA then contracted with several other vendors to provide models for testing. This testing produced mixed results until Bruntingthorpe, when another JAYCOR container with a usable door successfully contained one of the bombs detonated in this test. The FAA has contracted with JAYCOR and with Galaxy Corporation to produce further containers for explosives testing. The tests are being carried out in close cooperation with several U.S. air carriers. Potential bag resistance container systems that are currently being marketed are reviewed below.

JAYCO HULD (http://www.jaycor.com/set/jayhuld.htm).   The JAYCOR Hardened Unit Load Device (HULD) is made of composite material fabric and is a high-strength, lightweight, unibody container for widebody aircraft designed to confine the blast and suppress the fires caused by internal explosives. Prototype testing has demonstrated that the HULD can contain explosives significantly above the threshold of current detection technology. The device not only conforms to the IATA specifications for these containers, but also promises to provide greater durability due to its extraordinary strength. When it is combined with existing detection technology, comprehensive protection for airline passengers can be provided. This technology is ready to be put into service.

TELAIR **HULD** (http://www.telair.com/cargo-containment-systems/index.html).   The Telair HULD is the only aircraft security container to have successfully passed the FAA's blast resistance test and to be fully certified for onboard use. This innovative baggage container relies on advanced composites technologies. The blast-resistant capability is the last line of defense against undetected explosives concealed in passenger baggage and stowed in the under-wing cargo holds.

The future success of hardened containers could radically change the detection capability requirements for explosives detection equipment for checked baggage screening. Of course, for any given container, a large enough bomb can be constructed to overcome it. However, a larger bomb is more susceptible to detection, and increasing the mass of explosives that need to be detected would relax the requirements on the detection equipment. If it is possible, eventually, to protect against a bomb of substantially greater mass than is now needed to destroy an aircraft in flight, this approach could conceivably lead to smaller, cheaper, and faster bomb detection equipment.

COCKPIT DOOR REINFORCEMENT

On January 10, 2002, the FAA published new standards to protect cockpits from intrusion and small-arms fire or fragmentation devices, such as grenades. The rule requires U.S. operators of as many as 7000 airplanes to install reinforced doors by April 9, 2003. The rule issued new requirements for all airlines operating aircraft into or out of the United States under FAR 129. Sample reinforcement materials that are currently on the market are reviewed below.

TELAIR (http://www.telair.com/info/news/pr_jul2402_1.html).   Telair, which was developed by DuPont's Advanced Fibers division, is a composite material that weighs significantly less when compared to other hardening alternatives and is resistant to bullets, blunt-force impacts, and fire. It meets the National Institute of Justice (NIJ) level IIIA threat test criteria, regarded by the FAA and the Transportation Safety Administration as a threshold standard. This test calls for the firing of either a 9-millimeter or .45-caliber Magnum bullet into a panel of the material from close range. Telair's composite panel repeatedly absorbed the test bullets, permitting no ricochet and no loss of panel integrity.

PERMAGLASS X (http://www.permali-gloucester.co.uk/doors.htm).   Permaglass X is a lightweight composite armor material that provides protection against handgun bullets per NIJ level IIIA criteria. Permaglass X provides high mechanical strength and stiffness, and it is fully compliant with all JAA/FAA flammability regulations for aircraft cabin interiors. A further benefit is that unlike for metallic materials, the composite construction of Permaglass X virtually eliminates the risk to passengers and crew from ricochets.

SPECTRA SHIELD (http://www.flightsafetyzone.com/downloads/(Spectra)%20HASS%20 DataSheets.pdf).   Panels made with Spectra Shield Pre-Consolidated Roll (PCR) materials can be designed to meet the ballistic requirements set forth by NIJ level IIIA standards. Spectra Shield PCR materials were developed specifically for hard armor applications where light weight is critical. They can be used in a wide range of armor applications in fixed and rotor wing aircraft, military vehicles, watercraft, VIP vehicles, military and law enforcement helmets, riot shields, and breast plate inserts rated to defeat specified rifle rounds for soft body armor.

COMPUTER-ASSISTED PASSENGER SCREENING

For a number of years, the FAA, in cooperation with Northwest Airlines, has been developing a *computer-assisted passenger screening* (CAPS) system, which permits the airline's computer reservation system to use information in the passenger name record to exclude most passengers from further security measures. The FAA arrived at the criteria and algorithms used to perform this function through consultations with a large number of security and terrorism experts, who gave their assessments of the likely patterns of behavior of individuals intending to attack civil aviation, as reflected in their passenger name records. These criteria do not involve the ethnic, gender, or religious characteristics of passengers. In the United States, we would not use such information in passenger screening. Such actions by the government or the air carriers would be unlawfully discriminatory.

The Department of Justice was given the task of independently examining the criteria and procedures that the FAA directs air carriers to use in screening passengers. The Department's review, conducted by its Civil Rights Division with assistance from the FBI and the Department's Criminal Division, covered both the manual security screening process (in use before the introduction of the automated system) and the CAPS system. The review found that neither procedure unlawfully discriminated against passengers based on their race, ethnicity, national origin, or religion. The Department of Justice did recommend certain follow-up actions that could be taken to ensure that the civil rights of the flying public are maintained in an air transportation environment, secure from terrorist threats. The Department of Transportation and the FAA have acted on all the recommendations.

In September 1996, a follow-on grant was awarded to Northwest Airlines to refine the CAPS program to achieve operational capability and to assist in adapting CAPS to other airlines' reservation systems. Northwest met with other air carriers in the fall, conducted preliminary system tests during the winter, and progressed to operational tests on selected flights in its system in April 1997. Northwest has completed the process of phasing in CAPS throughout its domestic system, with over 150 stations online today.

Seven major air carriers, covering all major airline reservation systems, began work in earnest on developing their CAPS systems in 1997. Several major carriers began field-testing CAPS in 1998. The FAA is helping to fund these efforts through cooperative agreements that will disburse to the carriers funding appropriated by Congress for CAPS.

It is anticipated that all major carriers will have phased in CAPS voluntarily before a new federal regulation mandates its use as the method of determining which passengers' bags must be subjected to additional security measures, such as bag matching or screening with explosives detection devices.

NONTECHNOLOGICAL APPROACHES

High-technology detection methods can yield results, but it is clear that these results are expensive and are not perfect. The small added yield in security is arrived at only at some very great expense. Further, the cost is not just one of acquisition. There is also the matter of operations and maintenance and the opportunity costs to passengers standing in line, among others.

Although technology has its place in the overall security scheme, it is no panacea for air transportation security issues. Greater attention needs to be paid to security personnel. They play a critical role and are the ones who interpret the results of the technological analyses.

Until the TSA was established and took over security screening, most security personnel were low-paid workers who lacked the required training. The profession was plagued by questions of morale, turnover, stress, and ineffectiveness. FAA insiders had criticized security personnel as being inadequate. The ATA responded by developing standards that called for better selection, improved work conditions, and rewards for effective security personnel. In addition, many airlines upgraded their personnel screening. Nevertheless, despite new training requirements, the process remained uneven. The length of training was too short, and the subject matter was very loosely defined. However, when the TSA took over this function in 2001, conditions improved for security personnel in terms of pay, training, and other benefits. This has led to improved levels of security at airports.

A related problem is that of airport workers and staff. Airports are like cities; their personnel engage in a variety of activities and represent vocations ranging from salespeople to mechanics to cleaners. Because this diversity of workers provides terrorists with numerous opportunities, the FAA requires background checks of previous employment records and the wearing of badges. Despite these efforts, however, these measures do not guarantee security. Terrorists can forge badges, although forging computerized badges is a far more difficult task, requiring insider collaboration or access to the computerized database and badge process. Terrorists can follow a potentially riskier path, such as using threats to family members to pressure employees or simply bribing them to gain access.

There are alternative technological and tactical approaches to airport security. One might be able to divert some level of resources from hardware applications toward improved intelligence gathering to intercept terrorists long before they arrive at the airport. This suggestion, of course, raises the question of how intelligence gathering is to be improved, but this question is beyond the scope of this chapter. From an economic point of view, the use of resources to gather intelligence is more attractive than elaborate security technology. Although expensive hardware guarding the front door is very effective, better intelligence can be applied across the board to transportation, government, and business toward the improved security of them all, particularly since even the most effective technology remains vulnerable to changes in terrorist tactics. If we become experts in weapons and explosives detection, the threat could become biological (germs in the water supply) or chemical (e.g., the Tokyo subway system).

In addition, technology suffers from problems defined by two related axioms. The first is that security is very site-specific. What works well and inexpensively in one place is not necessarily a sound approach in some other place. The second axiom is political: Regulators seldom recognize the real-world needs of security. No matter how well conceived the legislation or regulation may be, it is not likely to have anticipated the kinds of problems described.

An important concluding point is that, in spite of all that has been accomplished to ensure the safety and security of the traveling public, the terrorist, in one sense, always has the upper hand. While we must protect every element of the transportation system at all times (not just aviation), the terrorist has the luxury of being the only one who knows the time, the place, and the method of the next attack.

KEY TERMS

Security versus safety

Federal Aviation Regulations (FARs) 107, 108, and 109

International Civil Aviation organization (ICAO) Annex 17

The International Security and Development Cooperation Act of 1985

Anti-hijacking or Air Transportation Security Act of 1974

The Air Carrier Standard Security Program (ACSSP)

Indirect air carrier security

The Air Marshal Program

Aviation Security Improvement Act of 1990

Antiterrorism Act of 1996

White House Commission on Aviation Safety and Security, 1996

Aviation and Transportation Security Act of 2001

Transportation Security Administration (TSA)

X-ray imaging technologies

Passive millimeter-wave imaging

Active millimeter-wave imaging

Trace detection technology

Explosives detection systems (EDSs)

Metal detectors

Biometric

Computer-assisted passenger screening (CAPS)

Body orifice security scanner (B.O.S.S.)

REVIEW QUESTIONS

1. What is the difference between security and safety?

2. Give three examples of attack categories on civil aviation.

3. Describe and distinguish between the important features of airport security, air carrier security, and indirect air carrier security.

4. What is the Air Marshal Program and why was it instituted?

5. Discuss the salient features of the Air Carrier Standard Security Program (ACSSP).

6. What caused the Aviation Security Improvement Act of 1990 to be enacted? Highlight the security measures that evolved as a result of this Act.

7. What was the reason for the White House Commission on Aviation Safety and Security, 1996? What did the commission produce?

8. What events led up to the Aviation and Transportation Security Act of 2001? What were the highlights of this Act?

9. Discuss some of the problems encountered in the gathering and analysis of intelligence for security purposes.

10. Give an example of four different security devices used to prevent weapons and explosives from entering an aircraft, and explain the workings of these devices.

11. What are some of the nontechnological approaches to airport security?

REFERENCES

Anderson, Theresa. 2000. "Airport Security." *Security Management*, pp. 73–74. February.

Dorey, F. C. 1983. *Aviation Security*. Van Nostrand Reinhold Company, New York.

FAR 107, Airport Security.

FAR 108, Airplane Operator Security.

FAR 109, Indirect Air Carrier Security.

ICAO Annex 17. *International Standards and Recommended Practices—Security, Air Carrier Standard Security Program.*

Moore, K. C. 1991. *Airport, Aircraft, and Airline Security.* 2d ed. Butterworth-Heinemann, Boston.

National Commission to Insure a Strong Competitive Airline Industry. 1993. Report to the President and the Congress. Washington, D.C., August.

Sweet, K. M. 2002. *Terrorism and Airport Security.* Symposium Series, vol. 8. The Edwin Mellen Press, Lewistown, N.Y.

WEB REFERENCES

http://www.boeing.com/ids/airport_security/equipment.html

http://books.nap.edu/books/0309054397/html/17.html#pagetop

http://transportationsec.com/ar/security_faas_working_group/

http://users.aber.ac.uk/scty34/50/civair.htm. Williams, Paul. "Securing the Skies: Threats to Civil Aviation from International Terrorism," *Interstate Online*, Spring 1997.

http://www.airsafe.com/index.html

http://www.carinex.hu/english/secsaf.htm

http://www.faa.gov

FAA Office of Civil Aviation Security web site:

http://www.fas.org/irp/threat/212fin~,1.html

http://www.gao.gov/airptsec.html

http://www.oceta.on.ca/profiles/barringer/ionscan.html

http://www.rangersecurity.com/wthru2/wtindex.html

http://www.tsa.gov/public/index.jsp

http://ww1.faa.gov/index.cfm/apa/1279/4322764F-434C-4415-A54FE6E9493AF9F

AIRLINE SAFETY

LEARNING OBJECTIVES

After completing this chapter, you should be able to

- List the elements of a successful safety program.
- Discuss ways in which management can demonstrate visible interest in safety.
- Explain why employees need to be involved in a safety program.
- List ways to involve employees in a safety program.
- Explain ways of holding employees accountable for their actions.
- Discuss the reasons for trends analysis and give examples of safety parameters that are trended.
- Explain why and how accidents are investigated.
- Explain the difference between the reactive and proactive approaches to safety.
- Define what an audit is and give examples of audits.
- List and discuss the basic elements of an internal evaluation program (IEP).
- List the three agencies that regulate safety for the airlines.
- Explain why communications is important for a successful safety program.
- Discuss the importance of safety training.
- Outline what it takes to deliver effective quality training.
- Explain the importance of documenting training efforts.
- Discuss the role of ALPA's committee structure and how accident investigations are conducted.
- Summarize the functions and importance of the Flight Safety Foundation (FSF).

INTRODUCTION

A commercial airline, like any other major business, has to integrate and manage several different functions/departments while delivering products and/or services to the customer and (hopefully) turn a profit. Safety is an major function, and it is important in that it permeates (or should permeate) through all departments. Safety must be built into every product, service, or operation and must be an integral part of every employee's job responsibility. The airline industry in particular has always placed great emphasis on safety and has moved aggressively to identify and control problems that cause accidents. Airlines have learned, sometimes the hard way, that active management of risk is an absolute requirement for a healthy company. This chapter brings together many of the concepts discussed in the previous chapters. This chapter will discuss the fundamental elements of a safety program that are essential for minimizing the occurrence of accidents and injuries. Also included in this chapter will be a sample airline organizational structure together with the functional descriptions of each position. This chapter will conclude by reviewing the role of the Air Line Pilots Association (ALPA) and Flight Safety Foundation in air safety.

SAFETY PROGRAM ELEMENTS

A safety program in essence is a tool used by management to assign responsibility and establish accountability for maintaining safe working and operating environments while maintaining (safety) regulatory compliance. Safety programs exist in various forms depending on the size and type of operation. However, all successful programs have the following elements:

- Management leadership and commitment
- Employee involvement
- Responsibility and accountability
- Trends analysis and risk assessment
- Accident investigation and auditing
- Hazard prevention and control programs
- Communications
- Training

MANAGEMENT LEADERSHIP AND COMMITMENT

Management must provide strong, *demonstrable, and visible* leadership if the safety program is to succeed. Management's accident prevention attitude and safety culture are typically reflected in the attitudes of the supervisors which in turn are reflected in attitudes of the operators and the rank-and-file employees. Hence, if top management does not demonstrate a genuine interest in safety, very few others in the organization will pay attention to safety.

Effective accident prevention can be linked inalterably to effective airline management. This precept is found in the earliest safety textbooks that were developed in the industrial safety field. It also can be found in the attitudes and practices of some airlines as early as the late 1930s. W. A. Patterson, United's board chairman, was frequently cited as putting particular emphasis on safety to his senior management staff. Some of the airline accident investigations in the 1930s identified so many air navigation problems and questions of investigation objectivity that the Air

Safety Board was formed in the United States along with the Civil Aeronautics Administration (CAA), the predecessors to today's National Transportation Safety Board (NTSB) and Federal Aviation Administration (FAA). These were government management corrective efforts.

The earliest teachings in accident investigation at the University of Southern California in the 1950s concentrated on human, machine, and environmental factors and began to discuss safety programs. The next decade saw the initial development of "Advanced Safety Management" and "Command" courses, first from the U.S. Air Force, then from the U.S. Navy, and later adopted by all U.S. military services. The classes were significant because they comprised higher-ranking officers than the safety officers who implemented safety programs at the working level (including the investigation specialists). These ranking officers then had access to very senior commanding officers, something which would be somewhat difficult for the lieutenants.

These new safety programs were also significant because they forced those who were teaching the courses to approach accident prevention more from management's point of view than had been done previously. The 5-M diagram discussed in Chap. 6 is an example of this. It portrays symbolically the overview role that management must play in accident prevention or, conversely, where causation may lie in the event of an accident. The man-machine-medium (environment) fundamentals were retained from the past to which another factor—mission—was added. This addition emphasized that in a military or civil endeavor, admitted to or not, one does not practice safety professionally just to prevent injuries or death. The mission, be it delivering ordnance or providing a viable air transportation system, is an important piece of the safety package.

Chapter 6 also shows the interrelationships of these various factors to stress and illustrates that one should not, for example, examine human or machine factors independently—the man-machine interface depicted by the overlapped areas between the two variables must be studied, too. Similarly, when one asks what most influences the combined man-machine-medium-mission, the answer most logically is management, hence its top position.

As mentioned earlier, management must demonstrate strong leadership and interest in safety if the safety program is to succeed. Some of the ways in which management can show visible interest in safety are as follows:

- Establish a written safety policy that clearly defines management's philosophy on safety. This policy should clearly state the priority that management places on safety and health protection in relation to other organizational values. Management should clearly communicate the policy to all employees, especially to personnel with site responsibilities, so that no confusion will exist when a conflict surfaces between two or more of these values, such as productivity and safety. The policy should also be realistic. For instance, a company's safety policy that simply states "safety is our total priority" is not facing reality. It may do so with a shock when the hard facts of compromises come to light in an investigation—compromises deemed necessary during day-to-day operations. The policy should be signed by the highest-ranking corporate official and be posted for all to see and in management offices. A specific policy that outlines clear goals and objectives for the safety program is more effective than a general generic policy statement. Employees should be able to relate their activities toward achieving these set goals and objectives.

- Consistently enforce policies, procedures, and programs.
- Emphasize that operations and safety are of *equal* (not greater) importance for an efficient operation.
- Set good examples by scrupulously following established worksite rules and practices. For example, if personal protection (earplugs, gloves, hard hats, etc.) is required in certain areas by company rules, then managers too must wear such protection when they are present in these areas.
- Regularly review department safety records and workers' compensation data with department heads.
- Publicly recognize good safety practices and results.
- Participate in site safety audits and inspections.
- Attend safety meetings.
- Participate in safety awards and other events that support the safety effort.
- Maintain awareness of safety throughout the organization through bulletin board announcements, e-mails, and employee meetings.
- Review incident and accident reports and promptly act upon them.
- Personally stop activities that are hazardous until the hazards can be corrected.
- Consistently hold employees and managers accountable for their actions.

A policy shortcoming frequently seen is the chief executive officer saying that he or she is the chief safety officer without acknowledging delegation of accident prevention tasks, which must surely be done to some degree. The CEO must create additional accountability with the intended result that all people in the organization realize their responsibility in safety is not just to themselves but to others as well.

EMPLOYEE INVOLVEMENT

Involving employees in programs that directly affect their safety and health, in addition to having a moral connotation, makes good business sense for the following reasons:

- Employees who are most in contact with the potential safety and health hazards will have a vested interest in effective protection programs.

- Line workers who have intimate knowledge of the job have proved to be very effective problem solvers.
- Group decisions draw from a broader database of experiences.
- Employees support programs that include their input.
- Encouraging employee involvement coupled with taking their contributions and ideas seriously increases job satisfaction, leading to increased employee productivity.

Employee involvement can take place through several activities related to safety. Employees can

- Participate in joint labor-management committees, employee safety committees, union committees, brainstorming quality circles, or other types of advisory or specific function committees.
- Conduct site and area inspections.
- Take part in accident or incident investigations.
- Perform job analyses and develop safe work practices.
- Train existing and new employees.
- Develop and/or revise safety and health programs.
- Make presentations at safety and health meetings.

RESPONSIBILITY AND ACCOUNTABILITY

While management has the ultimate responsibility for safety, it is management that sets up the organizational structure, assigns authority, and allocates resources to carry out the safety program. In this context, an organizational question worth asking is, What is the appropriate organization for an airline safety program, and where in the corporation should it report? Although there have been various approaches to these issues, it is generally agreed that, for optimum effectiveness, there must be a designated safety officer who serves as the principal focus for safety efforts in an organization. It is further generally recognized that to effect the necessary changes, it is important that the safety officer report to the highest levels of a corporation, ideally to the chair or chief executive officer. There are two reasons that this level of reporting is important. The first is the implicit message that safety is a high-priority corporate goal. The second is that the safety office can have direct access to all senior management in each of the operating departments. Because it is necessary to cross organizational lines to solve many safety issues, placing the safety office organizationally where it can easily do so is one way to ensure that effective corrective and preventive actions take place.

To have an effective program, management must assign responsibility and establish accountability through all management levels. While there will be varying (decreasing) levels of responsibility for safety as one moves from senior management through the level of rank-and-file employees, everyone in an organization should have some responsibility for safety. Responsibilities must cascade logically from objectives that were established for the overall program goal. One way of assigning responsibility for safety is to clearly spell it out in the employee job descriptions. Each position should be examined to determine what part it should play in the overall safety program. When viewed collectively, a company's job descriptions should reflect the overall organization's safety management system (in addition to other management systems that are needed for the business).

Once assigned, these safety responsibilities must be clearly communicated so that managers, supervisors, and employees throughout the organization understand what performance is expected of them and that they will be evaluated against these responsibilities.

Assigning responsibility, however, is meaningless unless it is also accompanied by adequate authority and sufficient resources to carry out this responsibility. Resources include the availability of well-trained and adequately equipped employees and funds for operations and capital expenditures. Once individuals have been assigned the responsibility and provided with appropriate authority and resources, they must be held accountable for what is expected of them. Accountability is crucial for success of any program, and it is a process in which individuals are held personally responsible for their performance. A portion of an employee's job evaluation (which could lead to pay increases or bonuses) should include performance on safety responsibilities.

TRENDS ANALYSIS AND RISK ASSESSMENT

To protect employees from workplace hazards, it is important to identify and understand what those hazards are. Also, as mentioned in Chap. 6, the risk management process requires that hazards be prioritized by levels of severity so that resources can be allocated for their control or elimination. Recording and trending accident, incident, injury, and other safety data (e.g., safety violations) support the hazard identification and prioritization process. The regulatory requirements for recording and/or reporting certain types of accidents, incidents, illnesses and injuries (AIII), and their causes is one way of forcing companies to identify and record hazards. Recording and reporting requirements were covered in earlier chapters. Each agency (EPA, FAA, or OSHA) has its own recording and reporting requirements. Sample safety parameters that are typically recorded and analyzed for trends include:

SITE/HARDWARE PARAMETERS

- Engine in-flight failures

- Hydraulic system failures
- Depressurizations
- Takeoff aborts
- Flight control failures
- Tire failures and aircraft fire incidents
- Violations from site audits and accident investigations
- Environmental spills

INJURY/INCIDENT PARAMETERS (AIII)

- Flight-related injuries
- OSHA recordable injuries and illnesses
- NTSB incidents

The data may be extracted from a variety of sources, including crew reports, maintenance information, flight operations quality assurance (FOQA) data, manufacturer data, accident/incident reports, OSHA 300 logs, and so forth. Data trending could reveal patterns of common causes not apparent when AIII were first reported. Risk assessment and reduction is a fundamental component of an effective accident prevention program.

SAFETY PERFORMANCE MONITORING

Trends analysis can also be used to monitor safety performance. Successful overview of technical safety performance requires the employment of proper statistical methods to determine safety trends. Once the data collection process is in place to determine incident levels, analytical normalization of the data is necessary. For example, to account for differences between aircraft with different numbers of engines, it is necessary to normalize shutdown rates in relation to engine hours logged by each fleet. Power plant failure rates are typically depicted in a rate per 1000 engine hours. Thus, one in-flight shutdown in a fleet logging 50,000 engine hours would be presented as 0.02 failure per 1000 engine hours. In this manner, the average shutdown rate is comparable between aircraft types independent of the number of engines per aircraft. A similar normalization process can be used for other system failures and other safety statistics. For example, 100,000 hours is used to standardize runway incursion rates, and OSHA accident rates are calculated per 200,000 human-hours. Following identification and selection of the desired indices representing safety performance, routine review of the data and comparison with past performance are essential for the identification of developing problem areas.

When one is developing an AIII trending and analysis system, it is important to do the following:

- Establish the number and type of safety parameters to be recorded.
- Assign responsibility for reporting such data.
- Assign responsibility for managing the data and developing reports.
- Establish procedures for acting upon the reports.
- Develop a system to disseminate the information.

ACCIDENT INVESTIGATION AND AUDITING

ACCIDENT INVESTIGATION

An important component of the safety effort at any organization is an incident and accident investigation system. By investigating accidents, organizations learn how to prevent future accidents. Some progressive companies investigate all unintended events, irrespective of whether these events led to injuries, illnesses, property damage, or equipment damage. Most companies, however, require some criteria to be met before an investigation is triggered. Investigations are initiated, for example, if accidents

- Require reporting on the OSHA 300 log
- Generate workers' compensation claims
- Cause lost workdays
- Meet NTSB definitions of accidents/incidents
- Result in an environmental spill
- Cause property and/or equipment damage over a preset dollar value

Once an investigation is triggered, the processes used to investigate accidents, whether they are FAA/NTSB, OSHA, or EPA related, are essentially the same. A formal notification system should be in place to provide timely notice of such events to the safety department. A typical investigation process is now described.

When the safety department is informed of an event, an immediate determination is made about whether the event meets the company's criteria for an investigation. When an investigation begins, depending on the circumstances, one or more individuals may be assigned to conduct the investigation, especially when it is clear that there may be more than one operating department involved, such as for incidents occurring during pushback. The manner in which the investigation is conducted is a function of the type of event and the circumstances under which it occurred. Typically, all key personnel involved in an incident or accident are asked to submit written statements describing the facts and circumstances as they saw them. In addition, provisions are made immediately for conducting interviews with the key individuals involved in the incident or accident. These interviews can be done in person or over the telephone, depending on the specific circumstances. In-person interviews are preferred. Similarly, interviews can be done individually or with a part or all of an entire crew. The people being interviewed are told that the purpose of the interview is safety only, that the information will not be used for any disciplinary or other purpose, and that it will remain confidential, to the extent permitted by law. Information learned from these interviews, along with written statements, forms an important element of these safety investigations.

In addition to the interviews and personnel statements, other records and documents may be obtained and reviewed. Such material might include training records, training manuals and syllabi, aircraft and procedures manuals, information bulletins, and similar material. Another important source of information is the flight data recorder, which is read out as part of the investigation of many in-flight events. Other relevant information is identified and obtained as necessary, including accident reports, technical reports, and any other documentation that may contain information relevant to the incident or that can provide useful information for formulating recommended practices and corrective actions.

Following the collection of basic information, the investigator assembles and issues an accident report. These are brief, synoptic reports that describe the basic facts and circumstances of the event under investigation, including a history of the flight, a summary of damage and injuries, an analysis, a list of findings, and, most importantly, recommendations for corrective or future preventive action. Also included is a brief summary of safety actions taken—operating departments are not required to wait for a formal recommendation from the safety department prior to initiating corrective or preventive action.

Findings from safety investigations are derived from the facts and circumstances associated with the event and are based on the investigator's analysis. There is no effort to determine a cause or probable cause of an event. Findings are a list of factors related to the event under investigation.

The recommendation process is probably the most important part of this safety investigation program, and it is important to understand how it is handled. The following points should be noted when recommendations are formulated:

1. Recommendations should be derived from the findings of an investigation. In general, there will be a direct link between a finding and one or more recommendations.

2. Recommendations should be coordinated with the appropriate departments prior to issuance; however, the safety department should retain sole responsibility for the decisions to issue recommendations.

3. The receiving department should respond in writing as to the disposition of each recommendation. In the event that the recommendation is not adopted or an alternative safety action is taken instead, the rationale should be set forth.

4. Open recommendations should be tracked on a regular basis (30, 60, and 90 days) until completed.

The accident investigation process described applies to investigations of minor accidents. In the case of a major accident (hull loss, occupational death, or a major environmental spill), coordination will be required with external agencies (NTSB, FAA, OSHA, EPA). For example, in the event of an aircraft crash, it is important that an airline have a detailed emergency response plan that contains, among other things, detailed plans for a go-team who will participate in the accident investigation. The go-team should be headed by a senior manager, who serves as the primary coordinator for all company activities related to an accident investigation. Appropriate technical personnel are designated as potential members of a go-team, the actual makeup of which is determined on the basis of known facts at the time of notification, most importantly, aircraft type and location. With regard to the latter, for any accident occurring outside of U.S. airspace, the company go-team reports to the U.S. accredited representative, in full accordance with International Civil Aviation Organization (ICAO) Annex 13.

To support the go-team, all necessary equipment and supplies, such as personal protective equipment, communications gear, and other material that may be necessary to conduct a major accident investigation, must be ready for instant shipment to the scene of an accident. All members of the go-team should be formally trained in accident investigation and must have the required (by OSHA) blood-borne pathogen training. The go-team roster should also identify other personnel who are tasked with providing administrative support to its technical members. The emergency response plan specifies the staffing of a command post by designated personnel from the affected airline. These people are responsible for the coordination of all activities related to the accident investigation, including the assembly of records, manuals, bulletins, and other necessary materials. Oversight and leadership of the entire effort should be under the direction of senior management, including the corporate safety officer. An annual emergency response drill is required to ensure that the plan is adequate and that the designated personnel are appropriately trained.

Although absolutely necessary, an emergency response plan and associated elements of the safety function are quite obviously something that no airline wants to activate for real. To meet this challenge, every airline company must have some form of independent, proactive accident prevention effort. The best means to accomplish this is to seek to identify hazards and risks consistently and then to eliminate these risks through accident prevention measures.

On a final note, virtually all operational incidents will require a certain level of technical investigation and analysis to fully understand and identify the underlying cause factors. Within the airline corporate structure, investigative responsibility for flight safety incidents must be clearly assigned. Similarly, professional investigative methods must be consistently employed in the technical area. Use of investigative tools such as the digital flight data recorder (DFDR) requires a consistent objective and confidential method of analysis. Also, since analysis methods require complex transcription methods, it is likely that DFDR analysis will occur at the maintenance/engineering department. However, DFDR information must be maintained in a strict confidential status with operational DFDR analysis performed by personnel familiar with current operational procedures.

AUDITING

Analyzing and addressing hazards after they have materialized (caused accidents and injuries) are commonly referred to as a *reactive approach* to safety. The preferred *proactive approach* to safety requires that hazards and hazardous conditions be identified before they cause accidents and injuries. One way of accomplishing this is through *audits*. The words *audit* and *inspection* are often used interchangeably with some confusion about their definitions. The problem is compounded when the word *survey* is thrown into the mix. Most schools of thought (including the authors of this text), however, consider an audit to be more comprehensive and have a greater scope of work than an inspection. While inspections usually identify workplace hazards, audits are designed to evaluate programs and management issues that have resulted (or could result) in workplace hazards. Safety audits are an integral part of a safety program, and in addition to uncovering hazards, audits reveal the level of compliance with regulatory standards as well as measure the effectiveness of the safety program. Some of the common approaches to auditing a facility include these:

- *Comprehensive audits*. These are extensive, detailed safety inspections/surveys (site conditions and programs) of the entire organization (an air carrier), a facility within an organization (an air carrier hub or maintenance facility), or an operation within a facility (fuel handling at the ramp). These audits are usually conducted on an annual basis. These audits are conducted by a team of safety professionals from within and outside the organization, facility, or site, and they are usually led by the company safety director or the divisional or site safety manager. This type of audit results in a formal written report of findings and recommendations and usually requires a response within an agreed time frame.

- *Self-audits*. These are informal daily, weekly, or monthly audits (more appropriately inspections) and form the crux of the *internal inspection program* (IEP). Additional information on IEP is presented in the next section. The frequency of checks depends on the application. For example, certain areas of airports have to be checked daily (or more often), and some operations in a hangar or on the ramp may require daily inspections. These audits are conducted by line management at the site or facility and usually result in a checklist of items that either are in acceptable condition or need fixing. No reports are normally generated as part of these audits. One of the functions of the director of safety is to help set up and provide expertise to local internal inspection teams.

* *Status audits.* These audits are intended to determine the status of compliance at the time of observation. These are usually subject-oriented audits and are conducted in areas of high risk and/or areas of known or perceived concern. A spot inspection of refueling for a randomly selected aircraft (or a group of aircrafts) and evaluating a welding operation are examples of status audits. These audits may generate reports if the findings are serious enough to require management's attention.

HAZARD PREVENTION AND CONTROL PROGRAMS

The airline safety department is normally responsible for corporatewide safety and compliance functions. Safety programs are developed to satisfy the regulatory requirements of the following three agencies:

* Federal Aviation Administration
* Occupational Safety and Health Administration
* Environmental Protection Agency

All through this text, excerpts were discussed of the above agencies' regulations together with the operational areas within aviation (airports and airlines) where they applied. In addition, each of these agencies' organizational structure, geographic location and territorial coverage, and (holistic) regulatory requirements were reviewed. While a detailed review of each agency's regulations is beyond the scope of this text, references to study these regulations in greater detail were included.

There are two points of a safety program that deserve emphasis. First, the department is not the compliance police department—it does not enforce compliance; it only provides corporatewide services to assist the operating organizations in maintaining compliance with applicable regulations. Responsibility for compliance and enforcement resides, as it must, with those who are responsible for the operation. Second, a critical element of any airline safety (and compliance) program is an internal evaluation program. The IEP has been demonstrated to be among the most powerful tools available to management to ensure that all aspects of the operation are being conducted in accordance with applicable regulations, internal policy, and standard practice, as defined in the company manuals. The FAA has officially recognized the importance of such programs and has published an Advisory Circular (AC-120-59) describing how these might be structured. Some important features of this Advisory Circular and other applicable orders are now reviewed.

INTERNAL EVALUATION PROGRAM

Advisory Circular 120-59, Air Carrier Internal Evaluation Programs (IEP), was issued in October 1992. In September 1995, the suggested IEP was extended to the repair stations when Advisory Circular 145-5, Repair Station Internal Evaluation Programs, was issued. This program gives management a tool to systematically evaluate the effectiveness of its management systems and preempt noncompliance before it occurs. There is no regulatory requirement for an IEP, and participation in the program is voluntary and left solely to the discretion of the certificate holder. The FAA encourages Parts 121 and 135 carriers and each repair station (Part 145) to develop and implement an IEP in accordance with the advisory material outlined in the circular. Such a program is designed to increase the awareness of management and employees of their responsibilities to promote compliance with all regulatory requirements and best operating practices. The program will help certificate holders develop formal compliance monitoring and verification programs. This approach to safety surveillance is consistent with the FAA's policy encouraging certificate holders to identify, correct, and voluntarily reveal occurrences of noncompliance. Internal evaluation should be

- An ongoing process that uses inspections, audits, and evaluations to assess the adequacy of management controls
- A review of systems and company operating practices for more than just regulatory compliance to discover and preempt deficiencies
- A continuous process that identifies deficiencies, develops corrective-action plans to correct these deficiencies, and performs follow-up evaluations
- An independent process that organizationally has autonomous and direct reporting responsibility to top management

The internal evaluation program does not replace existing regulatory auditing requirements, such as those listed under FAR Sections 121.373 and 135.431 (continuing analysis and surveillance). The internal evaluation concept stresses self-audit responsibilities of individual employees as well as top management to ensure that company policies and procedures permit safety compliance and allow employees to work safely. Those who implement an IEP remain eligible to disclose under the Voluntary Disclosure Reporting Program (AC 00-58). The basic elements of an IEP include

- Independence/defined responsibility
- Top management review
- A continual process
- An internal evaluation schedule

- Corrective-action plans
- Records

These elements are now briefly summarized from the original Advisory Circular.

INDEPENDENCE AND DEFINED RESPONSIBILITY

- A certificate holder's internal evaluation program should identify the person and/or group within the organization who has the responsibility and authority to
  Perform evaluations, audits, and inspections as a part of an ongoing IEP.
  Identify and record evidence necessary to substantiate findings or concerns.
  Initiate and recommend solutions to concerns through designated channels.
  Verify and implement solutions within a set time frame.
  Communicate and coordinate activities with the FAA on a regular basis.
- A top management representative who has direct contact with the chief executive officer or equivalent should be given the responsibility for establishing and maintaining an IEP.
- A certificate holder should allocate and identify resources and personnel dedicated to the IEP and should specify her or his organizational independence within the company. Individuals conducting internal evaluations should not be responsible for managing work in the areas being evaluated or the tasks being reviewed.
- When full-time, dedicated resources and personnel are not practical, the certificate holder should develop procedures that preclude persons directly responsible for the areas to be evaluated from participating in the selection of the evaluation team.
- An IEP for a very small operation might consist of developing checklists and a monthly (quarterly, semiannual, or annual) schedule for accomplishing items on the checklist. Even in this case, the review should include a signed written statement of a senior management official acknowledging the completion of the checklist.
- Certificate holders using outside resources for an IEP should demonstrate that the use of outside resources is independent and has direct access to top management.

TOP MANAGEMENT REVIEW

- Senior management (chief executive officer or a person in an equivalent position who has the authority to make things happen) should review internal evaluation results to verify that satisfactory corrective actions have been implemented. Senior management should be aware of the plans, findings, concerns, and follow-up actions undertaken in an IEP.

- The review of IEP information by senior management should be documented. The frequency, format, and structure for informing top management of internal evaluations are left up to the certificate holder. A functional diagram, which depicts the independence of personnel who perform or supervise internal evaluations and their reporting authority to top management, should exist and become part of the IEP.

CONTINUAL PROCESS

- The IEP should be a continual program, not just spot-check inspections. Stand-alone spot-check inspections will merely identify symptoms of potential problems.
- An ongoing process establishes whether findings are isolated instances or actual symptoms of policy, procedural, or managerial problems. The IEP should include scheduled evaluations, follow-up evaluations as necessary, and directed evaluations when trends are identified.

INTERNAL EVALUATION SCHEDULE

- The IEP should consist of structured activities with a defined schedule. This planned schedule will serve to verify that the internal evaluation process is complete and thorough, directed, credible, and recognized by senior management.
- A proper internal evaluation schedule should include

  A planned schedule of events that will be performed during a set calendar period

  A periodic review cycle for specific areas covered by the IEP

  Enough flexibility for resources to be committed as needed to special evaluations or follow-up evaluations based on concerns or priorities identified by top management, review of industry trends, FAA concerns, or identified internal trends

  A schedule for follow-up evaluations to ensure and verify that corrective-action commitments were met and to perform follow-up evaluations in response to FAA surveillance findings

CORRECTIVE-ACTION PLANS

- An IEP should include procedures to ensure that corrective-action plans are developed in response to findings and for monitoring and verifying timely completion of corrective-action plans. Responsibility and accountability for corrective-action plans should reside with the department that was cited with the concern. A proper corrective-action plan should include

  1. The identified violation or concern
  2. An analysis of the root cause(s) of the concern or violation

3. Planned corrective actions to ensure that the violation does not recur
4. An implementation schedule with time frames for corrective actions
5. Individuals/departments responsible for
   - Implementing the corrective actions
   - Ensuring corrective-action plans are developed in response to findings
   - Verifying corrective-action plans
   - Monitoring the implementation and completion of corrective-action plans
   - Providing top management with an independent assessment of the status of corrective-action plans
   - Initiating scheduled and/or unannounced follow-up evaluations

RECORDS

- Records are considered to be the principal form of evidence documenting the workings of an IEP and should be maintained by the certificate holder. Documented evidence is essential in determining the root causes of violations so that potential areas of noncompliance can be identified. Accurate records that document the activities and results of an internal evaluation must be maintained. The FAA suggests keeping files on the following:
  - Scheduled evaluation reports
  - Special evaluation reports, including the trends or other reasons associated with scheduling a special evaluation
  - Follow-up evaluation reports
  - Responses to findings or concerns contained in reports
  - Corrective-action plans submitted in response to findings
- IEP records that could be proprietary in nature are protected in accordance with applicable laws and regulations. Records should be maintained and secured on the premises. Information on voluntary disclosures is provided in Advisory Circular 120-56. Records related to findings that are voluntarily disclosed to the FAA should be made available to the FAA for review.

An extended explanation of each of these elements can be found in Advisory Circular 120-59 (Part 121/135 operators) or Advisory Circular 145-5 (repair stations). The advisory material and supporting documents will explain the program in detail and provide suggestions on how an air carrier/repair station can develop its own program. The advisory material also includes examples (models) of various IEP organizational structures to accommodate different air carrier needs.

Certificate holders customize elements of the IEP to fit their own unique program needs. The FAA partners with airlines to offer assistance and guidance using FAA internal guidance material, industry advisory material, supporting documents, best practices, and more. Air carriers should develop a program to include maintenance, operations, and security. Other areas (e.g., corporate staff operations) may also be included in the program. Initial air carrier programs may start with maintenance, operations, or security and then add others when appropriate.

Repair stations should include the entire repair station processes from incoming inspection to approval for return to service, including the use of contractors. A repair station IEP will strengthen the Joint Aviation Authorities' (JAA) requirements for a quality monitoring system. The quality monitoring function and process include independent, regular audits and emphasize the role of an accountable manager—both basic characteristics of an IEP. In addition, an IEP will fulfill the recent Part 145 requirement for a repair station capability list self-evaluation as outlined under 14 CFR Part 145.209 and Part 145.215.

While the FAA strongly encourages and assists in the development and implementation of such a program, it cannot approve or accept it. However, this methodical and systematic self-surveillance and self-evaluation program will benefit the air carriers and the repair stations and result in increased safety for the public.

ORDERS: 8400.10 AND 8300.10

In December 1994, the Secretary of Transportation initiated a safety conference in which senior U.S. aviation officials, the FAA Administrator, and more than 1000 industry, government, and aviation officials participated in various working sessions to address aviation safety issues. The conference ended in an agreement in which the participants agreed to voluntarily establish flight safety departments within their companies (commercial carriers) and undertake various other safety initiatives. In a final rule (titled "The Commuter Operations and General Certification and Operations Requirements") that followed in December 1995, the FAA required that each certificate holder that conducts operations under Part 121 have a Director of Safety. This individual would be responsible for keeping the highest levels of management of the certificate holder fully informed about the safety status of the entire organization. However, the FAA did make concessions for smaller operations, giving them the flexibility to permit the Director of Safety to have other responsibilities in addition to heading safety. Finally, on November 30, 1999, Orders 8400.10 and 8300.10 were issued. The purpose of this bulletin was to provide guidance to Part 121 and Part 135 air carriers on the development of a comprehensive and effective safety department, together with recommendations on functions, qualifications, and responsibilities of the Director of Safety position. Some highlights on the functions, reporting structure, and qualifications of the Safety Director's position are now reviewed.

SAFETY DIRECTOR FUNCTIONS.   The Safety Director is responsible for keeping the highest management officials of the certificate holder fully informed about flight, maintenance, and ground safety practices, procedures, and programs of the certificate holder's entire operation. The Director of Safety should provide safety concerns and findings to appropriate senior operations managers for appropriate corrective actions. In addition, she or he should be a primary participant in the development of an internal evaluation program. Specifically, the Safety Director should do the following:

1. Develop and implement a comprehensive safety program that emphasizes operational safety, including all aspects of flight and ground operations, maintenance, and passenger safety.
2. Implement a safety structure and staff appropriate to the size of the operator.
3. Ensure that the following essential safety program elements are developed, integrated, and coordinated throughout the air carrier:
   a. A safety incident/accident reporting system that includes
      • Monitoring and reporting to senior management on all air carrier activities that may have an impact on safety
      • Establishing a reporting system which provides for a timely and free flow of safety-related information
      • Developing and maintaining a database of incident/accident information to monitor and analyze trends
      • Monitoring and evaluating the various safety and malfunction reporting systems to ensure appropriate integration and evaluation of data
   b. Incident/accident investigation procedures with recommendations to preclude a recurrence
   c. Safety audits and inspections
   d. Internal evaluation program
   e. Operational risk assessment program
   f. Open reporting systems
   g. Routine monitoring and trend analysis programs
   h. Review of external evaluation programs
   i. Safety committee(s)
4. Solicit and process safety improvement suggestions
5. Develop and maintain a safety awareness program
6. Review and evaluate the adequacy of the emergency response plan
7. Monitor industry safety concerns that may have an impact on operations
8. Maintain close liaison with the FAA, NTSB, and industry safety organizations and associations.
9. Discharge his or her duties to meet applicable legal requirements and to maintain safe operations in accordance with section 119.65.
10. Ensure that the safety program has been disseminated to all appropriate personnel and that a detailed description of the safety program is incorporated in the appropriate manuals as described in Part 121, sections 121.133 and 121.135.
11. Ensure that adequate safety program management is maintained.

SAFETY DIRECTOR REPORTING STRUCTURE

• For Part 121 operations, the Director of Safety position should be a full-time position. While this position is not required for Part 135 operations, these operators are still encouraged to assign a company management official to monitor and evaluate flight maintenance and ground safety practices and programs.

- As far as possible, the Safety Director should be autonomous and separate from other departments and report directly to the company's chief executive.

- The Safety Director should have direct access to the most senior levels of management and to all managers/supervisors on safety issues.

SAFETY DIRECTOR QUALIFICATIONS

TRAINING AND EDUCATION.   The Director of Safety should be educated in aviation safety and its management and should participate in industry safety meetings, technical conferences, or school/university education programs (degree and/or certificate), as an essential part of continuing education consistent with the position's responsibilities. Training should include the following subjects:

- Corporate safety culture
- The role of the Safety Director as adviser to senior management officials
- Safety philosophy
- Safety data collection and analysis programs
- Risk management
- Incident/accident prevention and investigation
- Human factors

EXPERIENCE.   The Director of Safety should have extensive operational experience and professional qualifications in aviation safety standards, operating practices, and programs.

EXPERTISE.   The person assigned as the Director of Safety should have established professional qualifications that may include any of the following:

- An FAA commercial pilot or airline transport pilot certificate
- An FAA mechanics certificate
- An FAA aircraft dispatcher certificate
- Three years' experience in a supervisory position with a Part 121 or a scheduled Part 135 air carrier.
- Three years' experience in U.S. military aviation operations
- Three years' experience in a supervisory position with a U.S. government department, board, or agency that deals directly with aviation matters

KNOWLEDGE.   The Director of Safety should have a full understanding of the certificate holder's operations specifications, the manual required by section 121.133, and all appropriate maintenance and airworthiness requirements of 14 CFR chapter I (Parts 1 through 199).

COMMUNICATIONS

Communication takes on many forms. The one addressed earlier in Chap. 7 concentrated on verbal/voice radio communications within the cockpit and between the pilot and the air traffic controller. It was further discussed how ineffective (inadequate or incorrect) communications in addition to diction, accents, and unique nuances (double meaning and confusing words) in the spoken (English) language can lead to aviation accidents and disasters. Other avenues of communication are now explored.

SAFETY, MAINTENANCE, AND FLIGHT COMMUNICATION

The corporate safety, maintenance, and flight operations organizations are often mistakenly viewed as separate entities with little or no shared mutual interests, when in actuality the three organizations are closely interlinked in myriad complex relationships and objectives. While the maintenance/engineering department provides virtually all the technological expertise necessary to maintain the aircraft fleet, the flight and in-flight (flight attendant) organizations are considered the end user of the technical product and, therefore, must maintain a user's level of technical knowledge. The development and maintenance of this user requirement, whether for access of a complex onboard database or the operation of an evacuation system, define the first link in the flight/maintenance relationship.

The second link of the relationship is forged by the legal airworthiness concept. While the captain is responsible for ensuring the final airworthiness and safety of the aircraft, it is the maintenance engineering department that maintains or returns an aircraft to an airworthiness condition. Therefore, an active and formalized communication link between the two groups is necessary for mutual satisfaction of the end product, a flyable aircraft. A fundamental component of this link is the aircraft logbook, which serves to document the degradation and restoration of the aircraft between variable levels of serviceability, or airworthiness.

The third component of the flight-maintenance-safety relationship is the regulatory-procedural link. Both procedural and technical regulatory issues must be coordinated between the three departments. While regulatory requirements emanate from the FAA, they are often precipitated by the NTSB investigative findings, thereby necessitating direct communication with both agencies for effective implementation of evolving safety requirements.

In all the above-defined levels of interdivisional relationships, the concept of safety is prevalent, and dependent upon uninhibited access of information and communication between the flight, maintenance, and safety organizations (Fig. 12-1). To enhance and facilitate the communication of time-critical safety information between the maintenance, flight operations, and safety departments, dedicated formal communicative links must be established.

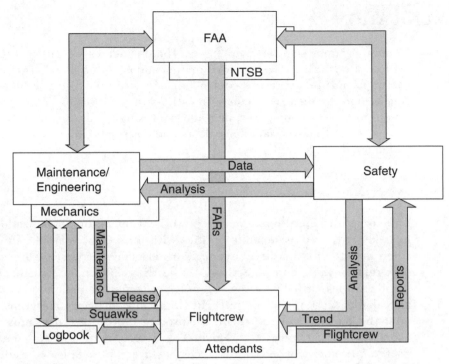

FIGURE 12-1 This schematic shows the primary operations departments at a major carrier and the basic data flow for safety-related issues.

FAA and NTSB Communication

To ensure that complete and consistent support of NTSB and FAA technical-based investigations is maintained, it is contingent upon the safety department to forge strong communication links with the NTSB and FAA. Included in this process is the requirement for consistent analysis of the technical aspects of NTSB and FAA investigations. Because of the highly complex nature of modern aircraft, a means to quickly analyze and disseminate critical safety information is required. In addition, conduct of both major and minor technically oriented investigations will require in-depth and thorough coordination of NTSB recommendations and FAA requirements. This is also true with foreign regulatory agencies; however, the specific processes vary widely between nations.

Flightcrew Communication

Critical to the safety process is the ability to quickly communicate technical information with both pilot and flight attendant groups. Safety information must be continually reviewed for information pertinent to each airline's operations. For example, developments in technical detail or aircraft operational requirements/

procedures must be provided to crewmembers in a timely manner. Included in the rapid communication process are issues related to cabin safety.

BRITISH AIRWAYS SAFETY INFORMATION SYSTEM (BASIS)

BASIS is a two-way safety communications system that includes a telephone hot line, where pilots can report safety issues, and a computer-based system for distributing safety-related material throughout the airline. This system involves all employees, including pilots' reports of irregular crew operations. It forms the backbone of the program's safety information collection and dissemination efforts. The irregular operations reports are routinely entered into a computer database that most airlines have adapted from the widely known *British Airways Safety Information System* (*BASIS*). The BASIS program is an excellent model of risk management feedback. It was designed and developed by safety professionals to provide support in capturing, investigating, and analyzing safety data from incidents and accidents. Its human-factors aspects facilitate investigative research into human errors throughout the system. It is a decision support tool for all levels of management in managing risk. It incorporates automatic alerting of problems and assists in setting priorities for preventive action.

BASIS promotes uninhibited reporting and open exchange of all safety information inside and outside the company. BASIS provides technical and operations managers with instant access to shared safety data within the company so that maintenance or ramp services can access safety information supplied by the flight operations department, and vice versa. This has bonded the usually disparate divisions into a cohesive unit that is dedicated to reducing risk in every element of the operation. There are many benefits of this program because it is

- Compatible with a personal computer
- Easy to use by non-computer-literate people
- Efficient (data can be entered promptly)
- An aid to fast investigation and report processing
- A tool to optimize safety department resources
- Accessible companywide through password control
- Compatible for an industrywide data exchange
- An important key in harnessing corporate operational expertise, while encouraging crew and individual feedback without fear of punitive action

COMMUNICATION MEDIA

Safety publications are another important component of an airline safety program. Most airlines publish regular internal safety documents to maintain high individual awareness of safety and risk management. Videos are also used to disseminate safety information to all employees. Communication and education are critical elements of any proactive airline safety program; there must be an effective mechanism in place to ensure the flow of critical safety information within the company.

COMMUNICATION FOR RISK REDUCTION

In conjunction with the risk assessment process, the communication of pertinent information necessary for reduction of the risk must be provided to the safety user. Depending on the type of risk involved, the user may be a pilot, flight attendant, mechanic, and/or related support person. The primary recipient of risk identity depends on the type of failure cause. Recurring material failure causes (e.g., fuel pump failure) are generally the responsibility of the technical services or engineering organization, while human-factor-related causes are best communicated to employee groups (e.g., pilots, flight attendants, mechanics) as appropriate.

TRAINING

Training in general is a very important component of any business plan, and safety training is no exception. Safety training can help employees develop the knowledge and skills they need to comprehend workplace hazards and protect themselves. In fact, at least one major regulatory agency (OSHA) considers training to be one of the four major elements of its recommended safety and health program guidelines. Safety training is most effective when it is integrated into a company's performance and job practices requirements. The importance of training has been discussed previously in Chaps. 6 and 7. While each regulatory agency sets different safety training requirements, the following important common issues should be noted and adhered to:

- The company's policy should clearly state the company's commitment to safety.
- The training should be conducted on paid work time.
- Training must be in the language that the employee understands and should be delivered at an appropriate pace and comprehension level.
- Employees and management should be involved in developing training programs.
- Training should be conducted based on identified needs.
- Training should have clearly defined goals.
- Training should follow good teaching pedagogies and should incorporate the following basic principles:
  - The attendee must understand the purpose of the instruction.
  - The information should be organized to maximize understanding.
  - Embed "hands-on" experiences whenever possible.

- Design the training for different learning styles such as written instruction, audiovisual instruction, hands-on exercises, and group learning assignments. Straight lectures should be no more than 20 minutes long.
- Repeat concepts as often as practical.

- There should be an evaluation component to training. Both the attendees and the training itself should be evaluated. Attendees can be evaluated through tests, quizzes, assignments, etc. Course evaluation questionnaires/surveys can be used to evaluate course content and its ability to achieve the stated goals.

- All employee training records should be documented and maintained. This will help ensure that every employee who needs training receives it, that refresher courses are provided when needed, that documentation is available when required to prove that training was conducted (a regulatory requirement is some cases), and that the training was appropriate. Documentation should include
  - The name and signature of the trained employee
  - The training date
  - The topic, including a brief lesson plan
  - Evidence of the employee's successful completion
  - The name and signature of the trainer

ROLE OF ALPA IN AIR SAFETY

Air safety is the primary responsibility of every airline pilot. As the oldest and largest airline pilots' union, the main role of the *Air Line Pilots Association's* (*ALPA's*) air safety structure is to provide channels of communication for line pilots to report air safety problems. An additional role is to stimulate safety awareness among individual pilots, to enable flightcrew members to be constructive critics of the airspace system. Finally, the air safety structure helps investigate airline accidents. Over the years, ALPA's air safety structure has contributed significantly to air safety.

ALPA's air safety structure can be compared to a three-sided pyramid, each side helping to support the others. That is, the air safety structure handles air safety problems from three perspectives:

1. By airline
2. By geographic area
3. By subject

LOCAL STRUCTURE

The basic unit of ALPA is the local council—the pilots of a single airline at a particular domicile. Each local council normally has an air safety committee, which processes local air safety problems. The committee is composed of not more than three members, headed by the local council air safety chairperson (LASC), who are appointed to 2-year terms by the local executive council (LEC) or the LEC chairperson.

An airline's central air safety committee is made up of each of the LASCs from that airline's various councils. The central air safety chairperson (CASC), appointed to a 2-year term by the master executive council (MEC) or MEC chairperson, presides over the committee, which handles problems unresolved at the local council level and those broader in scope than local issues.

Area safety coordinators (ASCs) domiciled near Federal Aviation Administration regional offices serve as the association's liaison with those offices. ALPA has 15 ASCs, including one in Berlin; their geographic areas of responsibility correspond roughly with the old (pre-1981) FAA regions. In a few cases, ARTCC boundaries are used to further subdivide particularly congested areas (e.g., Florida and California). The executive CASC, with the approval of ALPA's president, appoints ASCs to terms without time limits. At large hub airports, ASCs also serve as chairpersons of airport safety committees made up of the chairpersons of the various local air safety councils in that domicile.

TECHNICAL COMMITTEES

ALPA's air safety structure deals with air safety problems not only by airline and by locale but also by topic. The association has 14 technical committees to supply line pilot input to industry and government on a formal and continual basis. Each committee, established and dissolved by ALPA's president, is expected to stay abreast of developments in its assigned field and to represent the association by drafting positions, participating in rulemaking, detecting areas for regulatory or operational improvement, and maintaining a liaison with government and industry. ALPA's technical committees, whose chairpersons report directly to the executive central air safety chairperson, deal with the following areas of concern:

- Accident investigation
- Accident survival
- Airport standards
- Air traffic control
- Airworthiness and performance
- All-weather flying
- Aviation weather
- Charting and instrument procedures
- Hazardous materials
- Human performance

- New aircraft evaluation and certification
- Noise abatement
- Regional airline operations
- Training

ACCIDENT INVESTIGATION

ALPA's Accident Investigation Board oversees investigation of all air carrier accidents by an appropriate ALPA subgroup. The board coordinates ALPA's participation in accident or incident investigations by the NTSB or the FAA and ensures that the appropriate ALPA subgroup determines the significant factors in each air carrier accident.

The 10 pilots on ALPA's Accident Investigation Board are scattered throughout the United States, so that at least one trained accident investigator can rapidly reach the scene of the accident no matter its location. ALPA's president appoints the board's chairperson. The executive central air safety chairperson, with the approval of ALPA's president, appoints the other 9 members, and all 10 serve terms without time limits.

Under ALPA policy, each airline pilot group establishes an accident investigation team to function under the direction and responsibility of either the central air safety chairperson or the chief accident investigator, whichever that airline's MEC chooses to designate. (In most cases, the CASC heads the accident investigation team for his or her airline.) Either the CASC or the chief accident investigator is in charge of the technical aspects of any investigation involving one of his or her airline's aircraft. The CASC may request, however, that complete direction and operational control of an association accident investigation effort be assigned to a member of the ALPA Accident Investigation Board.

SPECIAL PROJECT COMMITTEES

As the need arises, the executive central air safety chairperson may, with the concurrence of ALPA's president, appoint pilots or committees to investigate special air safety topics of interest, apart from operating in a manner similar to that of the technical committees. These special project committees include engineering technical representatives for each airline to help the central air safety chairperson with service bulletins, airworthiness directives, and proposed *minimum equipment list* (MEL) changes regarding aircraft used by their airline; master minimum equipment list (MMEL) coordinators who act as ALPA spokespeople at FAA MMEL meetings and otherwise represent the association's interest in matters dealing with MMEL changes for a particular aircraft type that they are currently flying or are in some other way qualified to discuss; executive chairperson for aeromedical resources who works closely with ALPA's aeromedical adviser and with the executive central air safety chairperson to coordinate various medical activities such as cardiovascular fitness and health problems, as well as support for ALPA involvement in cases dealing with the issuance of aircrew medical certificates; flight-time/duty-time committee, which articulates association policy on proposed changes to flight-time/duty-time rules and acts to effect rulemaking accommodating that policy. The committee also interprets flight-time/duty-time regulations for interested ALPA parties; and the flight security committee, which works closely with the FAA, the Federal Bureau of Investigation, local law enforcement, and other concerned authorities worldwide to maintain high levels of airport and airline operational security from the threat of hijackers and saboteurs.

To provide continuity and technical expertise for line pilot volunteers in ALPA's air safety structure, the association maintains an engineering and air safety department plus an accident investigation department, with full-time staff engineers and support personnel. The activities of this group of volunteers and support staff range over a wide spectrum.

ALPA has been represented at virtually all government and industry safety seminars, symposia, and congressional hearings of interest to airline pilots. ALPA air safety representatives provide line pilot input to ongoing government and industry research projects and programs that deal with airline technology development and safety.

LINE PILOT INPUT

In promoting aviation safety, ALPA represents line pilot viewpoints on many major industrywide committees. Participating in the government-industry Air Traffic Procedures Advisory Committee, ALPA representatives advance association objectives to improve air traffic control. Similarly, ALPA participates on both the executive committee and several special committees of the Radio Technical Commission for Aeronautics, an association of U.S. government and industry aeronautical organizations that establishes standards for aviation electronic systems. ALPA members also serve on several aviation committees of the Society of Automotive Engineers and two committees of the National Fire Protection Association.

On the international level, ALPA is a founder, member, and active participant in the International Federation of Air Line Pilots Associations (IFALPA), among whose chief concerns is the development of worldwide standards for the design and operation of transport aircraft. IFALPA is represented on the International Civil Aviation Organization (ICAO), a United Nations agency devoted to obtaining uniformity and safety in world aviation. IFALPA has the status of observer or technical adviser to ICAO and processes safety problems of an international nature through it. ALPA often enjoys the same status within the U.S. ICAO delegation.

ALPA itself hosts numerous air safety seminars, workshops, and other meetings sponsored by MECs, LECs, or individual pilot groups. At ALPA's annual Air Safety Forum, held each summer in Washington, D.C., ALPA's air safety representatives and industry representatives exchange information and suggest solutions for the daily operating problems that various pilot groups experience. They also learn about the latest air transport technology. An annual meeting of ALPA's technical committee chairpersons serves a similar purpose, as do LASC-CASC training seminars.

ALPA also maintains 24-hour air safety hot lines for pilots to report an accident, incident, or alleged violation of a federal aviation regulation or simply report a safety problem or airspace system deficiency or to make a recommendation regarding safety.

Working to maintain the highest levels of airline safety is an enormous challenge. The depth and breadth of ALPA's air safety structure, and the critical role it plays in meeting that challenge, are unique in the air transportation industry.

FLIGHT SAFETY FOUNDATION

Another organization that plays a significant role in air safety is the *Flight Safety Foundation* (*FSF*). It was founded in 1945 by then-leaders of the aviation industry, who recognized the need for an independent body that would promote safety in aviation, anticipate flight safety problems, act as a clearinghouse on safety matters, and disseminate aviation safety information. Through the years, FSF has been responsible for the development of many aviation safety improvements that are taken for granted.

FSF doctrine is to anticipate and study flight safety problems and to collect and disseminate safety information for the benefit of all who fly. The most safety-conscious airline shares the same airspace with the less-informed or even careless operator, so it is of benefit to invest in the education and awareness-raising of such operators. FSF, with more than 800 member organizations in more than 70 countries, provides an information collection and feedback function that many less-developed aviation industries rely on for aviation safety information.

As an apolitical, independent, nonprofit, and international organization, FSF benefits from a nonofficial status because it avoids a great many of the postured responses that many businesses are obliged to present to their peers, governments, and media. Because it has no enforcement authority, its task is friendly persuasion. Several aviation leaders have described FSF as the "safety conscience" for the industry. FSF has the support from major manufacturers and airlines (which have a sense of responsibility as well as an enlightened self-interest) to make the skies as safe as possible.

The agendas of FSF's annual safety seminars, held in locations throughout the world for the past 50 years, feature a strong program of accident prevention methodology presented by the best safety experts in industry, government, and academia. Their aim, of course, is to provide effective feedback to the aviation community about hazard identification, design, training, inspection, procedures, trend analysis, etc., to use collective knowledge for the prevention of accidents.

Feedback occurs in other forums, such as industry association meetings, industry-government committees dealing with specific safety topics, meetings with other independent associations focusing on specific areas of safety improvement, and computer-based data exchanges.

Another means of obtaining information for feedback to the airline industry is FSF's confidential safety audits of corporate and airline operations. This is a valuable method of gaining firsthand information about how companies comply with their own operating standards, how they value safety, and how they manage risk. FSF shares this information on a nonattributable basis with its members through the regular publications it produces as well as its safety seminars. In addition, it completes the feedback loop by special workshops and conferences that focus on specific safety problems in various regions of the world.

FSF has helped the former Soviet Union to establish a Flight Safety Foundation in what is now the Commonwealth of Independent States (CIS). FSF is actively working through FSF-CIS to inculcate a safety-conscious culture in Aeroflot and the more than 60 emerging airlines in the CIS. Coordination of risk management information is a real challenge. For their part, the agencies of the former Soviet Union have been quite generous in sharing safety and accident information they have developed for their aviation operations. FSF, in turn, has shared these data with its worldwide membership.

SAFETY INFRASTRUCTURE AT A MAJOR AIRLINE

Included in this section are the safety organizational structure and job descriptions of a major airline. This real-world application brings together all the information studied in this chapter and delivers an appropriate ending to this text. Please refer to Fig. 12-2 for the organization chart of the safety department. What follows are job descriptions of the various positions within the safety organizational structure of the airline.

DEPARTMENTAL AND EMPLOYEE RESPONSIBILITIES

VICE PRESIDENT OF SAFETY

The Vice President of Safety has the authority to establish and modify policies, procedures, and programs within the Safety Division. The responsibilities of this position are to

- Fulfill the Director of Safety position described in FAR 119.65.
- Develop procedures, forms, and reporting mechanisms to support the safety program.
- Monitor the daily operation to permit the continual identification of hazards and to ensure that ongoing comprehensive risk analysis is performed.

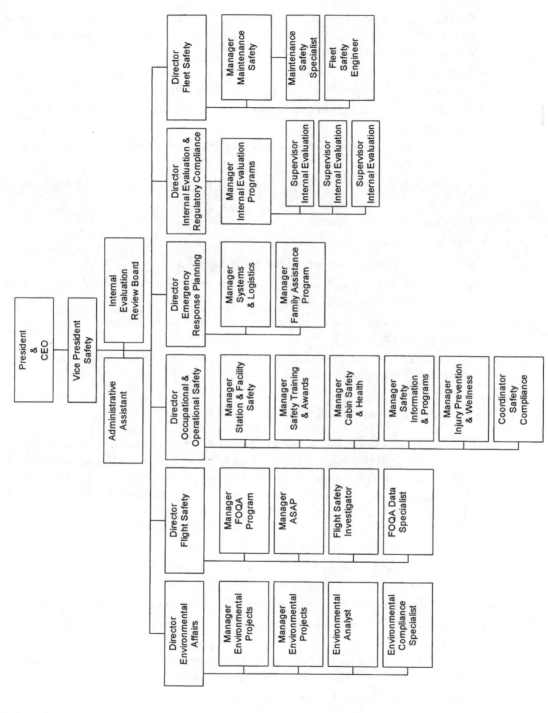

FIGURE 12-2 Organization chart of a major airline safety department.

- Provide recommendations and guidance to all operating departments in order to ensure that the company's vision for safety remains part of all departmental activities and processes.
- Serve as the subject matter expert on matters relating to aviation, ground, environmental, and occupational safety to the senior staff and senior management.
- Ensure that investigations of accidents and incidents are conducted where appropriate.
- Represent the airline regarding aviation safety matters dealing with government agencies and professional organizations.
- Provide timely advice and assistance on safety matters to line managers at all levels.
- Distribute safety information.
- Maintain an effective safety analysis program.
- Develop and manage the annual Safety Division budget.
- Continually monitor and assess operational risk and advise senior management of effective environmental affairs techniques.
- Ensure that adequate tools and methods are in place to effectively communicate safety to the workforce.
- Be responsible to the president and CEO for the development and management of the internal evaluation program.

DIRECTOR OF OCCUPATIONAL AND OPERATIONAL SAFETY

The Director of Occupational and Operational Safety is primarily responsible for ground safety at all locations. This position reports to the Vice President of Safety. Through audits, investigations, monitoring of records, and other tools, the Director of Occupational and Operational Safety will ensure that personnel, equipment, and facilities are in compliance with airline policies and procedures and with applicable local, state, and federal regulations. The responsibilities of this position are to

- Oversee and direct the efforts of the Department of Safety managers in the audit, inspection, and monitoring of all airline locations.
- Be the primary contact for state and federal OSHA complaints and responses.
- Ensure that deficiencies noted during audits and inspections are corrected in a timely manner.
- Ensure that ground handling procedures and processes are adequate to maintain a safe and healthy work environment for all personnel.
- Oversee development of general safety awareness materials for distribution companywide.

- Conduct accident/incident investigations and analysis to identify root causes and eliminate or mitigate hazards.

- Maintain damage statistics in the corporate safety database. Ensure that entries to the system are reviewed and analyzed. Produce recommendations for corrective action.

- Serve as airline representative with regard to ground safety issues to various trade associations and government agencies.

- Oversee cabin safety issues, and direct the efforts of the Manager of Cabin Safety and Health.

- Oversee publication of the quarterly safety magazine.

- Provide data analysis and reporting to the Vice President of Safety.

- Provide monthly statistics relating to damage to all stations.

- Provide monthly statistics relating to occupational injuries to all stations.

- Prepare reports on damages and incidents for the Director of Flight Operations Safety for distribution within flight operations.

MANAGER OF SAFETY INFORMATION AND PROGRAMS.   The Manager of Safety Information and Programs is primarily responsible for collecting, tracking, maintaining, analyzing, and reporting on all air, ground, and cabin incident and accident information. This position reports to the Director of Occupational and Operational Safety. Through the identification of trends and subsequent corrective actions identified by this position, the potential for cost avoidance as it pertains to aircraft ground damage, foreign object damage, and industrial injuries to the airline is substantial. This position is crucial to the fulfillment of the requirements as outlined in FAA order 8400.10, HBAT 99-19. The responsibilities of this position are to

- Receive, enter, and manage all incident and accident reports in the database.

- Control and administer the database system, including access rights, password assignment, system maintenance, and upgrades.

- Perform analysis of data contained in the database, identify and report on trends and root causes, and make recommendations for corrective action.

- Coordinate with all work groups the procedures for submitting safety information and data and identify nonreporting and underreporting areas.

- Prepare daily, weekly, monthly, quarterly, and annual reports on incidents and accidents. Create custom reports as required.

- Develop and maintain the safety intranet site. The site will serve as a repository for safety information and as a communication tool for employees with safety concerns.

- Serve as department subject matter expert on aircraft ground damage and foreign object damage (FOD).
- Represent the airline at industry meetings regarding ground damage and FOD.
- Develop and maintain a cost tracking system to accurately reflect charges (direct and indirect) related to incidents and accidents.

MANAGER OF STATION AND FACILITY SAFETY.   The Manager of Station and Facility Safety serves as the subject matter expert on ground safety and plays an important role in the program to reduce industrial injuries and accidents in the workplace. This position reports to the Director of Occupational and Operational Safety. Through audits, accident investigations, job hazard analysis, and the use of other tools, this position ensures that station personnel, equipment, and facilities are in compliance with company policy and procedures and all local, state, and federal regulations. The responsibilities of this position are to

- Assist with investigations of accidents and incidents involving airline personnel, property, or other individuals under the supervision or responsibility of the airline.
- Conduct, communicate, and publish hazard assessments for station operations.
- Ensure that station operation procedures are adequate to maintain a safe operation.
- Respond to ground operation audit deficiencies.
- Investigate select safety hazard reports and safety hotline calls.
- Provide guidance to station safety committees.
- Represent the airline at industry meetings regarding ground operation safety.
- Attend monthly Employee Injury Prevention Review Board meetings.

MANAGER OF SAFETY TRAINING AND AWARDS.   The Manager of Safety Training and Awards is responsible for resolving deficiencies in employee training as they relate to safety. This position ensures that safety training is in compliance with company policies and procedures and all local, state, and federal regulations and is a subject matter expert on both safety compliance and training issues. The Manager of Safety Training and Awards will recognize individuals and the groups they represent for their contribution to the airline safety program and will assemble and publish the quarterly airline safety magazine. This position reports to the Director of Occupational and Operational Safety. The responsibilities of this position are to

- Ensure the safety training programs throughout the company are in compliance with all local, state, and federal safety regulations and airline safety policies.

- Work with training functions of the airline to coordinate the implementation of new training material as it relates to safety requirements.

- Provide guidance and technical assistance to safety committees to ensure their effectiveness.

- Assemble, publish, and distribute the safety magazine each quarter.

- Conduct, participate, and review training audits, accident investigations, and trend analysis of safety information to detect training deficiencies related to safety.

- Revise and update safety manual sections pertaining to employee or ground safety as needed.

- Coordinate and maintain the Airline Safety Award Program. Compile data and select individual, flightcrew, safety committees, and operating locations for their outstanding contributions to the airline safety program.

- Assist with investigations of accidents and incidents involving airline personnel, property, or other individuals under the supervision or responsibility of the airline.

MANAGER OF CABIN SAFETY AND HEALTH.   The Manager of Cabin Safety and Health serves as the subject matter expert on cabin safety and health issues within the safety organization. This position will assist in reducing injuries to cabin crew, flight-deck crew, and ground crew while working in the aircraft's cabin. This position reports to the Director of Occupational and Operational Safety. Through audits, accident investigations, monitoring of Cabin Safety Reports, station visits, and other tools, the Manager of Cabin Safety and Health will ensure that personnel, aircraft, and equipment are in compliance with applicable regulations and airline policies and procedures. These policies and procedures will be evaluated for their effectiveness in supporting safety and compliance. The Manager of Cabin Safety and Health may make recommendations for change. The responsibilities of this position are to

- Attend airline and industry safety meetings and conferences.

- Review policies and procedures for compliance with FAA and applicable OSHA regulations.

- Perform risk assessments on concerns raised by employees or other groups in the company and industry, and make recommendations.

- Develop safety awareness campaigns to prevent injury and promote compliance.

- Attend annual flight attendant recurrent training.

- Assist internal evaluation with the annual compliance audit of the in-flight department.

- Hold regular office hours at each flight attendant base.

- Serve as interdepartment liaison in the coordination of policy and procedures as they relate to the aircraft cabin.
- Manage Cabin Safety Report input and trend analysis.

DIRECTOR OF ENVIRONMENTAL AFFAIRS

The Director of Environmental Affairs is responsible for corporate policy development, regulatory compliance, training, environmental affairs, and audits in the area of environmental protection. This position reports to the Vice President of Safety. The Director is the primary contact and liaison with related local, state, and federal agencies, e.g., EPA, Air Quality Management Districts, etc. The Director develops and prepares an annual financial plan, including expenditure controls and tracking for all environmental affairs issues and projects. He or she also provides direction and performs work plan control for all environmental aspects of projects initiated in other divisions. The Director must continually evaluate proposed federal, state, and local regulations to determine the impact upon the company in the area of environmental compliance, thereby formulating procedural and policy recommendations and implementation programs as required. The Director has primary responsibility for international environmental issues to ensure that adequate environmental programs are established in order to provide the safest possible workplace and to ensure compliance with all applicable international environmentally related directives. The responsibilities of this position are to

- Protect the company's long-term liability by assessing all potential risks associated with operations and property transactions and making appropriate recommendations to the environmental steering committee.
- Report significant environmental issues annually to the Board Audit Committee.
- Develop and maintain environmental portions of the safety manual.

MANAGER OF ENVIRONMENTAL PROJECTS. The Manager of Environmental Projects is responsible for administering the environmental audit program. This individual, reporting to the Director of Environmental Affairs, is responsible for developing and maintaining the curricula for all environmental training. The responsibilities of this position are to

- Coordinate the air compliance program.
- Oversee pollution prevention programs, including tracking and trending spills.
- Supervise storage tank compliance.
- Conduct periodic environmental audits and training.

SUPERVISOR OF ENVIRONMENTAL PERMITTING AND COMPLIANCE. The Supervisor of Environmental Permitting and Compliance is responsible for managing environmental remediation projects. This individual, reporting to the Director of Environmental Affairs, is also responsible for supervising the storm water program. The responsibilities of this position are to conduct periodic environmental audits and conduct periodic environmental training.

ENVIRONMENTAL ANALYST. The Environmental Analyst is responsible for overseeing compliance reporting. This individual, reporting to the Director of Environmental Affairs, is also responsible for supervising the hazardous waste program. The Environmental Analyst's responsibilities are to

- Manage chemical information databases.
- Conduct periodic environmental audits.
- Conduct periodic environmental training.

DIRECTOR OF FLIGHT SAFETY

The Director of Flight Safety is responsible for ensuring that appropriate standards are established and maintained to provide the utmost safety for airline flight operations. This position reports to the Vice President of Safety. The Director of Flight Safety will develop, coordinate, and implement preventive measures affecting present and potential problems relating to mandatory FAR and Advisory Circular directives. He or she will act as a technical correspondent to the FAA and industry groups. The Director of Flight Safety should be actively involved in policy and procedure development for aircraft operating standards. The Director of Flight Safety's responsibilities are to

- Serve as party coordinator during all National Transportation Safety Board accident or incident investigations.
- Serve as the company's authorized representative to all agencies and organizations seeking specific information concerning the company's Flight Safety Department.
- Attend air carrier and/or ATA meetings concerning flight safety issues.
- Maintain contact with other air carriers and trade organizations as directed to maintain an interchange of industry safety concerns.
- Regularly fly the line to maintain awareness of line-related issues.
- Oversee the flight safety incident database. This database is to be used to pinpoint areas where enhanced equipment, training, and procedures are needed.
- Oversee and maintain the proper functioning of the Flight Operations Quality Assurance (FOQA) Program.

- Facilitate the bimonthly Flight Safety Committee meetings.
- Ensure pilot awareness of all abnormal occurrences with review of accidents or incidents in the form of a published article in the company magazine *The Safe Approach*.
- Evaluate aircraft systems and procedures.
- Oversee the flight operations college internship program.
- Provide the Vice President of Safety with an ongoing evaluation to ensure an optimum level of safety is maintained for all the company's flight operations.
- Determine the schedule and budget for the Flight Safety Department, including task schedules, methods, staffing, and resources.
- Maintain the readiness of the airline NTSB accident investigation team.
- Provide input to the internal evaluation board (IEB) concerning matters of flight operations.
- Represent flight operations, when directed, at the monthly executive ground damage review board.
- Serve as an active airline crewmember and maintain flight currency by operating regularly scheduled flight and training events.
- Participate in the internal evaluation program oversight review of flight operations.
- Perform all other duties/projects as assigned.

AVIATION SAFETY ACTION PROGRAM (ASAP) MANAGER.   The Aviation Safety Action Program (ASAP) Manager reports to the Director of Flight Safety. This position is responsible for administering the Aviation Safety Action Program, ensuring that the daily tasks are handled in a timely manner and assisting the ASAP Event Review Committee in providing an open communication avenue among airline management, work groups, union representatives, and the FAA to identify and correct safety hazards. The duties of the ASAP Manager are to

- Act on behalf of the Director of Flight Safety in her or his absence.
- Act as the focal point for information about, and inquires concerning, the status of ASAP reports.
- Act as liaison between the Event Review Committee (ERC) and airline employees, management, union representatives, and the FAA.
- Schedule, and prepare agendas for, Event Review and Steering Committee meetings.
- Maintain the aviation quality database for all ASAP reports.
- Coordinate and track ERC recommendations.

- Publish a quarterly ASAP publication for dissemination to the appropriate work groups.
- Perform other duties as assigned.

FLIGHT OPERATIONS QUALITY ASSURANCE MANAGER.   The Flight Operations Quality Assurance Manager reports to the Director of Flight Safety. This position is responsible for coordinating the collection, transcription, analysis, and dissemination of digitally recorded flight data to appropriate parties, and will assist the Director of Flight Safety in facilitating the identification of abnormalities and enabling the appropriate departments in addressing and correcting any identified hazards. The duties and responsibilities of the FOQA Manager are to

- Be a technical liaison between flight operations and the Maintenance and Engineering Division.
- Be a technical liaison between flight operations and outside vendors to help continue the development of the FOQA and NASA Aviation Performance Measuring System (APMS) programs.
- Evaluate options and write specifications for new equipment required to implement current FOQA/APMS programs and new data-gathering programs in the future.
- Investigate mechanical problems reported by maintenance, engineering, or flight operations by reviewing the gathered data.
- Design software routines that will gather data required for special flight performance studies.
- Report to maintenance and engineering pertinent flight data recorder sensor problems found while reviewing the gathered data.
- Be a liaison between the Flight Operations Training Department and all matters concerning FOQA.
- Coordinate flight operations data retrieval and storage needs with Information and Communication Services (ICS).
- Act as the liaison between the company and the FAA in all matters concerning FOQA.
- Act as the liaison between the company and ALPA for all matters regarding FOQA.
- Define and develop an auditing and tracking system to ensure measurable benefit.

ADMINISTRATOR OF FLIGHT SAFETY.   The Administrator of Flight Safety reports to the Director of Flight Safety. This position ensures that the Flight Safety Department's daily tasks are handled in a timely manner and assists the Director of Flight Safety in providing an open communication avenue between upper management, the pilot group, and ALPA, to identify and correct hazards to flight safety. The duties and responsibilities of the Administrator of Flight Safety are to

- Act on behalf of the Director of Flight Safety in her or his absence.
- Liaison with other departments to schedule meetings and business-related trips.
- Perform daily administration of the flight operations college internship program. This includes interviewing and placement of candidates and maintaining relationships with college internship coordinators.
- Assist in maintaining the flight safety incident database.
- Attend meetings as directed by the Director of Flight Safety.
- Perform other duties as assigned.

DIRECTOR OF EMERGENCY RESPONSE PLANNING

The Director of Emergency Response Planning is responsible for all the organization's emergency response plans and procedures. This position reports to the Vice President of Safety. The Director of Emergency Response Planning plays a key role in the company's response in the event of an accident and is responsible for ensuring procedures are in compliance with legislative requirements. The duties and responsibilities of the Director of Emergency Response Planning are to

- Update, maintain, file, and distribute all documents and manuals related to emergency response (ER).
- Develop and administer the annual budget for training, supplies, equipment, and costs for all emergency response teams.
- Evaluate airline emergency procedures for consistency and viability.
- Coordinate airline response plans with code share partners.
- Direct development and implementation of initial and recurrent ER training for all public-contact employees.
- Liaison with airline, agency, and community constituents.
- Plan and implement departmental and companywide emergency drills, assess results, and recommend improvements.
- Determine emergency response program goals, and monitor progress and results.
- Promote increased companywide awareness of ER goals and requirements.
- Develop, implement, and ensure maintenance of internal and external ER communications systems.
- Report to airline officers on the status of the company's overall emergency response readiness.

- Supervise staff as required.

MANAGER OF EMERGENCY RESPONSE SYSTEMS AND LOGISTICS. The Manager of Emergency Response Systems and Logistics is responsible for coordinating procedures, equipment, and logistical support for the airline emergency response plan. The duties and responsibilities of this position are to

- Cooperate with emergency response team leaders and other division representatives, evaluate procedures, and modify as necessary to ensure a well-coordinated company response.
- Participate in the design and maintenance of the corporate emergency command center, including layout, equipment, and procedures.
- Determine and implement a strategy to meet logistical requirements of accident response, including equipment and facility needs.
- Establish procedures for timely setup of the Airline Air Accident Location Command Center and the Joint Family Support Operations Center.
- Evaluate communications requirements and recommend equipment, systems, and procedures to facilitate effective communications between accident site, command centers, and response teams.
- Develop and maintain comprehensive companywide emergency procedures manuals for the airline, integrating all response components.
- Assist in the development and execution of team and companywide emergency drills.
- Coordinate and maintain emergency phone trees and other callout activities.
- Serve as assistant to the emergency response director in the executive command center during accident response.
- Perform related duties as assigned or as the situation dictates.

MANAGER OF FAMILY ASSISTANCE PROGRAMS AND EMERGENCY RESPONSE. The Manager of Family Assistance Programs and Emergency Response is responsible for development, implementation, and coordination of air group family support functions as required by the Aviation Disaster Family Assistance Act of 1996. The duties and responsibilities of this position are to

- Provide support and direction to leaders of CARE, records and notification, coordination, and other family-services emergency response teams for the airline.
- Ensure ongoing training is provided to all public-contact employees as required under the airline aviation disaster family assistance plan.
- Assess training needs, assist in developing additional training curricula as necessary, and provide leadership to training personnel.

- Coordinate the airline family assistance procedures with code share partners.
- Assist in the development and execution of team and companywide emergency drills.
- Serve as leader of family assistance functions at the executive command center during accident response.
- Perform related duties as assigned or as the situation dictates.

Director of Internal Evaluation and Regulatory Compliance

The Director of Internal Evaluation and Regulatory Compliance reports to the Vice President of Safety and has overall responsibility for the day-to-day management of the internal evaluation program (IEP). This position serves as the chairperson of the Internal Evaluation Review Board (IERB) with regard to IEP management and control. He or she will develop and manage the IEP to ensure compliance with the written program as outlined in the safety manual and will seek to continually improve and enhance the program. The duties and responsibilities of the Director of Internal Evaluation and Regulatory Compliance are to

- Develop, implement, and maintain an IEP in accordance with the guidance contained in the Federal Aviation Administration Advisory Circular 120-59 (per its most current revision).
- Coordinate with the IERB to develop an annual evaluation schedule.
- Ensure that all evaluations are conducted on schedule.
- Review all evaluation information and prepare a summary of all evaluations for the IERB.
- Prepare finding reports for submission to each operating department with evaluation findings.
- Review departmental corrective-action plans for accuracy and effectiveness.
- Track both findings and corrective actions to ensure follow-up and completion.
- Prepare a monthly summary of evaluation findings and corrective-action plans for the Vice President of Safety.
- Prepare quarterly and annual reports of internal evaluation progress for senior management and the board of directors.
- Maintain the electronic database of evaluation information and findings.
- Assume the duties of the Director of Safety position in the event that the Vice President of Safety is unavailable to perform those duties.

Manager of Internal Evaluation Programs.   The Manager of Internal Evaluation Programs (IEP) reports to the Director of Internal Evaluation and Regulatory

Compliance. The Manager of Internal Evaluation Programs oversees and coordinates the efforts of the supervisors of internal evaluation in the performance of evaluations and audits of systems, processes, operations, stations, and facilities. The responsibilities of the Manager of Internal Evaluation Programs are to

- Schedule audits and evaluations in accordance with the IEP guidelines.
- Conduct audits and evaluations of stations and facilities.
- Conduct annual oversight evaluations of operating departments.
- Manage subordinate staff within the department.
- Compile and analyze audit and evaluation findings for root cause identification.
- Recommend corrective actions pursuant to findings.
- Track corrective actions to completion.
- Maintain databases of audit materials, checklists, findings, root causes, corrective actions, and analyses.
- Prepare reports for the Director of Internal Evaluation and Regulatory Compliance, the Vice President of Safety, the IERB, and senior management.

SUPERVISORS OF INTERNAL EVALUATION. The supervisors of internal evaluation report to the Manager of Internal Evaluation Programs. These positions perform the inspections, audits, and evaluations of airline stations, facilities, and operating departments in accordance with the IEP guidelines. The responsibilities of the supervisors of internal evaluation are to

- Conduct scheduled and unscheduled inspections, audits, and evaluations as directed by the Manager of Internal Evaluation Programs.
- Report findings related to inspections, audits, and evaluations.
- Track corrective actions related to findings to completion.
- Maintain files, records, and data related to the IEP.
- Analyze findings and deficiencies for causal factor and root cause identification.

DIRECTOR OF FLEET SAFETY

The Director of Fleet Safety is responsible for the safety and compliance of the airline's aircraft fleet. This position reports to the Vice President of Safety. Through audits, records reviews, reliability data analysis, and other tools, the Director of Fleet Safety ensures that the equipment is safe and compliant with airline policies and procedures as well as applicable federal regulations. The responsibilities of the Director of Fleet Safety are to

- Oversee and direct the activities of reporting personnel.

- Develop and administer an annual budget for the department.

- Act as internal coordinator for policies and procedures that affect fleet safety.

- Prepare *fleet safety recommendations* (FSRs) to be distributed to operating departments.

- Ensure that FSRs are acted upon in a timely manner.

- Identify potential unsafe conditions, and coordinate with operating departments to plan fleet action accordingly. Review corrective-action plans for effectiveness.

- Investigate select safety hazard reports and safety hotline calls.

- Conduct accident and incident investigations as well as root cause analysis to eliminate or manage safety hazards. Coordinate with the Director of Flight Safety in all National Transportation Safety Board accident and incident investigations.

- Serve as the airline focal point with regard to fleet safety issues with government agencies, other airlines, and trade associations. Attend pertinent meetings.

- Provide input to the Internal Evaluation Board regarding fleet safety matters.

- Perform other duties and special projects as assigned by the Vice President of Safety.

MANAGER OF MAINTENANCE SAFETY.   The Manager of Maintenance Safety serves as the subject matter expert on aircraft maintenance practices and procedures as they relate to safety. This position reports to the Director of Fleet Safety. Through audits, accident and incident investigations, monitoring of safety reports, station visits, and other tools, the Manager of Maintenance Safety ensures that aircraft maintenance is performed in compliance with applicable regulations, manufacturers' recommendations, and airline policies and procedures. The Manager of Maintenance Safety's responsibilities are to

- Assist with investigations of accidents and incidents involving company aircraft or other aircraft maintained by the company.

- Ensure that maintenance procedures are performed in accordance with applicable regulations, manufacturers' recommendations, and airline policies and procedures.

- Investigate select safety hazard reports and safety hotline calls, and make recommendations where warranted.

- Attend monthly Executive Ground Damage Review Board meetings.

- Make recommendations to ensure compliance and improve safety wherever the need for improvement is identified.

- Keep abreast of the day-to-day operation of the maintenance organization in order to identify any adverse trends before they become hazards.

- Oversee and direct the efforts of the maintenance safety specialists in the audit, inspection, and investigation at all airline maintenance locations.

- Perform audits of select maintenance organizational units and procedures, and report findings of those audits.

MAINTENANCE SAFETY SPECIALIST.   The Maintenance Safety Specialist reports to the Manager of Maintenance Safety. Through audits, accident and incident investigations, monitoring of safety reports, station visits, and other tools, the Manager of Maintenance Safety ensures aircraft maintenance is performed in compliance with applicable regulations, manufacturers' recommendations, and airline policies and procedures. The Manager of Maintenance Safety's responsibilities are to

- Assist with investigations of accidents and incidents involving airline aircraft or other aircraft maintained by the airline.

- Ensure that maintenance procedures are performed in accordance with applicable regulations, manufacturers' recommendations, and airline policies and procedures.

- Investigate select safety hazard reports and safety hotline calls, and make recommendations where warranted.

- Make recommendations to ensure compliance and improve safety wherever the need for improvement is identified.

- Perform audits of select maintenance organizational units and procedures, and report findings of those audits.

KEY TERMS

Audits

Inspections

Comprehensive audits

Status audits

Proactive approach (to safety)

Reactive approach (to safety)

Internal evaluation program (IEP)

British Airways Safety Information System (BASIS)

Air Line Pilots Association (ALPA)

Flight Safety Foundation (FSF)

REVIEW QUESTIONS

1. List the elements of a successful safety program.

2. List ways in which management can demonstrate visible interest in safety.

3. Why is it important that the chief safety officer report to the highest levels of management in the corporation?

4. Why is it important for employees to be involved in a safety program?

5. List ways to involve employees in a safety program.

6. Describe ways of holding employees accountable for their actions.

7. Give examples of safety parameters that are measured and monitored for trends. Why do we trend safety data?

8. Explain why and how accidents are investigated.

9. How are recommendations from an accident investigation handled?

10. Describe some of the elements included in an emergency response plan involving a major accident.

11. Explain the difference between the reactive and proactive approaches to safety.

12. What is the difference between a safety inspection and an audit?

13. Define what an audit is and describe the different types of audits.

14. Describe the functions of an airline safety department.

15. What role does the safety department play in safety?

16. List and discuss the basic elements of an internal evaluation program.

17. List the three agencies that regulate safety for the airlines.

18. Explain why communication is important for a successful safety program.

19. What is the British Airways Safety Information System? List some of the benefits of this system.

20. Discuss the importance of safety training.

21. Outline what it takes to deliver effective quality training.

22. Explain the importance of documenting training efforts.

23. What is the role of the Air Line Pilots Association (ALPA) in safety?

24. How is an air carrier's local council organized?

25. Identify some of ALPA's technical committees.

26. What are the roles of ALPA's Accident Investigation Board and Special Project Committees?

27. What is the primary function of the Flight Safety Foundation?

28. Why is FSF referred to as the *safety conscience* of the industry?

29. What is the purpose of FSF's annual safety seminars and safety audits of corporate and airline operations?

REFERENCES

Arbon, Capt. E. R., Capt. L. Homer Mouden, and R. A. Feeler. 1990. The Practice of Aviation Safety: *Observations from Flight Safety Foundation Audits*. Arlington, Va.: Flight Safety Foundation. June.

IATA. 1989. *Technical Policy Manual OPS-20*, Amendment No. 37. Montreal, Canada: July.

Lederer, Jerome F., and John H. Enders. 1987. *Aviation Safety—The Global Conditions and Prospects*. Arlington, Va.: Flight Safety Foundation. June.

Mouden, Capt. L. Homer. 1991. *Airline Accident Prevention Management Factors*. Herndon, Va.: Aviation Research and Education Foundation. January.

National Safety Council. 2000. *Aviation Ground Operation—Safety Handbook*. 5th ed. National Safety Council, Itasca, Ill.

U.S. Department of Labor. 1982. *Managing Worker Safety and Health—Draft*. Office of Cooperative Programs, Occupational Safety and Health Administration, U.S. DOL. Washington, D.C.

Wood, R. H. 1997. *Aviation Safety Programs—A Management Handbook*. Englewood, Colo.: Jeppesen Sanderson.

WEB REFERENCES

http://www.ncpa.org/pd/regulat/reg-1.html
http://www1.faa.gov/avr/afs/hbat/hbat9919.doc/
http://www2.faa.gov/AVR/afs/news/archive/Sep_Oct2002/Eval.htm
http://www1.faa.gov/fsdo/orl/files/advcir/AC120-59.TXT Air Carrier Internal Evaluation Model Program Guide

INDEX

Note: Boldface numbers indicate illustrations; italic *t* indicates a table.

ABOUT THE AUTHORS

ALEXANDER T. WELLS, ED.D., is a consultant, Emeritus Professor of Aviation in the College of Business at Embry-Riddle Aeronautical University in Daytona Beach, Florida, and the author of such successful texts as *Airport Planning & Management*, Fifth Edition, also from McGraw-Hill. During his 43 years in aviation, he held many positions with United Airlines and several major aviation insurance firms. In 2000, Dr. Wells was presented the University Aviation Association annual Dr. Frank E. Sorenson Award for outstanding achievements of excellence in aviation research scholarship.

CLARENCE C. RODRIGUES, PH.D., P.E., CSP, CPE, is an associate professor of applied aviation sciences in the College of Aviation and is the program director for the BS in Safety degree at Embry-Riddle Aeronautical University (ERAU) in Daytona Beach, Florida. Before joining ERAU, Dr. Rodrigues was on the safety sciences' faculty at the Indiana University of Pennsylvania (IUP) and was also an OSHA consultant for the State of Pennsylvania. Prior to joining IUP, Dr. Rodrigues was a worldwide engineering manager for the Campbell Soup Company for seven years. While at Campbell Soup, he also held an adjunct faculty appointment at The University of Pennsylvania's systems engineering department. Dr. Rodrigues has consulted for industry and for government, has made several national and international presentations, and has authored or coauthored numerous technical publications. Dr. Rodrigues holds a Ph.D. in industrial engineering and a master's degree in civil engineering from Texas A&M University. He also holds a bachelor's degree in civil engineering, and his master's degree is in environmental engineering from the Indian Institute of Technology (IIT), Bombay, India.